Polymers Handbook

Polymers Handbook

Edited by **Stacy Spoon**

NY RESEARCH PRESS

New York

Published by NY Research Press,
23 West, 55th Street, Suite 816,
New York, NY 10019, USA
www.nyresearchpress.com

Polymers Handbook
Edited by Stacy Spoon

International Standard Book Number: 978-1-63238-365-5 (Hardback)

Printed in China.

Contents

Preface

In the last few decades, advancements in polymer science has gained considerable rise. Earlier, it was considered a specialization subject for engineers. However, it has now been converted into a specific field of interest. It was Jöns Jakob Berzelius in 1833, who coined the term 'polymer'. Also known as macromolecular science, the origins of the field of polymer science can be traced to the rubber industries in pre-Columbian Mexico. Polymer science is a fundamental discipline for numerous engineering branches and scientific approaches.

This book covers different varieties of polymers, including their applications and production processes. This book is suitable for researchers, as advanced level scientific approaches have been discussed in detail. Polymer science is about mechanisms of organic compounds, physico-chemical characterization and processing, chemo-enzymatic polymerization, electrokinetic potential method, emulsion polymerizations, polymer surface patterning, biopolymers in the environment, and carbon nanostructure. These are some of the aspects our readers will get to learn while going through the chapters in this book.

Finally, I am most happy to have received contributions from internationally renowned contributors from different parts of the world. They joined hands with us to report on traditional and innovative approaches and reviewed some of the most relevant and impacting aspects of polymers in this book. This effort would not have been possible without the kind cooperation of our contributors who patiently went through revisions of their chapters.

Instead of organizing the book into a pre-formatted table of contents with chapters, sections and then asking the authors to submit their respective chapters based on this frame, the authors were encouraged by the publisher to submit their chapters based on their area of expertise. The editor was then commissioned to examine the reading material and put it together as a book. I convey my heartfelt thanks to all the contributors and to the team at the publishing house for their encouragement and excellent technical assistance as and when required.

Editor

1

Investigation of Cure Reaction, Rheology, Volume Shrinkage and Thermomechanical Properties of Nano-TiO_2 Filled Epoxy/DDS Composites

Jyotishkumar Parameswaranpillai,[1] Abhilash George,[1] Jürgen Pionteck,[2] and Sabu Thomas[3,4]

[1] *Department of Polymer Science and Rubber Technology, Cochin University of Science and Technology, Cochin, Kerala 682022, India*
[2] *Leibniz-Institute for Polymer Research Dresden, Hohe Strare 6, 01069 Dresden, Germany*
[3] *School of Chemical Sciences, Mahatma Gandhi University, Priyadarshini Hills, Kottayam, Kerala 686560, India*
[4] *Centre for Nanoscience & Nanotechnology, Mahatma Gandhi University, Priyadarshini Hills, Kottayam, Kerala 686560, India*

Correspondence should be addressed to Jyotishkumar Parameswaranpillai; jyotishkumarp@gmail.com

Academic Editor: Jae-Woon Nah

The cure reaction, rheology, volume shrinkage, and thermomechanical behavior of epoxy-TiO_2 nanocomposites based on diglycidyl ether of bisphenol A cured with 4,4′-diaminodiphenylsulfone have been investigated. The FTIR results show that, at the initial curing stage, TiO_2 acts as a catalyst and facilitates the curing. The catalytic effect of TiO_2 was further confirmed by the decrease in maximum exothermal peak temperature (DSC results); however, it was also found that the addition of TiO_2 decreases the overall degree of cure, as evidenced by lower total heat of reaction of the cured composites compared to neat epoxy. The importance of cure rheology in the microstructure formation during curing was explored by using rheometry. From the PVT studies, it was found that TiO_2 decreases the volume shrinkage behavior of the epoxy matrix. The mechanical properties of the cured epoxy composites, such as tensile strength, tensile modulus, flexural strength, flexural modulus, impact strength, and fracture toughness of the polymer composites, were examined. The nanocomposites exhibited good improvement in dimensional, thermal, and mechanical properties with respect to neat cross-linked epoxy system. FESEM micrographs of fractured surfaces were examined to understand the toughening mechanism.

1. Introduction

Epoxy resins are widely used in industrial applications, such as matrices in composite materials, and for aerospace applications, adhesives, coating, electronic circuit board-laminates, and so forth due to their excellent mechanical and thermal properties, low cost, ease of processing, fine adhesion to many substrates, good chemical resistance, good weathering resistance, low specific weight, and long pot life period [1]. The major problems with epoxy are low stiffness and low strength compared to metals and also cured epoxy resins are highly brittle, which greatly limits their application in some areas [2]. Toughness reinforcement of epoxy resins has attracted considerable attention for over the last 30 years. A new approach aiming to overcome this basic problem is related to nanotechnology and uses fillers in the nanometer scale. Nanoparticles embedded in polymer matrix have attracted increasing interest because of the unique mechanical, optical, electrical, and magnetic properties displayed by the nanocomposites [3]. Nanoparticles can significantly alter the mechanical properties of the polymer close to a particle surface due to the changes in polymer chain mobility. The enhancement in toughness is attributed to the size of the nanoparticles, since they can provide high specific surface area which helps to develop new material behaviour and is determined by interfacial interactions, resulting in unique properties and completely new classes of materials [4, 5]. Recently many researchers have shown that the addition of nanofillers not only increases the toughness but also strengthens the obtained composites [6–8].

The cure kinetics and reaction mechanisms of various epoxy monomer and its nanocomposites with nanofillers have been studied using different methods including FTIR spectroscopy [9, 10], and DSC [11, 12]. FTIR and DSC have been a useful technique for studying cure kinetics of cross-linking reactions in thermosetting epoxy. FTIR spectral analysis is based on the band intensity change of the functional group during the reaction time at a given temperature. In DSC analysis, both the isothermal and the dynamic heating experimental modes have been used extensively to determine the cure kinetic models and their parameters for the neat epoxy and epoxy/nanocomposites, which are mostly cured with amine curing agent. The basic assumption is that the heat evolution monitored and recorded by DSC is proportional to the extent of consumption of the functional groups, such as the epoxide groups in the epoxy resin or amine groups in the curing agent [13].

The processing of the thermoset matrix composites is more complicated and less controlled because of their reactivity. In this processes, polymer synthesis and shaping take place in a single operation; this involves the conversion of liquid monomers or polymers into solid crosslinked polymers [14, 15]. In addition, tendency of polymers to expand or contract during processing is a problem that has been widely recognized, especially for epoxy systems [16, 17]. Thus, the knowledge of rheology and volume shrinkage behavior of the epoxy resin system is particularly necessary to have a correct control of the processing mechanism and the subsequent engineering design.

The very high specific surface area of nanoparticles, however, causes the nanoparticles to attract each other due to van der Waals forces and to form agglomerates with dimensions of several micrometers [6]. It is reported that a considerable amount of improvement in mechanical properties can be achieved using very low amounts of nanofiller loadings [7, 8]. To extract maximum benefit from their high specific surface, nanoparticles have to be homogenously dispersed in the polymer matrix to increase the effective interfacial surface between the fillers and the matrix. The mechanical properties significantly depend on the dispersion of the particles in the matrix. This interfacial surface allows a good transfer of the applied load from the matrix to the nanoparticles. Good dispersion also results in a more uniform stress distribution and minimizes stress concentration which should increase the general strength and modulus of the composites [18, 19]. Nanoparticles are generally introduced into the polymer matrix by various techniques which include application of high shear forces during mechanical stirring [7, 20–22], pulsed ultrasound vibrations [23], and direct incorporation with chemical methods [24, 25]. This avoids the agglomerated state.

Nanoscaled TiO_2 has gained great interests in the last decades because of its unique properties and good performance. The TiO_2 is extensively used in industry, such as additives for epoxies, plastics, and rubbers [6, 26–30]. Moreover, TiO_2 filled polymers are well-known antimicrobial materials [31–35]. Since epoxy is a very common material for aerospace and automotive industry, addition of TiO_2 may act as substitute for pollution treatment (in presence of UV light, TiO_2 generates hydroxyl radicals, which can degrade or oxidize the pollutants into more environmentally friendly products) and disinfectant (photo catalytic degradation of bacteria and grime) [35] which is a major industrial issue. TiO_2 is well known for photo chemical degradation of toxic chemicals, UV shielding, waste water purification, and so forth. Apart from that, it is widely used in piezoelectric capacitors, solid oxide fuel cells, electrode materials in lithium batteries, energy converter in solar cells, and so forth [36].

Very few studies have been reported on the epoxy/TiO_2 nanocomposites, most of them are focused on the final mechanical properties [37, 38]. However, the effect of TiO_2 on cure, rheology, and volume shrinkage of epoxy systems has never been investigated. In this context, a detailed study in this topic is very important. In this paper, we present an experimental investigation on the effect of mechanically dispersed TiO_2 on the network formation of an epoxy/amine system. The effect of TiO_2 on the volume shrinkage was also examined. Finally, we have investigated the thermomechanical properties of the cured epoxy/TiO_2 nanocomposites.

2. Experimental Part

2.1. Materials Used. The epoxy resin, diglycidyl ether of bisphenol-A (DGEBA) (Lapox L-12, Atul Ltd, India.) was used as the matrix material. The amine hardener, diamino diphenyl sulfone (DDS) (Lapox K-10, Atul Ltd., India.), was used as a cross-linker for the epoxy. Nano-TiO_2, with size around 100 nm, was obtained from Riedel-de Haen (product code: 14027), Germany.

2.2. Preparation of Epoxy/TiO_2 Nanocomposites. TiO_2 nanoparticles were dispersed in DGEBA (epoxy) and mixed at 180°C using a high speed magnetic stirrer for two hours; then a stoichiometric amount of DDS with an epoxy: amine ratio of 2 : 1 was added as the curing agent and mixed well. The solution was transferred to an open mold. The composites were cured in the open mold at 180°C for three hours and then postcured at 200°C for 2 hours. Composites with TiO_2 contents of 0, 1, 5, and 10 g in epoxy-hardener mixture (100 g DGEBA + 35 g DDS) were prepared, and the samples were named as neat epoxy, 0.7 wt% TiO_2, 3.6 wt% TiO_2, and 6.9 wt% TiO_2, respectively.

For fourier transform infrared spectrometer (FTIR), differential scanning calorimetry (DSC), rheology and pressure-volume-temperature (PVT) observations, the freshly prepared mixtures were immediately used or stored before use in a freezer at −20°C. For other experiments, the freshly prepared mixtures were cured in the air oven at 180°C for 3 h and then postcured at 200°C for further 2 h. The resultant composites were then allowed to cool slowly to room temperature.

3. Characterization

3.1. Fourier Transform Infrared Spectroscopy. In situ curing studies were carried out using a fourier transform infrared-spectrometer EQUINOX 55 (Bruker Optik GmbH). A few milligrams of sample were placed in a sample holder and *in-situ* FTIR measurements were performed at 150°C for three hours. Please note that at 180°C the reaction between epoxy and amine is too fast to be followed by *in-situ* FTIR analysis.

3.2. Differential Scanning Calorimetry . The noncured samples were analyzed using a DSC Q1000 of TA-Instruments. Samples of about 6 mg were encapsulated in aluminum pans. Modulated DSC measurements were carried out in the temperature range from −60°C (5 min.) to 300°C (0.5 min.) at a heating rate of 2 K/min. in N_2 atmosphere with an amplitude of ±0.31 K and a period of 40 s. Please note that at higher heating rates the curing reaction was not complete up to 300°C. Above this temperature beginning degradation (proven by TGA investigations, which are not shown here) overlaps the curing reaction. Thus, to ensure complete curing below 300°C the low heating rate had to be used in this work. The heat of the reaction (ΔH) was determined from the total heat flow of the 1st heating scan. The curing enthalpy is proportional to the extent of the reaction. In the composites, the measured curing enthalpy ΔH is normalized to the epoxy content ($\Delta H_{(corr)}$).

3.3. Oscillatory Shear Rheology. Rheological properties of the TiO_2/epoxy/amine mixtures during isothermal curing were studied by the oscillatory shear measurements using a stress-controlled ARG2 rheometer (TA Instruments). Due to the complete vitrification of the samples, a disposable parallel plate system was utilized. The system consisted of two aluminium plates of 25 mm, one standard upper plate and one lower plate with drip channel, which were replaced at the end of each measurement. Temperature control was achieved by a sealed environmental testing chamber (ETC system, TA Instruments). All experiments were carried out with 5% strain and an angular frequency of 1 Hz and were used to determine the storage modulus (G', Pa), loss modulus (G'', Pa), tan δ, and complex viscosity (η^*).

3.4. Pressure-Volume-Temperature Analysis. The PVT measurements were done using a fully automated GNOMIX high pressure mercury dilatometer. Below 200°C the absolute accuracy of the instrument is of 0.002 cm^3/g. In practice, the change in specific volume as small as 0.0002 cm^3/g can be resolved reliably. The cross-linking reaction was characterized in the so-called data acquisition mode (DAQ) at 10 MPa and 180°C by following the volume shrinkage of the samples as a function of time. In order to check whether any cross-linking has taken place during the sample preparation stages, the experimental initial specific volumes were compared with those calculated initial specific volumes from the densities of the individual components by assuming an additive behavior (Table 1). The experimental specific volumes and calculated specific volumes are comparable, or in other words, the deviation was within an acceptable range (1%), this means that no curing or little curing takes place during sampling.

3.5. Scanning Electron Microscopy. Each sample was cooled and fractured in liquid nitrogen so that the morphology of the composites will not be effected. After that the samples were sputtered with gold in the Leica EM SCD050 sputter coater (Germany) to avoid the charging effect. Finally the sputtered samples were analyzed by using a Hitachi S-4800 high-resolution scanning electron microscope (Japan) with an accelerating voltage of 1 kV.

Table 1: Experimental and calculated initial specific volume epoxy/TiO_2 composites at 180°C.

Samples	Specific volume (experimental) cm^3/g V_{E0}	Specific volume (calculated) cm^3/g V_{T0}
Neat epoxy	0.893	0.900
3.6 wt% TiO_2 nanocomposite	0.827	0.827

3.6. Transmission Electron Microscopy. Due to the low penetration power of electrons, in order to be able to see the nanoparticle filler distribution under the sample surface, it is necessary to mount objects for examination in the electron microscope as very thin films. Therefore, samples were cut by using an ultramicrotome (ULTRACUT E. from Reichert-Jung (Germany)) into 50 to 80 nm thick ultra thin films. The prepared samples were examined in an EM 902 transmission electron microscope (Zeiss, Germany) with an accelerating voltage of 80 kV.

3.7. Thermogravimetric Analysis. Thermal stability of the blends was analyzed by thermogravimetric analyser (TGA). TA Instrument Mettler Toledo TGA/SDTA/851 (USA) was used to monitor the samples. The measurements were performed on 3–5 mg of the samples from room temperature to 700°C at a heating rate of 20°C/min and under nitrogen atmosphere.

3.8. Thermomechanical Analysis. The thermomechanical properties of neat epoxy and epoxy/TiO_2 composites were measured using a TA Instruments, Co., Q 400 thermomechanical analyser (USA). The samples were scanned from 50 to 250°C at a heating rate of 1°C/min. Rectangular specimens of 20 × 10 × 3 mm^3 were used for the analysis.

3.9. Dynamic Mechanical Analysis. The investigation of the viscoelastic properties was performed using dynamic mechanical analyzer (DMA 2980, TA Instruments, Co.), USA. Rectangular specimens of 40 × 10 × 3 mm^3 were used. The analysis was done in single cantilever mode at a frequency of 1 Hz, from −100 to 300°C and at a heating rate of 1°C/min.

3.10. Tensile Properties. Specimens for mechanical testing were machined to the required dimensions from the cast panels with a cutting machine. Tensile measurements were performed according to ASTM D 638. The measurements were taken with a universal testing machine (Tinius Olsen, Co.) Model H 50 KT (UK) at a cross-head speed of 10 mm/min. Rectangular specimens of 100 × 10 × 3 mm^3

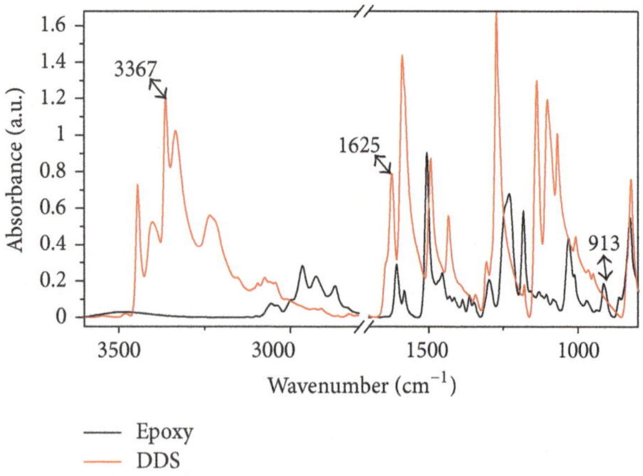

FIGURE 1: FTIR spectra of epoxy monomer and DDS.

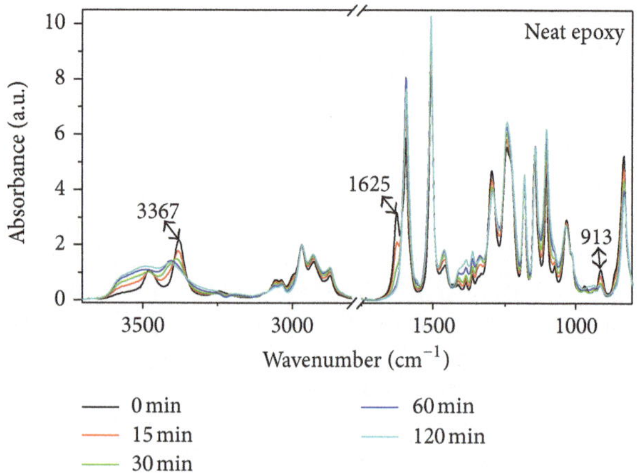

FIGURE 2: FTIR spectra of neat epoxy/DDS system us a function of reaction time (T = 150°C).

were used for determining the tensile strength. The tests were performed on 6 different specimens of the same sample and the mean values were calculated.

3.11. Flexural Properties. The flexural measurements were performed according to ASTM D 790. The cross-head speed was 10 mm/minute. The measurements were taken with a universal testing machine (Tinius Olsen Co) Model H 50 KT (UK) at a cross-head speed of 10 mm/min. Rectangular specimens of 100 × 10 × 3 mm^3 were used. The flexural modulus was calculated from the slope of the initial portion of the flexural stress-strain curve. The tests were performed on 6 different specimens of the same sample and the mean values were calculated.

3.12. Impact Strength. Charpy impact strength of the unmodified and modified epoxy resin was measured by means of a Charpy impact test following the specifications ISO 179/1eA. Impact tests were performed on a Zorn Stendal impact testing machine (Germany). The dimensions of the specimens were approximately 40 × 8 × 4 mm^3. At least 10 successful measurements were used to obtain the mean values.

3.13. Fracture Toughness. Fracture toughness of the specimens was determined according to ASTM D 5045-99. The measurements were taken with a universal testing machine Zwick (UPM—Z010) (Germany). Rectangular specimens of 60 mm length, 10 mm width, and 4 mm thickness were used for fracture toughness measurements. A notch of 5 mm was made at one edge of the specimen. A natural crack was made by a fresh razor blade at the base of the notch. The analysis was done in bending mode at room temperature. At least six successful measurements were used to obtain the mean values.

3.14. Field Emission Scanning Electron Microscopy. The morphology of the fractured surface of crosslinked epoxy as well as epoxy nanocomposites, after the K_{Ic} test, was examined using a ULTRA FESEM, (model-ultra plus) Nano Technology Systems Division Carl Zeiss SMT AG, Germany. The samples were coated with platinum by vapor deposition using a SCD 500 Sputter Coater (BAL-TEC AG, Liechtenstein).

4. Results and Discussion

4.1. Spectroscopic Analysis. Figure 1 shows the FTIR spectra of epoxy monomer and DDS. From the FTIR spectra, the most remarkable peaks are those of the amine region of DDS. We have strong peaks for the NH_2 groups from the DDS at 1625 and 3367 cm^{-1} [39]. The intense signal of the epoxy ring is centered at 913 cm^{-1} due to the asymmetrical ring stretching, in which the C–C bond stretches during the contraction of the C–O bond [39].

Figure 2 reveals the FTIR spectra of pure epoxy/DDS mixture at different stages of cure. The epoxide band at 913 cm^{-1} is very evident at the beginning of the experiment (t = 0). The intensity of epoxide band reduces with curing time and is almost not detectable after 90 minutes, indicating the time for the consumption of epoxy groups. On the other hand, the strong absorbances at 1625 and 3367 cm^{-1} of the NH_2 groups from the DDS also reduce with curing time and almost disappear after 90 minutes, again indicating the time for complete epoxy/amine conversion. However, an evaluation of the curing kinetics is difficult since during polymerization the spectrum between 3100 and 3600 becomes complex; the unreacted amines and hydroxyl groups overlap to a broad band and the absorbance at 1625 cm^{-1} forms after short curing time a shoulder to the strong absorption at 1594 cm^{-1} caused by phenyl ring [40]. Therefore, the rate of epoxy/amine polymerization was estimated only by following the loss of epoxide band intensity with respect to cure time.

Figure 3 shows the FTIR spectra of the 3.6 wt% TiO_2 containing epoxy nanocomposite. As in the pure epoxy/DDS system the band at 913 cm^{-1} is very evident at the beginning of the experiment (t = 0) and reduces with curing time. Again, after 90 min. it almost disappears. The rate of epoxy/amine polymerization can be estimated by following the loss of epoxide band intensity with respect to cure time. In Figure 4,

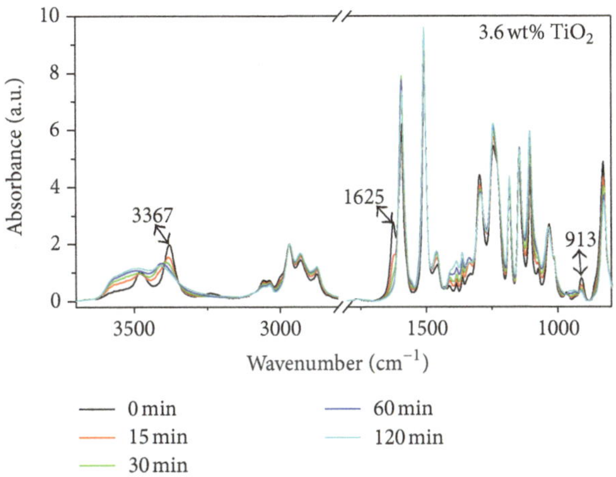

FIGURE 3: FTIR spectra of 3.6 wt% TiO_2 modified epoxy/DDS system using a function of reaction time (T = 150°C).

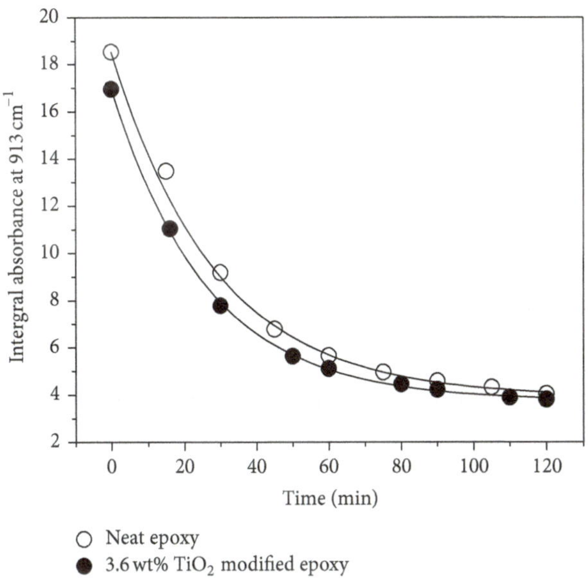

FIGURE 4: Exponential decay behavior of epoxy resin in epoxy/amine mixture during curing (dot is corresponding to the experiment data, and line corresponding to the result simulated by $y = y_0 + A_1 \exp(-x/t_1)$) ($T$ = 150°C).

we compare the loss of epoxy band with respect to cure time for the pure epoxy/DDS system and 3.6 wt% TiO_2 containing epoxy nanocomposite studied. We observe similar behavior for both systems that is a rapid decrease of epoxide band intensity within the first 45 min, followed by a slower decrease in the next hour. However, the band does not completely disappear and the intensity levels off.

To evaluate the cure reaction in more detail, the loss of epoxide band with respect to cure time was simulated with Maxwell's decay equation [41] (Figure 4):

$$y = y_0 + A_1 \exp\left(-\frac{x}{t_1}\right). \qquad (1)$$

TABLE 2: First-order decay constants for nanocomposites.

Specimen	t_1 (min)	y_0	A_1	R^2
Neat epoxy	30.42 ± 1.81	3.71 ± 0.24	15.04 ± 0.34	0.997
3.6 wt% TiO_2 nanocomposite	26.08 ± 0.55	3.75 ± 0.06	13.26 ± 0.11	0.999

TABLE 3: $\Delta H_{t\,(\mathrm{Corr})}$ and T_p for the TiO_2/epoxy composites.

Samples	ΔH_t (J/g)	$\Delta H_{t\,(\mathrm{Corr})}$ (J/g)	T_p (°C)
Neat epoxy	−394	−394	179
0.7 wt%	−353	−355	176
3.6 wt%	−354	−367	176
6.9 wt%	−328	−352	177

As $t \to \infty$, $y \to y_0$. Here A_1 is a constant and t_1 is the relaxation time of the cure reaction and indicates the rate of the epoxy/amine reaction. As shown in Figure 4, the simulation results gave a good fit to the experimental data. The fit parameters t_1, A_1, y_0, and R^2, to assess the quality of fit, are given in Table 2. The relaxation time t_1 decreases with filler addition due to the acceleration of the epoxy/amine reaction by the presence of TiO_2 particles. The increase in reaction rate was supposed to be from the high thermal conductivity and high specific surface area of the TiO_2 particles.

4.2. Nonisothermal Cure Behavior. Figure 5 shows differential scanning calorimetry thermograms during the curing of composites as obtained during first heating in modulated DSC scan. For any given specimen, only one (exothermic) DSC peak was observed due to the polymerization of DGEBA/amine, owing to the heat evolved when epoxy rings are opened during their reaction with amine functionalities. The peak area under the baseline extrapolated at the end of the reaction was used to calculate the total heat of reaction (ΔH). The maximum exothermal peak temperature (T_p) and the ΔH_{corr} values, normalized to the epoxy weight content, as a function of TiO_2 concentration, are reported in Table 3. From the Table it is clear that TiO_2 act as catalyst and facilitate curing at the initial stage by lowering T_p; on the other hand, ΔH_{corr} tends to be smaller as higher the TiO_2 content. This is because the addition of TiO_2 dilutes the epoxy-DDS reaction volume and also the presence of TiO_2-epoxy interactions reduces the crosslinking degree attained by the polymer matrix.

4.3. Rheological Analysis. For the in-depth understanding of cure process, oscillatory shear flow measurements were performed under isothermal condition at three different temperatures (150, 165, and 180°C) to investigate the rheological parameters of epoxy nanocomposites. The rheological profile remains the same irrespective of the temperature. For avoiding overlapping of the results, we are giving only the representative rheological profile obtained at 180°C. In the rheological characterization of thermosets, the oscillatory shear flow measurements are preferred to those of the steady

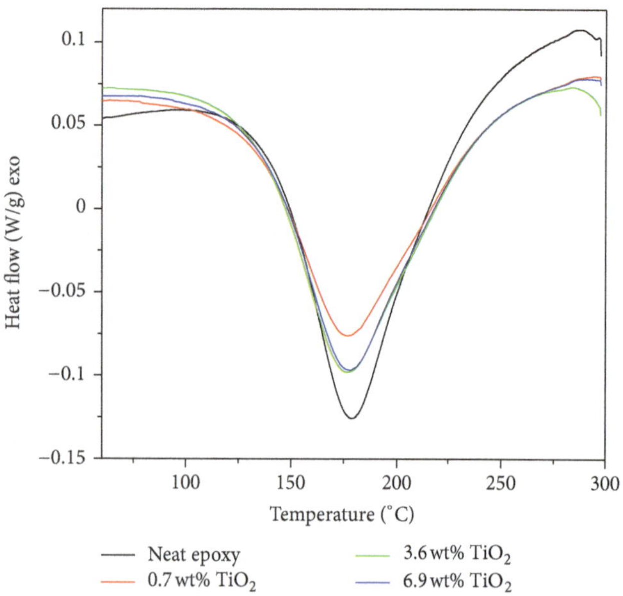

FIGURE 5: Variation of heat flow with respect to temperature.

shear because they can be applied to a material not only in the liquid state but also in rubbery and glassy.

The profile of change in complex viscosity as a function of cure time at 180°C is shown in Figure 6(a). At the beginning of the curing, the viscosity slowly increases with time, and then, at a certain point a rapid increase in viscosity is observed. The reaction between epoxy and amine induces various chemical and physical phenomena; the cross-densification advances and hence resin viscosity increases. Moreover, this increase in complex viscosity curves shifted to shorter times with TiO_2 addition due to better epoxy/amine reaction by the acceleration effect caused by TiO_2 particles. The growth of the complex viscosity follows first-order exponential equation:

$$y = y_0 + A_1 \exp\left(\frac{x}{t_1}\right). \tag{2}$$

Figure 6(b) reveals the variation of rheological parameters (tan δ, G', and G'') as a function of real time/conversion scale in case of neat epoxy system at 180°C. The tan δ value during the initial stages of cure is around 90, indicating the liquid nature of the material [42]. As the curing progresses, a rapid drop in the tan δ is observed corresponding to the chemical gelation of the material. This behavior is typical for epoxy systems [43].

From Figure 6(b), during the early stages of the reaction, G' is below G'' indicating the liquid nature of the material, after some time both G' and G'' increase with the linear growth of the epoxy chains. This is followed by a crossover point (gelation point) where G' equals G'' that means the system acts as both elastic and viscous, storing and dissipating an identical amount of energy at this point [14, 42]. Epoxy composites with, 0.7, 3.6, and 6.9 wt% TiO_2 follow a similar trend. Figure 6(c) shows the rheological profile of the 3.6 wt% TiO_2 containing epoxy composites, which is representative for all other nanocomposites studied.

In the present study, three criteria for gelation time where applied to the dynamic mechanical data. First criterion is time at $G' = G''$, second criterion is the time at the decrease in tan δ [44, 45]. A third criterion for gel point is to locate the time at which the viscosity reaches a value of 10^3 Pa·s [46]. For accuracy, we make use of all the criteria to evaluate the gel point and taken the average of all to determine the precise gelation point to calculate the activation energy. On the other hand, two criteria for vitrification time where applied, for accuracy. First criteria is time at minimum tan δ and the second criteria is time at maximum G''. The gelation and vitrification times calculated from the rheological profiles at different isothermal temperatures are listed in Table 4. The gelation and vitrification time of the epoxy/amine system decreases with the addition of TiO_2 particles. The cure acceleration effect caused by TiO_2 could bring positive effect on the processing of composite since it needs shorter pre-cure time and postcure time.

In order to characterize gelation and vitrification as a function of temperature and time, a TTT diagram is proposed. The concept of a time-temperature-transformation (TTT) cure diagram has been developed by Gillham and Babayevsky [47] for polymer systems and represents the time required to reach gelation and vitrification during an isothermal cure. The TTT diagram for the 0.7 wt% TiO_2 is proposed in Figure 6(d). The transition from viscous to elastic gel is produced by the gelation transition. On the other hand, transition from elastic gel to completely crosslinked glass produced by the vitrification transition. From the TTT diagram, it is clear that the behavior of the epoxy monomer depends on the cure temperature.

For better evaluation of the effect of TiO_2 content on rheology, activation energies for polymerization were obtained from gelation times, by using the following differential equation [48, 49]:

$$\frac{dx}{dt} = A \exp\left(-\frac{E_a}{RT}\right) f(x), \tag{3}$$

where A is the preexponential constant, E_a is the activation energy, R is the gas constant, and T is the absolute temperature. The extent of the reaction is independent of the cure temperature.

Integration of this equation from $x = 0$ to $x = x_{gel}$ by using natural logarithm leads to

$$\ln \int_0^{x_{gel}} \frac{dx}{f(x)} = \ln A + \ln\left(t_{gel}\right) - \left(\frac{E_a}{RT}\right). \tag{4}$$

Based on Flory theory [50] the extent of reaction at gel point is constant, so the previous equation can be expressed as

$$\ln\left(t_{gel}\right) = \text{constant} + \frac{E_a}{RT}. \tag{5}$$

Thus, from the slope of the plot regarding the linear relationship between ln (t_{gel}) and the inverse of temperature the apparent activation energy can be calculated.

The values of ln (t_{gel}) obtained by the rheological tests are plotted against $1/T$ in Figure 6(e). The activation energies

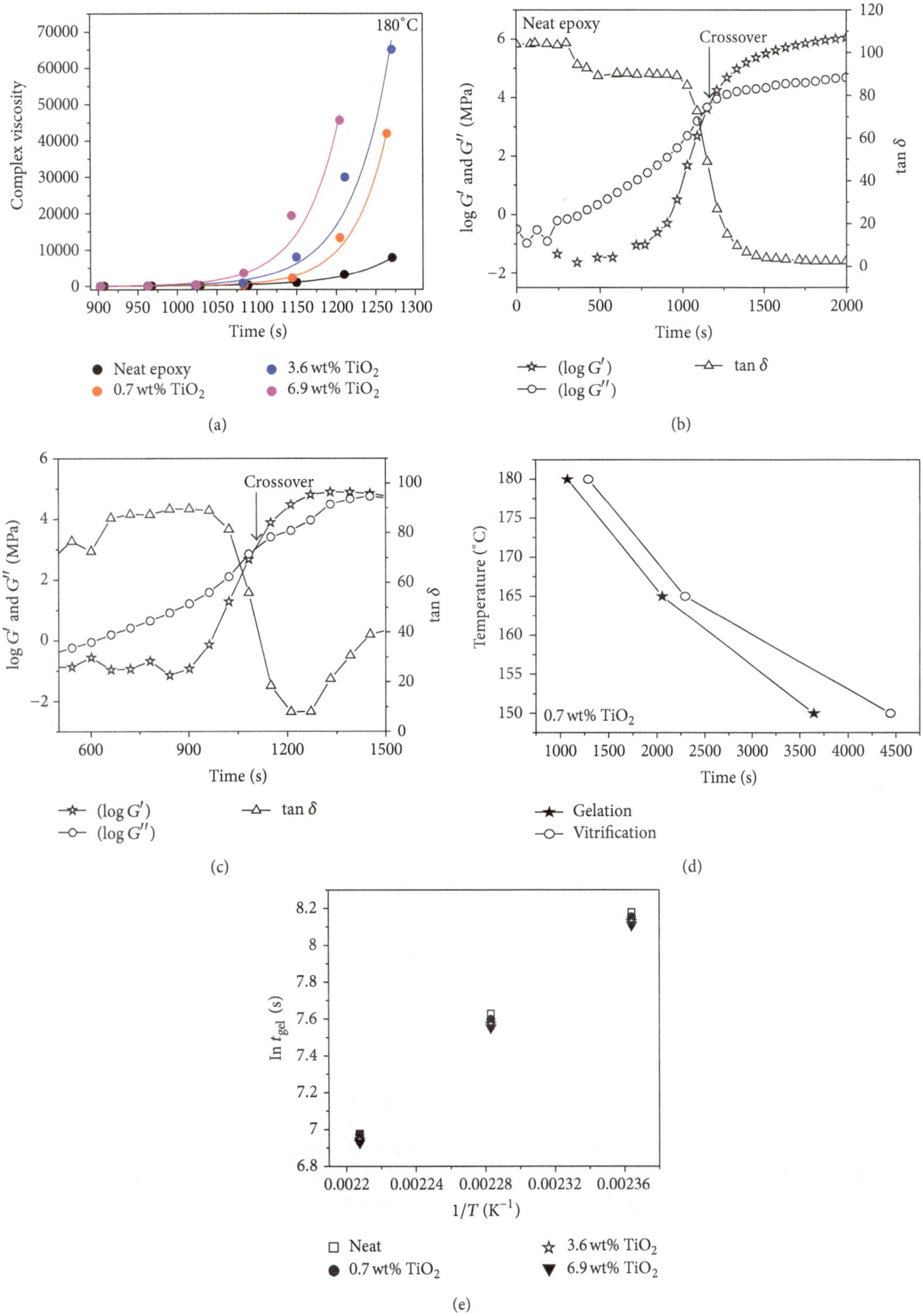

FIGURE 6: (a) Exponential growth behaviors of complex viscosity during curing of epoxy/TiO_2 nanocomposites at 180°C (dot is corresponding to the experiment data and line is corresponding to the result simulated by $y = y_0 + A_1 \exp(x/t_1)$). (b) Rheological profile of neat epoxy at 180°C. (c) Rheological profile of 3.6 wt% TiO_2/epoxy nanocomposites at 180°C. (d)Times to gelation and vitrification for 0.7 wt% TiO_2 modified epoxy system. (e) Variation of ln (t_{gel}) from complex viscosity with cure temperature for the neat matrix and epoxy nanocomposites.

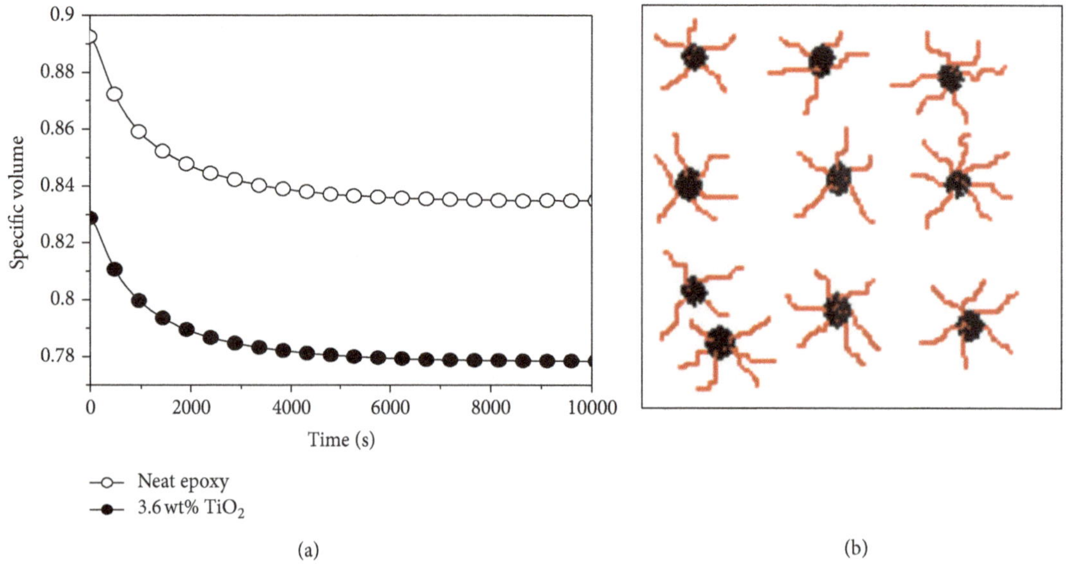

FIGURE 7: (a) Specific volume versus curing time from PVT test (epoxy/TiO_2 composites) at 180°C, 10 MPa. (b) A schematic representation of constrained layer of polymer chains around the TiO_2 nanoparticles dispersed in the epoxy matrix (white-epoxy matrix, black-TiO_2. particles, red-constrained polymer chains).

have been established from the plot. The activation energy values are 65 ± 5, 63 ± 3, 63 ± 2, and 63 ± 2 for neat epoxy, 0.7 wt% TiO_2/epoxy, 3.6 wt% TiO_2/epoxy, and 6.9 wt% TiO_2/epoxy composite, respectively. In comparison to amine cured DGEBA epoxy, the amine cured DGEBA/TiO_2 composites exhibit lower activation energies. These results validate our FTIR and DSC studies.

4.4. Pressure Volume Temperature Studies. The typical isothermal curing of thermoset/TiO_2 based composites is accompanied by the polymerization shrinkage and vitrification of the thermoset-rich phase. The material shrinkage that occurs during curing can adversely affect the desired tolerance of the final product and hence a PVT study is necessary for thermosetting composites. Pressure-volume-temperature characterization is a well-known tool for the investigation of change in specific volume with respect to cure time for epoxy systems. Figure 7(a) gives the change in specific volume data obtained for neat epoxy/DDS and 3.6 wt% TiO_2/epoxy/DDS mixtures with respect to cure time. The absolute specific volume decreases with TiO_2 content due to the higher dense TiO_2 nano filler in the epoxy matrix. A temperature of 180°C and a pressure of 10 MPa were applied for 13 hours, during which a sharp decrease in the specific volume was observed for both neat epoxy and composite due to the *in-situ* epoxy-amine reaction (volume shrinkage) [51–53]. After heating for three hours, the specific volume reached a constant value for composites. The beginning of the final plateau region corresponds to the vitrification process, although at this stage the curing reaction could continue slowly.

The change in specific volume is 0.05847 cm^3/g and 0.05061 cm^3/g for pure epoxy and epoxy composites, respectively, and is a indirect measure of volume shrinkage. Decrease in volume shrinkage for epoxy/TiO_2 composites can be explained as follows. The addition of TiO_2 dilutes the epoxy-DDS reaction volume. Moreover, presence of dense TiO_2 nano filler may increase the viscosity of the system and hence progressively decrease the relative volume shrinkage. Moreover, due to the high surface to volume ratio of nano TiO_2, epoxy polymer/oligomer get attached to the filler surface, which generates a constrained polymer around the filler surface as depicted in Figure 7(b). These constrained layer of polymer chains does not or only slightly undergo shrinkage due the presence of rigid TiO_2 particles.

4.5. Micro- and Nanostructure of the Nanocomposites

4.5.1. Scanning Electron Microscopy (SEM). The surface morphology of the cured nanocomposites was investigated by SEM. SEM micrographs were taken for all the composites. Figure 8(a) shows the SEM micrograph of the neat epoxy system which reveals a single phase. Even though, we broke the sample in liquid nitrogen, the surface morphology of neat epoxy look fractured because of the brittle nature of the control epoxy. On the other hand SEM micrograph with 0.7 wt% TiO_2 filler content (Figure 8(b)) is flat and smooth and quiet nicely form line boundaries at the interface between the pure epoxy matrix. The filler catalyses the epoxy polymerization and crosslinking and the interfaces are the contact areas of the approaching polymerization front. The apparent roughness of the surface increases with TiO_2 content. Figure 8(c) shows a SEM micrograph of the 3.6 wt% TiO_2 modified epoxy which reveals a similar morphology as observed for 0.7 wt% TiO_2 filler content. When the concentration was increased to 6.9 wt% TiO_2, particles are more agglomerated, (Figure 8(d)), [24]. As mentioned before even though the neat epoxy was fractured in liquid nitrogen there was observed fracture step edges due to the highly brittle nature of the material. On the other hand, fracture step edges are absent in the case of composites with 0.7 wt% TiO_2 and 3.6 wt% TiO_2 due to increased ductility of the epoxy matrix.

4.5.2. Transmission Electron Microscopy. For better understanding of the dispersion of the TiO_2 in the epoxy matrix, TEM micrographs were taken. TEM is a straightforward technique to visualize quality of the dispersion of nanoparticles within epoxy matrix. The micrographs obtained from the sections of the nanocomposites reveal the presence of individual TiO_2 particles embedded in the epoxy matrix. Figure 9(a) shows a TEM image of the neat epoxy; it is homogeneous with a single phase; Figure 9(b) shows a TEM image of epoxy modified with 0.7 wt% TiO_2; there were a few agglomerates. The size of the particles varies around 100 nm. A similar morphology was observed for 3.6 wt% TiO_2 and 6.9 wt% TiO_2 modified epoxy system (Figures 9(c) and 9(d)) but agglomerates of particles appear. The particle agglomerates are due to weak van der Waals attraction forces, which are higher for smaller particles leading them to poor dispersion.

4.6. Thermal Stability. Thermal stability of the neat epoxy and TiO_2 nanocomposites was studied by TGA. TGA and DTG (derivative thermogram) curves for all the cross-linked composites are given in Figures 10(a) and 10(b), respectively. From the thermograms (Figure 10(a)) nano TiO_2 filled composites show relatively less thermal decomposition than the neat crosslinked epoxy. In general, the decomposition started at around 350°C and was completed at around 500°C. The average weight loss of around 1-2% up to 300°C is due to the release of moisture if any. On the other hand, the weight loss above 350°C is related to the decomposition of the polymer. The main degradation peak in the composites occurred between 390 and 470°C, where 70%–80% of the degradation occurred. Beyond the main degradation stage, all the volatile materials were driven off from the sample resulting in the residual char. The % residual of the composites was better than the neat epoxy due to the presence of the thermally stable TiO_2 particles. For better understanding the thermal stability of the nanocomposites, their weight percentage at different temperatures were taken and are given in Table 5; it would immediately point out that for temperatures up to 400°C (the first 4 rows) little difference appears for different loaded composites. On the contrary, above 400°C (500 to 700°C) the stabilizing effect of the filler becomes clear, each column value is greater than that of the column on the left. The final (at 700°C) mass percent lefts from Table 5 are 12.5, 17.3, 19.2, and 23.7 for the loadings of 0, 0.7, 3.6 and 6.9 wt%, respectively. It means that, in the first case, still a left mass from neat matrix is 12.5: if we add the mass left to the different loading, in the next values, such as to obtain, roughly, $12.5 + 0.7 = 13.2$, $12.5 + 3.6 = 16.1$, $12.5 + 6.9 = 19.4$. If we substrate the final (at 700°C) mass percent left with these calculated values, then we have $17.3 - 13.2 = 4.1$, $19.2 - 16.1 = 3.1$, $23.7 - 19.4 = 4.3\%$; this suggest that % residual of the composites was better that neat epoxy system. The temperature at which decomposition rate has its maximum can be calculated from the DTG curves, shown in Figure 10(b). The thermal stability of composites is more evident from DTG curves where the main degradation for composites is comparatively smaller than that of the neat epoxy, while the main degradation temperature remains unaffected.

4.7. Thermomechanical Analysis. Thermal expansion behaviour of the nanocomposites was measured using a TMA to investigate the effect of TiO_2 on the dimensional change of the epoxy nanocomposite samples. The TMA thermograms of cured epoxy nanocomposites are shown in Figure 11. From the thermograms, the dimensions increase with increase in temperature is visible. During heating, the nanocomposites expand and then undergo a glassy-to-rubbery transformation (T_g) (at around 200°C); it then continues to expand in the rubbery phase with a greater liner expansion coefficient, showing the normal expansion behaviour of rubber state. In neat epoxy system drop in the dimension change near to T_g was observed. The observed drop in the dimension change near to T_g is due to relaxation of below T_g frozen nonequilibrium states of the cured samples [53]. It is important to mention that the inserted NPs or the associated mixing avoided the freezing of such local sites in the nanocomposites. From the plot, it can be concluded that the dimensional stability of crosslinked epoxy remains unaltered by the addition of nanoparticles.

4.8. Dynamic Mechanical Analysis. The storage modulus demonstrates strength and the stiffness. Figure 12(a) shows the storage modulus curves of the epoxy composites. From the profile, the storage modulus values of the epoxy nanocomposites were higher than the neat crosslinked epoxy for all temperatures. The rigid TiO_2 particles improved the stiffness of the epoxy matrix composites, especially for 6.9 wt% TiO_2 modified epoxy nanocomposites. There is an inflection point at around 200°C which indicates the transition from the solid state to the rubbery state, that is, the glass transition (T_g) of the system. Loss modulus curves (Figure 12(b)) indicate the energy dissipation and can be used to measure the viscous component of the material. The loss modulus is greater for the composites, the dispersed TiO_2 dissipates energy due to resistance against viscoelastic deformation of the surrounding epoxy matrix. Only one peak is observed for the entire system, the peak corresponds to the T_g of the crosslinked epoxy phase. It is important to mention that the peak is observed at lower temperature for the composites, which indicates slightly reduced crosslinking density for the composites.

4.9. Mechanical Tests. Tensile tests were conducted to measure the stress–strain behavior of different wt% nanocomposites under uniaxial tension. The stress-strain behavior from the tensile tests is given in Table 6. It can be seen that addition of filler shows slight improvement in the tensile properties of the epoxy system. The modulus, strength, and ductility were approximately stable, with fluctuations for the ductility

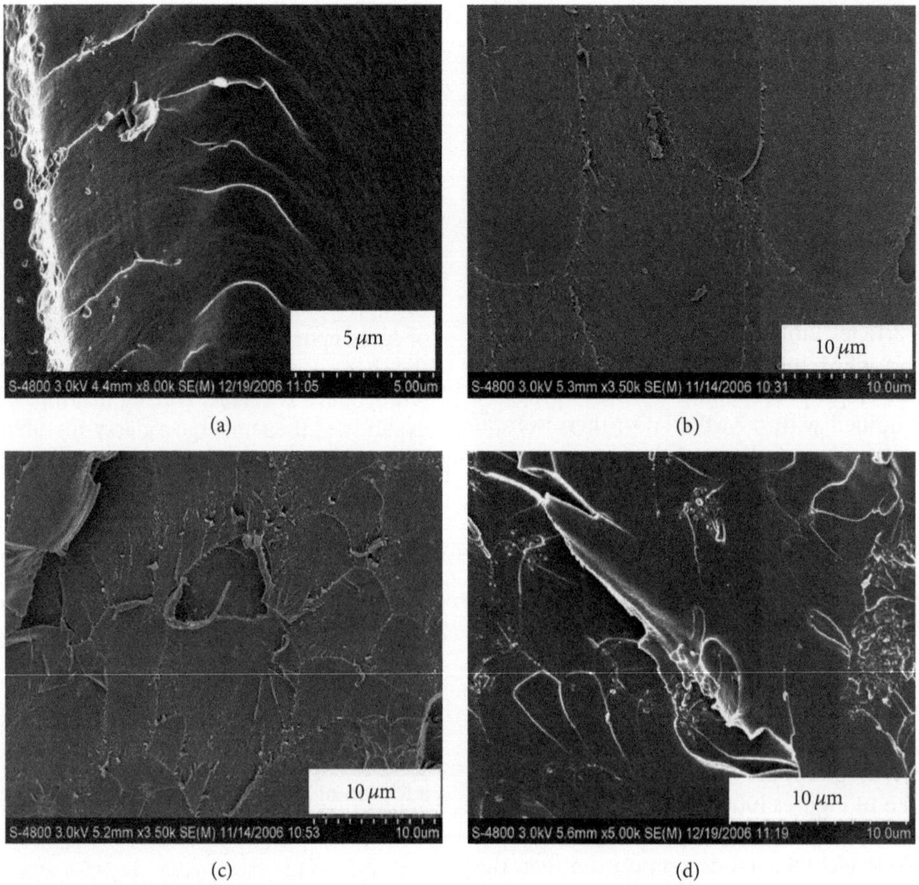

FIGURE 8: SEM micrographs of epoxy nanocomposites; (a) neat epoxy, (b) 0.7 wt% TiO_2, (c) 3.6 wt% TiO_2, and (d) 6.9 wt% TiO_2.

TABLE 4: Gelation and vitrification times for the TiO_2/epoxy composites.

T (°C)	Gelation time (sec) of the epoxy phase $(G' = G'')$ (s)	tan δ Max (s)	η^* (s)	Average of all the three parameters (s)	Vitrification time (sec) of the epoxy phase tan δ (s)	Max G'' (s)
			Neat epoxy			
150	4015	3580	3568	3721	4653	4853
165	2266	2048	2139	2151	2476	2536
180	1146	1000	1147	1097	1571	1452
			0.7 wt% TiO_2 composites			
150	3939	3534	3470	3648	4415	4475
165	2174	2008	1992	2058	2295	2295
180	1145	1024	1066	1078	1265	1323
			3.6 wt% TiO_2 composites			
150	3906	3587	3397	3630	4235	4186
165	2137	1986	1943	2022	2415	2475
180	1101	1000	1039	1047	1211	1271
			6.9 wt% TiO_2 composites			
150	3954	3558	3328	3613	4355	4417
165	2115	1941	1905	1987	2356	2356
180	1057	952	1021	1010	1144	1205

Figure 9: TEM micrographs of epoxy nanocomposites; (a) neat epoxy (b) 0.7 wt% TiO_2, (c) 3.6 wt% TiO_2, and (d) 6.9 wt% TiO_2.

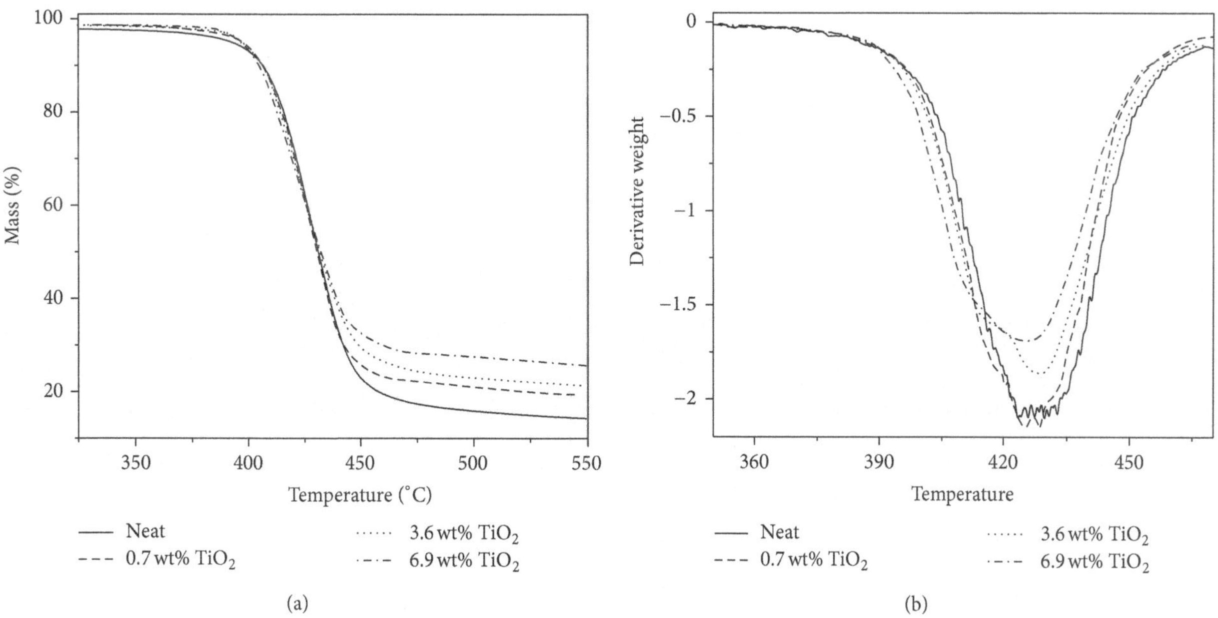

Figure 10: (a) TGA curves for epoxy nanocomposites. (b) DTG curves for epoxy nanocomposites.

TABLE 5: TGA data of epoxy composites at different temperatures.

Temperature (°C)	% residue			
	Neat	0.7 wt% TiO_2	3.6 wt% TiO_2	6.9 wt% TiO_2
100	99.5	99.7	99.7	99.8
200	98.2	98.9	98.9	99.1
300	98.0	98.7	98.7	98.7
400	93.1	94.2	94.1	93.5
500	15.9	20.8	22.9	27.2
600	13.4	18.3	20.2	24.8
700	12.5	17.3	19.2	23.7

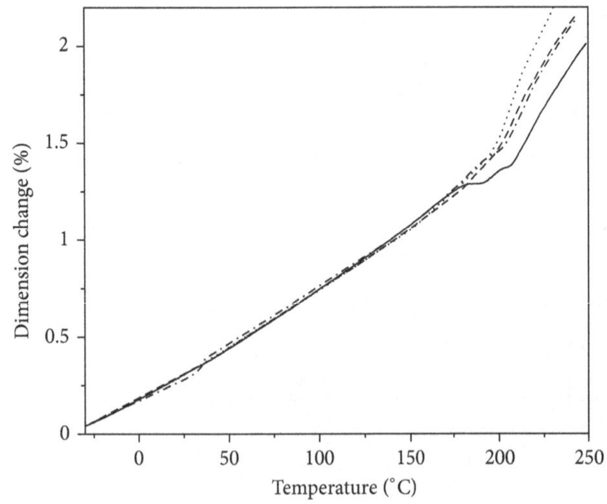

FIGURE 11: TMA curves for epoxy nanocomposites.

and an apparent moderate increase (10%) for modulus and strength.

The flexural properties of the epoxy composites are given in Table 7. With filler addition, the modulus increased; particularly with 6.9 wt% TiO_2 addition, on the other hand, the % strain and strength decreased. The flexural load is a combination of tensile and compressive load, and therefore compressive modulus should be increased by addition of stiff fillers; one could expect that the flexural modulus would rather increase with TiO_2 addition. When the composites were subjected to flexural bending, the TiO_2 particles are not able to transfer the applied load but hinder the movement of epoxy resins molecular chains. This leads to stress within the materials, so that the bend fracture strength of material decreases [36].

The impact toughness of a material describes the energy required to break the specimen. The magnitude of the impact strength reflects the ability of the material to resist impact. The impact strength of the cross-linked epoxy and epoxy composites is shown in Figure 13. Addition of TiO_2 particles enhanced the impact strength of the neat epoxy slightly, but not for 6.9 wt% TiO_2. The increase in the impact strength can be attributed to the increased toughness of the matrix upon filler addition. The nanometer-sized TiO_2 particles with good dispersion are able to block the propagation of cracks through the polymer. This is assigned to the reinforcing effect of nano-TiO_2. The impact properties of polymers are mainly enhanced by small particles with low aspect ratio, since large particles can act as crack initiation sites. However, large filler content (6.9 wt%) made it more difficult for dispersion and easier for nanosized particle to "agglomerate." Since agglomerated particles make it possible to generate defects, stress concentration will likely occur within the epoxy due to external force, resulting in decreased impact strength.

Nanocomposites simultaneously improve the fracture toughness as well as the stiffness of a neat polymer. The resistance of a material to crack initiation and propagation can be explained in terms of fracture toughness. We express the fracture toughness of the cured epoxy resin and its composites in terms of the stress intensity factor K_{IC}. The value of the stress intensity factor (K_{IC}) was calculated using (1) [41]:

$$\text{Stress intensity factor, } K_{IC} = \frac{QPa^{1/2}}{bd}, \tag{6}$$

where P is the load at the initiation of a crack, is the crack length, b is the breadth of the specimen, d is the thickness of the specimen, and Q is the geometry constant.

Q is calculated using the following equation:

$$Q = 1.99 - 0.41\left(\frac{a}{b}\right) + 18.7\left(\frac{a}{b}\right)^2 - 38.48\left(\frac{a}{b}\right)^3 + 53.85\left(\frac{a}{b}\right)^4. \tag{7}$$

The variation of fracture toughness with filler content is also given in Figure 13. The fracture toughness increases negligibly (by 8%) with the addition of TiO_2 at intermediate concentrations. However, when the TiO_2 content was increased to 6.9 wt% there was a dramatic decrease in K_{IC}, even lower than that of the control neat crosslinked epoxy system due to the agglomeration of TiO_2 particles.

FESEM analysis of the fracture surfaces of the TiO_2-filled crosslinked epoxy network composites can give an insight into the cause and location of failure as well as the dispersion state of the particles within the epoxy matrix. Figure 14 shows FESEM micrographs of the samples after the K_{IC} fracture tests. The FESEM fracture micrographs after the tensile, flexural, and impact tests were similar to those of the K_{IC} fractured samples. It is well known that the presence of rigid fillers may induce several toughening mechanisms in epoxy matrices; this includes crack deflection [54, 55], plastic deformation [56, 57], and crack front pinning [58]. Some of these toughening mechanisms was observed the for epoxy/TiO_2 composites. Figure 14(a) shows the fracture surface of the neat, crosslinked epoxy; it reveals a brittle behavior characterized by large smooth areas, typically flat, and the crack propagated uninterrupted in the direction of the applied load. The nanoparticle reinforced composites and on the other hand, exhibited rougher fracture surfaces, as shown in Figures 14(b)–14(d). Nanoparticle agglomerates of

TABLE 6: Tensile properties of the epoxy modified nanocomposites.

Sample	Tensile strength MPa	Tensile modulus GPa	Tensile elongation %	Ductility (area under the stress/strain curves)
Neat epoxy	51 ± 2	2.3 ± 0.1	3.99 ± 0.2	121
0.7 wt% TiO_2	55 ± 3	2.5 ± 0.1	4.23 ± 0.3	136
3.6 wt% TiO_2	56 ± 3	2.6 ± 0.1	3.88 ± 0.3	124
6.9 wt% TiO_2	57 ± 2	2.7 ± 0.1	4.12 ± 0.2	135

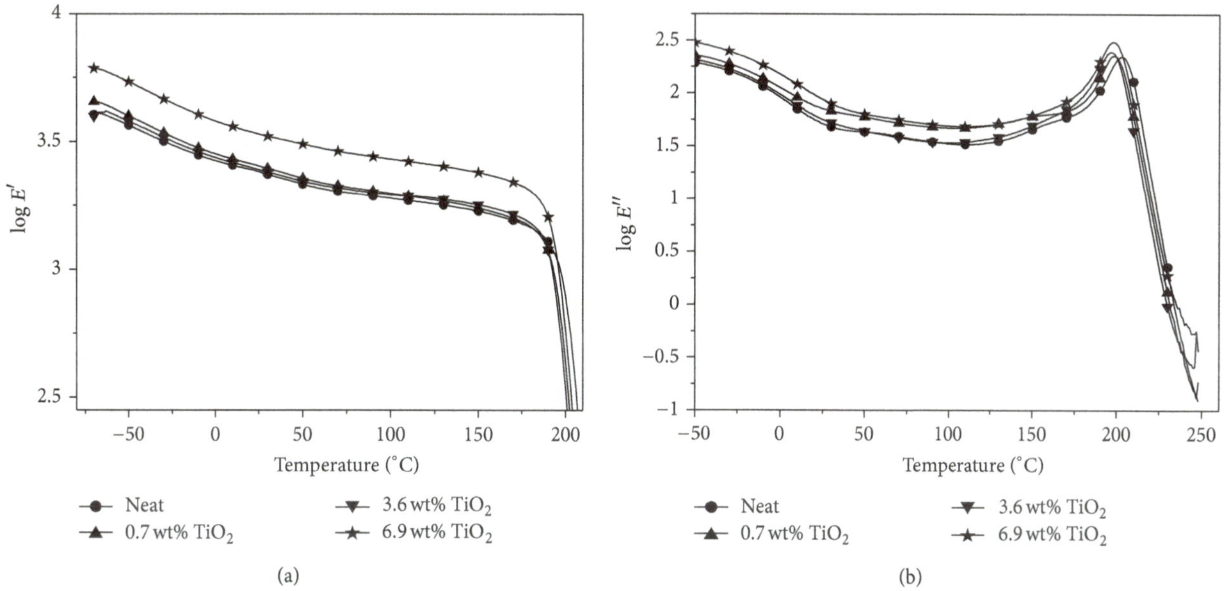

FIGURE 12: (a) Storage modulus curves for epoxy nanocomposites. (b) Loss modulus curves for epoxy nanocomposites.

TABLE 7: Flexural properties of the epoxy modified nanocomposites.

Sample	Flexural strength MPa	Flexural modulus MPa	Flexural elongation %
Neat epoxy	110 ± 5.5	3.3 ± 0.2	5.45 ± 0.3
0.7 wt% TiO_2	91 ± 4	3.1 ± 0.1	3.74 ± 0.2
3.6 wt% TiO_2	64 ± 3.7	3.5 ± 0.1	2.03 ± 0.1
6.9 wt% TiO_2	77 ± 3.8	4.2 ± 0.2	2.14 ± 0.1

a few micrometers size can be seen for all the samples along with individual nanoparticles. The FESEM micrographs of the 0.7 wt% filler content reveal the prevention of crack propagation by the agglomerates of TiO_2 and the crack seems to be deviated from the original path and hence more energy is needed for the crack propagation. Some of the TiO_2 particles were pulled out from the epoxy matrix by the application of the external load, so that smooth black holes remain visible (Figure 14(b)). The latter indicates a breakdown of the filler/matrix interface which would reduce interactions between crack and particle and subsequently force the crack to breakaway at lower stresses. This could explain why the measured fracture toughness improvement was low [59].

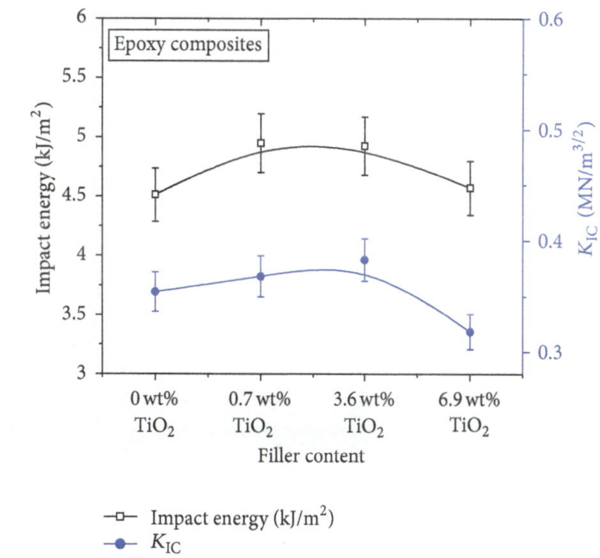

FIGURE 13: Impact strength and fracture toughness of epoxy nanocomposites.

With increase in particle loading, the number of particles increased; this causes particle-particle interaction rather than the particle-matrix interaction. Hence the particles begin to agglomerate and form lumps which eventually affected the

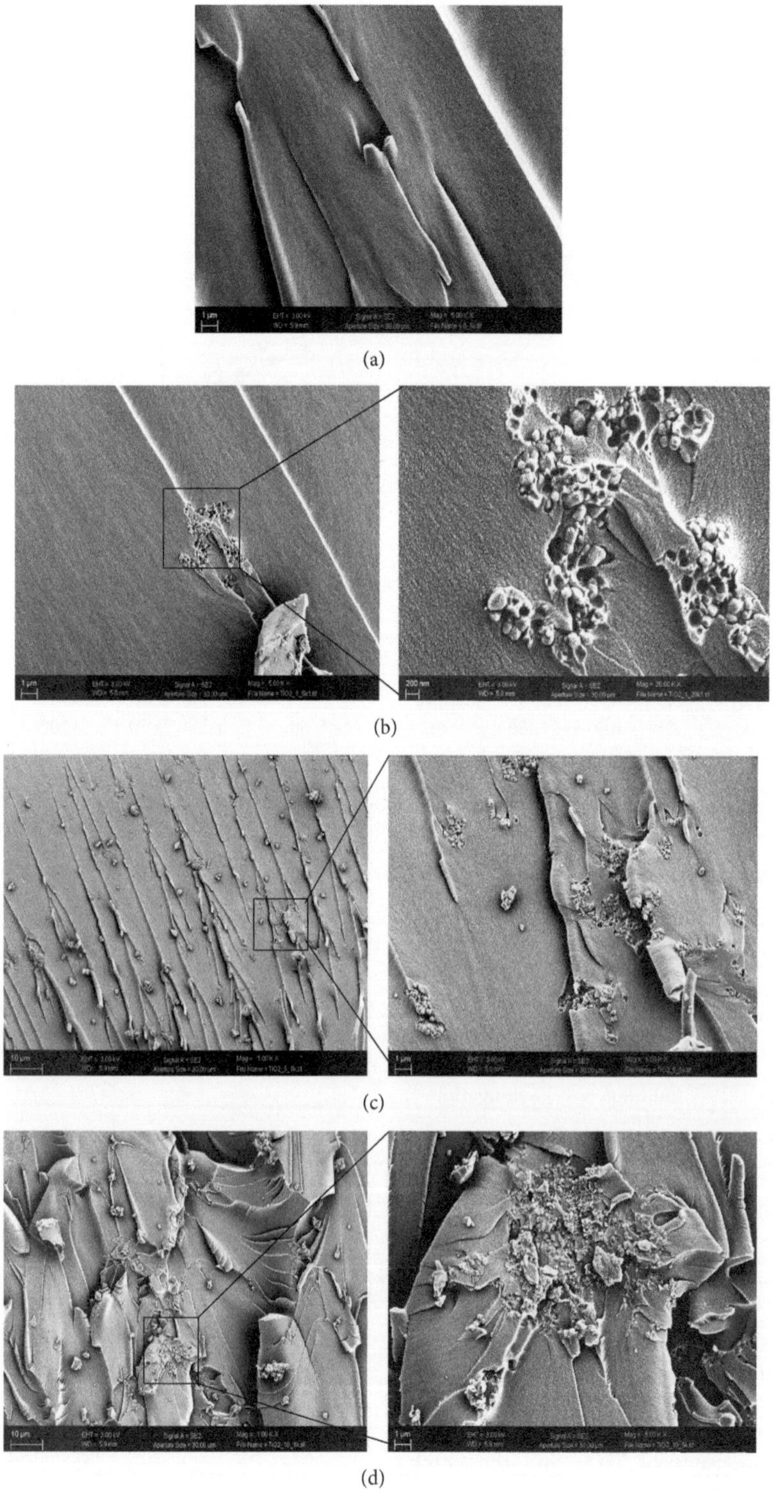

FIGURE 14: Scanning electron micrographs of failed surfaces of nanocomposites; (a) neat epoxy, (b) 0.7 wt% TiO_2, (c) 3.6 wt% TiO_2, and (d) 6.9 wt% TiO_2.

van der Waals interaction between the polymer chains and that may cause a decrease in properties. The agglomerates of TiO_2 are believed to cause high stress concentrations in the polymer matrix near the particle edges, which will facilitate failure during the fracture tests. In the case of 3.6 wt% TiO_2 (Figure 14(c)) a morphology similar to 0.7 wt% TiO_2 was observed; the particles are more agglomerated but these agglomerates not act as failure sites and hence better properties compared to neat epoxy. The particle agglomerates remaining in the matrix should be small enough to not induce a brittle and detrimental failure in the way like "large" particles would. On the other hand, when the concentration was increased to 6.9 wt% TiO_2 (Figure 14(d)), the fracture morphology was quite different from the rest of the samples, with larger particle agglomerates. However, the particle agglomerates were large enough to induce a brittle and detrimental failure in the way "large" particles would do and hence a decline in mechanical properties.

5. Conclusion

The catalytic effect of TiO_2 nanoparticle on epoxy/amine reaction was carefully analyzed using FTIR and DSC studies. This was further confirmed by rheological studies. The gelation time was decreased, with the addition of TiO_2 nanofiller due to the acceleration effect caused by the nanoparticles on the epoxy/amine reaction. PVT studies reveal that small amount of TiO_2 can potentially reduce the volume shrinkage behavior of epoxy. The effect of filler loading on the thermal and mechanical properties and the morphology were investigated. TiO_2 is an effective reinforcement for epoxy system for high performance applications. The thermal expansion behaviour was not influenced by the addition of TiO_2. Similarly, thermal stability was enhanced negligibly by the addition of TiO_2 nanoparticles. The DDS-cured epoxy nanocomposites have a single T_g corresponding to that of the epoxy phase, which shows only minor decrease with TiO_2 content and was close to T_g of the pure cross-linked epoxy system. Storage modulus results reveal better stiffness (40% increase (MPa)) for the 6.9 wt% TiO_2 composites. The rigid TiO_2 particles have no effect or little effect on the mechanical properties of the epoxy system; however, modulus increases at some point. Morphological investigation of the K_{IC} fracture surface indicates that the slight improvement in properties of the composites can be ascribed to the presence of the TiO_2 nanoparticles in the epoxy, which leads to increased surface roughness and hence energy dissipated during the propagation of the crack. However, when the concentration of filler is more than a critical value the particles were more agglomerated and the particle agglomerates are large enough to induce a brittle and detrimental failure.

Conflict of Interests

The authors do not have any direct financial relation with the commercial identities.

Acknowledgments

Jyotishkumar Parameswaranpillai acknowledges the Department of Science and Technology, Government of India, for the financial support under INSPIRE Faculty Fellowship (IFA-CH-16).

References

[1] K. Varma and V. B. Gupta, "Thermosetting resin—properties," *Comprehensive Composite Materials*, vol. 2, pp. 1–56, 2003.

[2] P. Jyotishkumar, J. Pionteck, R. Häßler, S. M. George, U. Cvelbar, and S. Thomas, "Studies on stress relaxation and thermomechanical properties of poly(acrylonitrile-butadiene-styrene) modified epoxy-amine systems," *Industrial & Engineering Chemistry Research*, vol. 50, pp. 4432–4440, 2011.

[3] J. S. Tsuji, A. D. Maynard, P. C. Howard et al., "Research strategies for safety evaluation of nanomaterials, part IV: risk assessment of nanoparticles," *Toxicological Sciences*, vol. 89, no. 1, pp. 42–50, 2006.

[4] M. Antonietti and C. Goltner, "Superstructures of functional colloids: chemistry on the nanometer scale," *Angewandte Chemie*, vol. 36, no. 9, pp. 910–928, 1997.

[5] G. Schmid, "Large clusters and colloids. Metals in the embryonic state," *Chemical Reviews*, vol. 92, no. 8, pp. 1709–1727, 1992.

[6] P. Carballeira and F. Haupert, "Toughening effects of titanium dioxide nanoparticles on TiO_2/epoxy resin nanocomposites," *Polymer Composites*, vol. 31, no. 7, pp. 1241–1246, 2010.

[7] B. Wetzel, F. Haupert, and F. Zhang, "Epoxy nanocomposites with high mechanical and tribological performance," *Composites Science and Technology*, vol. 63, no. 14, pp. 2055–2067, 2003.

[8] R. P. Singh, M. Zhang, and D. Chan, "Toughening of a brittle thermosetting polymer: effects of reinforcement particle size and volume fraction," *Journal of Materials Science*, vol. 37, no. 4, pp. 781–788, 2002.

[9] L. Guadagno, L. Vertuccio, A. Sorrentino et al., "Mechanical and barrier properties of epoxy resin filled with multi-walled carbon nanotubes," *Carbon*, vol. 47, no. 10, pp. 2419–2430, 2009.

[10] G. C. Huang and J. K. Lee, "Isothermal cure characterization of fumed silica/epoxy nanocomposites: the glass transition temperature and conversion," *Composites A*, vol. 41, no. 4, pp. 473–479, 2010.

[11] W. K. Goertzen, X. Sheng, M. Akinc, and M. R. Kessler, "Rheology and curing kinetics of fumed silica/cyanate ester nanocomposites," *Polymer Engineering and Science*, vol. 48, no. 5, pp. 875–883, 2008.

[12] M. Harsch, J. Karger-Kocsis, and M. Holst, "Influence of fillers and additives on the cure kinetics of an epoxy/anhydride resin," *European Polymer Journal*, vol. 43, no. 4, pp. 1168–1178, 2007.

[13] P. Jyotishkumar, J. Pionteck, L. Häußler, G. Adam, and S. Thomas, "A novel hydroxyapatite/ethylene-vinyl acetate/copolymer 66 composite for hard tissue regeneration," *Journal of Macromolecular Science B*, vol. 51, no. 1, pp. 1–11, 2012.

[14] P. Jyotishkumar, S. M. George, P. Moldenaers, and S. Thomas, "Dynamics of phase separation in poly(acrylonitrile-butadiene-styrene)-modified epoxy/DDS system: kinetics and viscoelastic effects," *Journal of Physical Chemistry B*, vol. 114, no. 42, pp. 13271–13281, 2010.

[15] Y. Liu, X. Zhong, and Y. Yu, "Gelation behavior of thermoplastic-modified epoxy systems during polymerization-induced phase separation," *Colloid and Polymer Science*, vol. 288, no. 16-17, pp. 1561–1570, 2010.

[16] M. Holst, K. Schänzlin, M. Wenzel, J. Xu, D. Lellinger, and I. Alig, "Time-resolved method for the measurement of volume changes during polymerization," *Journal of Polymer Science B*, vol. 43, no. 17, pp. 2314–2325, 2005.

[17] J. Jose, K. Joseph, J. Pionteck, and S. Thomas, "PVT behavior of thermoplastic poly(styrene-co-acrylonitrile)-modified epoxy systems: relating polymerization-induced viscoelastic phase separation with the cure shrinkage performance," *Journal of Physical Chemistry B*, vol. 112, no. 47, pp. 14793–14803, 2008.

[18] M. Z. Rong, M. Q. Zhang, H. Liu, H. M. Zeng, B. Wetzel, and K. Friedrich, "Microstructure and tribological behavior of polymeric nanocomposites," *Industrial Lubrication and Tribology*, vol. 53, no. 2, pp. 72–77, 2001.

[19] M. A. L. Manchado, L. Valentini, J. Biagiotti, and J. M. Kenny, "Thermal and mechanical properties of single-walled carbon nanotubes-polypropylene composites prepared by melt processing," *Carbon*, vol. 43, no. 7, pp. 1499–1505, 2005.

[20] M. Avella, M. E. Errico, S. Martelli, and E. Martuscelli, "Preparation methodologies of polymer matrix nanocomposites," *Applied Organometallic Chemistry*, vol. 15, no. 5, pp. 435–439, 2001.

[21] H. R. Dennis, D. L. Hunter, D. Chang, S. Kim, J. L. White, and J. W. Cho, "Nanocomposites: the importance of processing," *Plastics Engineering*, vol. 57, no. 1, pp. 56–60, 2001.

[22] J. W. Cho and D. R. Paul, "Nylon 6 nanocomposites by melt compounding," *Polymer*, vol. 42, no. 3, pp. 1083–1094, 2001.

[23] Q. Wang, H. Xia, and C. Zhang, "Preparation of polymer/inorganic nanoparticles composites through ultrasonic irradiation," *Journal of Applied Polymer Science*, vol. 80, no. 9, pp. 1478–1488, 2001.

[24] T. Adebahr, C. Roscher, and J. Adam, "Reinforcing nanoparticles in reactive resins," *European Coatings Journal*, no. 4, pp. 144–149, 2001.

[25] S. Kang, S. I. Hong, C. R. Choe, M. Park, S. Rim, and J. Kim, "Preparation and characterization of epoxy composites filled with functionalized nanosilica particles obtained via sol-gel process," *Polymer*, vol. 42, no. 3, pp. 879–887, 2001.

[26] G. Xian, R. Walter, and F. Haupert, "A synergistic effect of nano-TiO_2 and graphite on the tribological performance of epoxy matrix composites," *Journal of Applied Polymer Science*, vol. 102, no. 3, pp. 2391–2400, 2006.

[27] M. Sangermano, G. Matucelli, E. Amerio et al., "Preparation and characterization of nanostructured TiO_2/epoxy polymeric films," *Macromolecular Materials and Engineering*, vol. 291, no. 5, pp. 517–523, 2006.

[28] R. J. Nussbaumer, W. R. Caseri, P. Smith, and T. Tervoort, "Polymer-TiO_2 nanocomposites: a route towards visually transparent broadband UV filters and high refractive index materials," *Macromolecular Materials and Engineering*, vol. 288, no. 1, pp. 44–49, 2003.

[29] M. Xiong, S. Zhou, B. O. You, G. U. Guangxin, and L. Wu, "Effect of preparation of titania sol on the structure and properties of acrylic resin/titania hybrid materials," *Journal of Polymer Science B*, vol. 42, no. 20, pp. 3682–3694, 2004.

[30] Y. Tong, Y. Li, F. Xie, and M. Ding, "Preparation and characteristics of polyimide-TiO_2 nanocomposite film," *Polymer International*, vol. 49, no. 11, pp. 1543–1547, 2000.

[31] H. Segawa, K. Tateishi, Y. Arai, K. Yoshida, and H. Kaji, "Patterning of hybrid titania film using photopolymerization," *Thin Solid Films*, vol. 466, no. 1-2, pp. 48–53, 2004.

[32] M. D. Soucek, A. H. Johnson, L. E. Meemken, and J. M. Wegner, "Preparation of nano-sized UV-absorbing titanium-oxo-clusters via a photo-curing ceramer process," *Polymers for Advanced Technologies*, vol. 16, no. 2-3, pp. 257–261, 2005.

[33] P. V. Kamat, "Photochemistry on nonreactive and reactive (semiconductor) surfaces," *Chemical Reviews*, vol. 93, no. 1, pp. 267–300, 1993.

[34] J. C. Yu, W. Ho, J. Yu, H. Yip, and P. K. Wong, "Efficient visible-light-induced photocatalytic disinfection on sulfur-doped nanocrystalline titania," *Environmental Science & Technology*, vol. 39, no. 4, pp. 1175–1179, 2005.

[35] J. C. Yu, W. Ho, J. Lin, H. Yip, and P. K. Wong, "Photocatalytic activity, antibacterial effect, and photoinduced hydrophilicity of TiO_2 films coated on a stainless steel substrate," *Environmental Science & Technology*, vol. 37, no. 10, pp. 2296–2301, 2003.

[36] A. Chatterjee and M. S. Islam, "Fabrication and characterization of TiO_2-epoxy nanocomposite," *Materials Science and Engineering A*, vol. 487, no. 1-2, pp. 574–585, 2008.

[37] K. S. Huang, Y. H. Nien, J. S. Chen, T. R. Shieh, and J. W. Chen, "Synthesis and properties of epoxy/TiO_2 composite materials," *Polymer Composites*, vol. 27, no. 2, pp. 195–200, 2006.

[38] A. A. Wazzan, H. A. Al-Turaif, and A. F. Abdwlkader, "Influence of submicron TiO_2 particles on the mechanical properties and fracture characteristics of cured epoxy resin," *Polymer-Plastics Technology and Engineering*, vol. 45, no. 10, pp. 1155–1161, 2006.

[39] A. Sabra, T. M. Lam, J. P. Pascault, M. F. Grenier-Loustalot, and P. Grenier, "Characterization and behaviour of epoxy-based diaminodiphenylsulphone networks," *Polymer*, vol. 28, no. 6, pp. 1030–1036, 1987.

[40] T. M. Don and J. P. Bell, "Fourier transform infrared analysis of polycarbonate/epoxy mixtures cured with an aromatic amine," *Journal of Applied Polymer Science*, vol. 69, no. 12, pp. 2395–2407, 1998.

[41] P. Jyotishkumar, E. Abraham, S. M. George et al., "Preparation and properties of MWCNTs/poly(acrylonitrile-styrene-butadiene)/epoxy hybrid composites," *Journal of Applied Polymer Science*, vol. 127, no. 4, pp. 3093–3103, 2013.

[42] P. Jyotishkumar, P. Moldenaers, S. M. George, and S. Thomas, "Viscoelastic effects in thermoplastic poly(styrene-acrylonitrile)-modified epoxy-DDM system during reaction induced phase separation," *Soft Matter*, vol. 8, no. 28, pp. 3452–7462, 2012.

[43] W. Gan, G. Zhan, M. Wang, Y. Yu, Y. Xu, and S. Li, "Rheological behaviors and structural transitions in a polyethersulfone-modified epoxy system during phase separation," *Colloid and Polymer Science*, vol. 285, no. 15, pp. 1727–1731, 2007.

[44] S. Poncet, G. Boiteux, J. P. Pascault et al., "Monitoring phase separation and reaction advancement in situ in thermoplastic/epoxy blends," *Polymer*, vol. 40, no. 24, pp. 6811–6820, 1999.

[45] S. Swier and B. van Mele, 1998, Polymer Network Preprints P12, Trondheim.

[46] I. Alig and W. Jenninger, "Curing kinetics of phase separating epoxy thermosets studied by dielectric and calorimetric investigations: a simple model for the complex dielectric permittivity," *Journal of Polymer Science B*, vol. 36, no. 13, pp. 2461–2470, 1998.

[47] J. K. Gillham and P. G. Babayevsky, "Epoxy thermosetting systems: dynamic mechanical analysis of the reactions of aromatic diamines with the diglycidyl ether of bisphenol A," *Journal of Applied Polymer Science*, vol. 17, no. 7, pp. 2067–2088, 1973.

[48] P. A. Oyanguren and R. J. J. Williams, "Cure of epoxy novolacs with aromatic diamines. I. Vitrification, gelation, and reaction kinetics," *Journal of Applied Polymer Science*, vol. 47, no. 8, pp. 1361–1371, 1993.

[49] J. M. Barton, D. C. L. Greenfield, and K. A. Hodd, "Some effects of structure on the cure of glycidylether epoxy resins," *Polymer*, vol. 33, no. 6, pp. 1177–1186, 1992.

[50] P. J. Flory, *Principles of Polymer Chemistry*, Universidad de Cornell, Ithaca, NY, USA, 1953.

[51] P. Jyotishkumar, J. Pionteck, C. Özdilek et al., "Rheology and pressure-volume-temperature behavior of the thermoplastic poly(acrylonitrile-butadiene-styrene)-modified epoxy-DDS system during reaction induced phase separation," *Soft Matter*, vol. 7, no. 16, pp. 7248–7256, 2011.

[52] J. A. Ramos, N. Pagani, C. C. Riccardi, J. Borrajo, S. N. Goyanes, and I. Mondragon, "Cure kinetics and shrinkage model for epoxy-amine systems," *Polymer*, vol. 46, no. 10, pp. 3323–3328, 2005.

[53] P. Jyotishkumar, E. Logakis, S. M. George et al., "Preparation and properties of multiwalled carbon nanotube/epoxy-amine composites," *Journal of Applied Polymer Science*, vol. 127, no. 4, pp. 3063–3073, 2013.

[54] K. T. Faber and A. G. Evans, "Crack deflection processes—I. Theory," *Acta Metallurgica*, vol. 31, no. 4, pp. 565–576, 1983.

[55] K. T. Faber and A. G. Evans, "Crack deflection processes—II. Experiment," *Acta Metallurgica*, vol. 31, no. 4, pp. 577–584, 1983.

[56] T. Kawaguchi and R. A. Pearson, "The moisture effect on the fatigue crack growth of glass particle and fiber reinforced epoxies with strong and weak bonding conditions, part 2: a microscopic study on toughening mechanism," *Composites Science and Technology*, vol. 64, no. 13-14, pp. 1991–2007, 2004.

[57] A. J. Kinloch, D. Maxwell, and R. J. Young, "Micromechanisms of crack propagation in hybrid-particulate composites," *Journal of Materials Science Letters*, vol. 4, no. 10, pp. 1276–1279, 1985.

[58] J. Spanoudakis and R. J. Young, "Crack propagation in a glass particle-filled epoxy resin," *Journal of Materials Science*, vol. 19, no. 2, pp. 473–486, 1984.

[59] B. Wetzel, P. Rosso, F. Haupert, and K. Friedrich, "Epoxy nanocomposites—fracture and toughening mechanisms," *Engineering Fracture Mechanics*, vol. 73, no. 16, pp. 2375–2398, 2006.

2

Condensation Polymers of Terephthalic Acid and 1,4-Diaminobutane and Their Schiff Base Complexes

Sandeep Rai,[1] Anjali Bajpai,[2] and Snehal Lokhandwala[1]

[1] *Shroff S. R. Rotary Institute of Chemical Technology, Block No. 402, At & Post Vataria, District Bharuch, Gujarat 393002, India*
[2] *Department of Chemistry, Government Autonomous Model Science College, Jabalpur 482001, India*

Correspondence should be addressed to Sandeep Rai; sadeep1964@yahoo.co.in

Academic Editor: Yves Grohens

Amino-terminated oligomeric poly(tetramethylene terephthalamide) (PTTA) was prepared by condensation of terephthalic acid and 1,4-diaminobutane using phosphorylation technique. Schiff base complexes of this polyamide were synthesized with salicylaldehyde and 2-hydroxy-1-naphthaldehyde complexes of Co(II), Ni(II), and Cu(II). The polyamide as well as Schiff base complexes were characterized by elemental analysis, IR spectroscopy, and magnetic susceptibility measurements. Thermal stabilities of ligand and its various complexes were compared by thermogravimetric analysis.

1. Introduction

Polyamides, the condensation polymers of dicarboxylic acids and diamines, are frequently referred to as nylons, an important group of polymers which can be synthesized using phosphorylation technique [1]. Oligomers based on these polyamides having reactive terminals are of great importance in order to yield materials with novel properties. Most of the polyamides do not withstand elevated temperature for a longer time. We have tried to get thermally resistant polyamide-metal complexes by inserting metal ions in ligands through the formation of Schiff's base complexes. We have reported Schiff base complexes with oligomeric poly(methylene diphenylene terephthalamide) [2]. In the present communication we are reporting Schiff base complexes of amino-terminated low molecular weight poly(tetramethylene terephthalamide), PTTA, with salicylaldehyde and 2-hydroxy-1-naphthaldehyde complexes of Co(II), Ni(II), and Cu(II).

Various Schiff bases involving aldehydes with aromatic diamines are extensively studied by numerous workers and are critically reviewed and reported for their biological properties, such as antibacterial and antifungal activities [3–5]. Few polymeric complexes with Schiff's base structures have been synthesized from aromatic amines and phenolic aldehydes. These metal complexes have been widely studied because they have anticancer and herbicidal applications [6–8].

2. Materials and Methods

2.1. Materials. The materials of analytical quality or equivalent grades were used. Salicylaldehyde was from S. Merck., India, and 2-hydroxy-1-naphthaldehyde (AG) was from Fluka. 1,4-diaminobutane of AR grade was obtained from E. Merck, India, and was used without further purification. Dimethyl formamide (DMF), dimethyl sulphoxide (DMSO), and pyridine were GR grade products of Qualigens, Glaxo, India, and were distilled before use. Acetates of cobalt(II), nickel(II), and copper(II) (all of AR grade from E. Merck, India) were used as such.

2.2. Synthesis of Amino-Terminated Oligomeric Polyamide (PTTA). The amino terminated oligomeric polyamide ligands were prepared by a known procedure. Terephthalic acid (0.1 moL) was treated with 0.12 moL of 1,4-diaminobutane, which was finely powdered in the presence of triphenyl phosphite. A mixture of DMF (150 mL) and pyridine (30 mL) was taken as solvent. The resulting mixture was maintained at 80°C for 1.5 hrs under nitrogen atmosphere. The amino group terminated oligomer thus formed was separated out by precipitating out from the reaction mixture by addition of excess of distilled methanol. The precipitated products were washed several times with methanol, filtered under suction, and then dried in air oven at 80°C for 20 hrs (see Scheme 1).

The PTTA oligomer was characterized by IR spectroscopy and elemental analysis. The mass average molecular mass $\overline{M}_w$ was calculated by viscometery using the following equation given by Aoki et al. [9]:

$$[\eta] = 9.9 \times 10^{-4} \overline{M}_w 0.8. \quad (1)$$

2.3. Synthesis of Metal-Aldehyde Complexes. The metal-aldehyde complexes were prepared by mixing aqueous solutions of acetates of cobalt(II), nickel(II), and copper(II) with ethanolic solutions of salicyladehyde or 2-hydroxy-1-naphthaldehyde in respective preparations (see Schemes 2 and 3). The reaction mixture was refluxed with continuous stirring for 2-3 hrs. The coloured solid complexes were filtered, washed with ethanol followed by ether, and then dried in an oven at 60°C. The complexes were characterized by elemental analysis and IR spectroscopy (see Tables 1 and 2).

2.4. Synthesis of Polyamide-Metal Complexes. Equimolar solutions of metal aldehyde and the amino-terminated oligomer PTTA in 50 mL hot DMF were mixed in respective preparations and refluxed on a heating mantle for 6–8 hrs to obtain solid coordination polymers. The coloured precipitates were filtered, washed repeatedly with hot DMF and alcohol, and then dried. The polyamide PTTA-metal complexes were coloured, powdery solids insoluble in water and common organic solvent. The reaction scheme for the synthesis of amide-metal complexes is given in Figure 1.

2.5. Characterization. Elemental analysis of polyamide ligand (PTTA) and its coordination polymers was carried out on Perkin Elmer Elemental Analyzer model 2400. The metal contents were determined by complexometric titration with EDTA after decomposing the coordination polymers with fuming nitric acid. Molecular weights of the coordination polymers could not be determined due to their insoluble nature. IR spectra of the ligand and the coordination polymers were recorded on a Perkin Elmer Infrared Spectrophotometer Model 1430 using CsI pellets in 4000–200 cm^{-1} range. The magnetic susceptibility measurements were done at room temperature with a guoy's balance using $HgCo(NCS)_4$ as a calibrant. Thermogravimetric analysis was performed on a Perkin-Elmer Thermal Analyzer TGA 7 with professional computer 7700 in an inert atmosphere at a heating rate of 20°C/min.

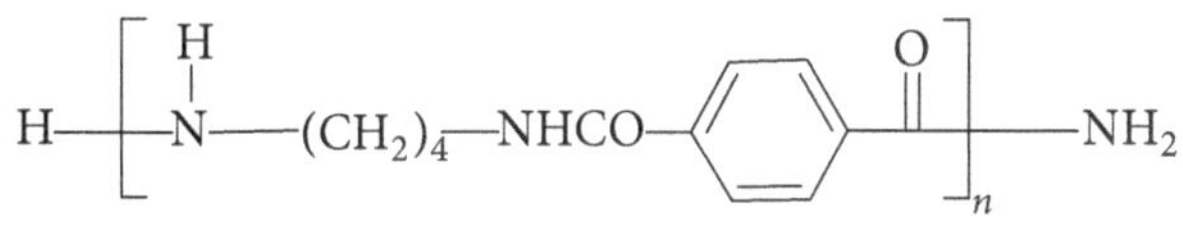

PTTA

SCHEME 1

TABLE 1: Analytical data of PTTA and its metal complexes.

No.	Compound	Elemental analysis*			
		%C	%H	%N	%M
1	PTTA	63.40 64.70	5.96 6.50	13.20 14.10	— —
2	PTTA-Co(II)-Sal′	60.36 62.40	6.19 5.90	9.69 11.80	3.00 2.80
3	PTTA-Ni(II)-Sal′	59.86 62.40	4.97 5.9	9.22 11.8	3.22 2.79
4	PTTA-Cu(II)-Sal′	60.14 62.20	5.02 5.88	10.04 11.79	3.78 3.00
5	PTTA-Co(II)-Naphthal′	62.25 64.60	4.97 5.77	9.98 11.03	3.28 2.68
6	PTTA-Ni(II)-Naphthal′	63.98 64.62	4.88 5.77	10.12 11.03	4.00 2.66
7	PTTA-Cu(II)-Naphthal′	62.46 64.41	5.16 5.76	10.06 10.99	4.58 2.88

*In every case the upper value represents the observed results, and the lower value represents the theoretically calculated values.

3. Results and Discussion

The analytical data for the ligand and their metal complexes are presented in Table 1. The compositions and theoretical analytical results are based on assumed structures. In case of PTTA the number of repeated units (n) is taken as 8 to 9 for the theoretical calculations. The average of molecular masses calculated on the basis of $n = 8$ is $\overline{M}_w = 1750$ which is close to the experimentally found value that is 1758.

It was observed that the deviations of the experimental values of elemental analysis from the theoretical ones were beyond experimental error. This discrepancy between the calculated and observed results may be due to the strong moisture pickup tendency of polyamide [10].

Puffer and Sebenda [11] have described sorption and desorption isotherms of water for several crystalline as well as amorphous polyamides and the model amides. They have shown that, in nylon 6, three molecules of water are sorbed on two neighbouring amide groups. The first molecule is firmly bound between the CO groups, and the other two molecules form H-bonds between the CO and the NH groups (loosely bound water). In nylons, two to six water molecules are sorbed on two amide groups, probably as firmly bound water. Nylon 7 stands somewhere between the two groups.

In our report on poly(methylene diphenylene terphthalamide) and its coordination polymers, we have also observed the moisture pickup by the polyamide as well as its coordination polymers. The number of water molecules sorbed to the amide groups has been considered as 1.5 per CONH group excluding the capillary condensation. Because of the polymeric nature, the possibility of structural defects is always there. Further, the exact amount of adsorbed and capillary condensation water cannot be calculated.

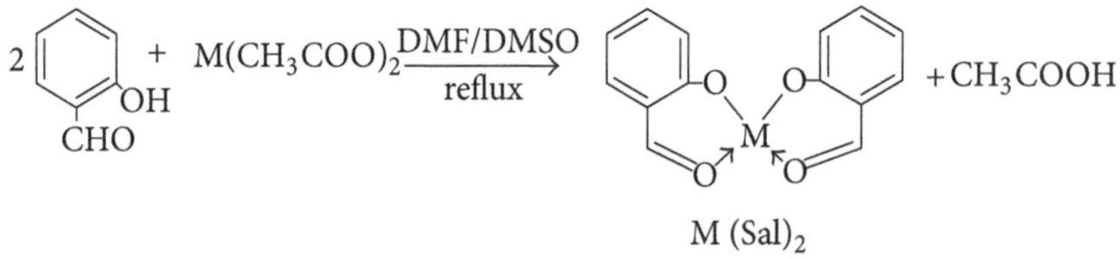

where M = Co(II), Cu(II), or Ni(II) and
Sal = salicylaldehyde

SCHEME 2

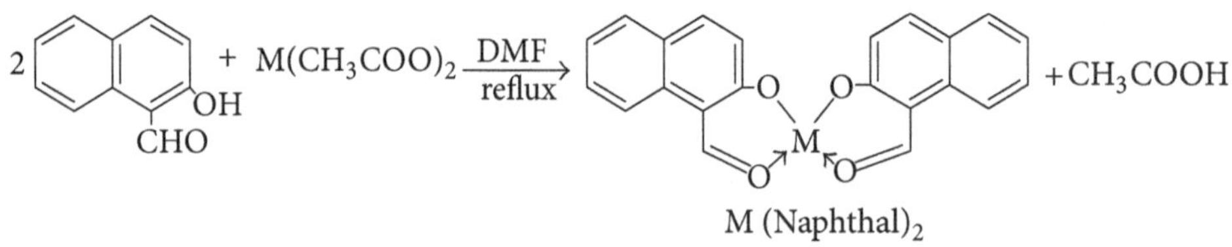

where M = Co(II), Cu(II), or Ni(II) and
Naphthal = 2-hydroxy-1-naphthaldehyde

SCHEME 3

3.1. IR Spectral Studies. The structurally important infrared spectral bands for the amino terminated oligomeric ligands and their metal complexes are depicted in Table 2. The assignments are based on the available literature [12, 13]. In the spectrum of PTTA ligand, there is a sharp and strong band at 3305 cm^{-1} attributed to C=O of amide group of polyamide chain. The band at 1630 cm^{-1} is attributed to carbonyl frequency of amide I. The absorption due to NH deformation appears at 1545 cm^{-1} or amide II. Another band due to C–N appears at 1285 cm^{-1}; p-substituted benzene ring vibration appears at 855 cm^{-1}. A band at 1470 cm^{-1} is assigned to-CH_2 bending mode of polyamide chain of PTTA.

The spectra of all the complexes show a strong and sharp band at 3305 cm^{-1} due to the NH of polyamide chain of PTTA. Appearance of bands in the region 1550–1555 cm^{-1}, as a result of splitting of amide or NH deformation in the spectra of all the complexes, indicates involvement of amino group in the complexation. Therefore, one band is ascribed to C=N–M while the other may be due to amide or NH deformation of amide group of PTTA ligand.

The appearance of new bands in the lower frequency region due to M–O at 560–520 cm^{-1} and due to M–N in the region of 480–420 cm^{-1} further supports and confirms the involvement of nitrogen and oxygen atoms in complex formation.

The other bands of the ligands appear uninfluenced upon coordination suggesting noninvolvement of amide groups of polyamide in coordination. The characteristic absorptions due to acetate group could not be observed in the IR spectra of metal complexes thereby suggesting absence of acetate ions. This observation supports the suggested formula without acetate ions on the basis of elemental analysis data. The appearance of new bands in the far IR region attributed to M–O and M–N further confirms the complexation through nitrogen and oxygen.

3.2. Magnetic Susceptibility Measurements. The diamagnetic behaviour of Ni(II) complexes indicates their square-planar geometry [14]. The square-planar geometry of Cu(II) complexes is indicated by the magnetic moment values in the range of 1.69–1.85 BM [15, 16].

Tetrahedral geometry of Co(II) complexes is inferred from the magnetic moment values of 4.59 and 4.44 for PTTA-Co(II)-Sal and PTTA-Co(II)-Naphthal, respectively. Diamagnetic behaviour of Ni(II) complexes indicates their square-planar behaviour. PTTA-Cu(II)-Sal and PTTA-Cu(II)-Naphthal complexes have magnetic moment values of 1.72 and 1.78, respectively, which were in the range predicted for square-planar complexes of Cu(II).

On the basis of all the aforementioned studies, a tentative structure as shown in Figure 1 can be proposed for the complexes. Magnetic moment values of amide-metal complexes are summarized in Table 3.

3.3. Thermogravimetric Analysis. Thermogravimetric data of polyamide ligand PTTA and its metal complexes are presented in Table 4. The TG curves of ligand PTTA and its complexes were compared in order to determine the influence of metal ion on the thermal stability. Thermal stabilities have been compared on the basis of initial decomposition temperature, residual percentage by weight at a particular temperature, and energy of activation (E) calculated by Fuoss method [17]. PTTA and its complexes with Ni(II)-Sal and Cu(II)-Sal show stepwise degradation whereas PTTA-Co(II)-Sal complex exhibits continuous degradation above 250°C, and the weight becomes constant above 550°C in all the cases. Residual percentage weights at 600°C for the complexes are

TABLE 2: IR spectral assignments of amides and their metal complexes.

Compounds	Assignments cm^{-1}					
	$^{\nu}$NH	Amide (I)	Amide (II)	$^{\nu}$C=N	$^{\nu}$M–O	$^{\nu}$M–N
PTTA	3305 s	1630 s	1545 s	—	—	—
PTTA-Co(II)-Sal′	3305 s	1630 s	1545 s	1550 s	560 s	480 m
PTTA-Co(II)-Naphthal′	3305 s	1630 s	1540 s	1550 s	560 s	485 w
PTTA-Ni(II)-Sal′	3305 s	1630 s	1540 s	1550 sh	540 s	485 s
PTTA-Ni(II)-Naphthal′	3305 s	1630 s	1535 s	1550 s	570 s	470 m
PTTA-Cu(II)-Sal′	3305 s	1630 s	1530 s	1555 s	520 s	490 s
PTTA-Cu(II)-Naphthal′	3305 s	1630 s	1535 s	1560 s	560 m	480 s

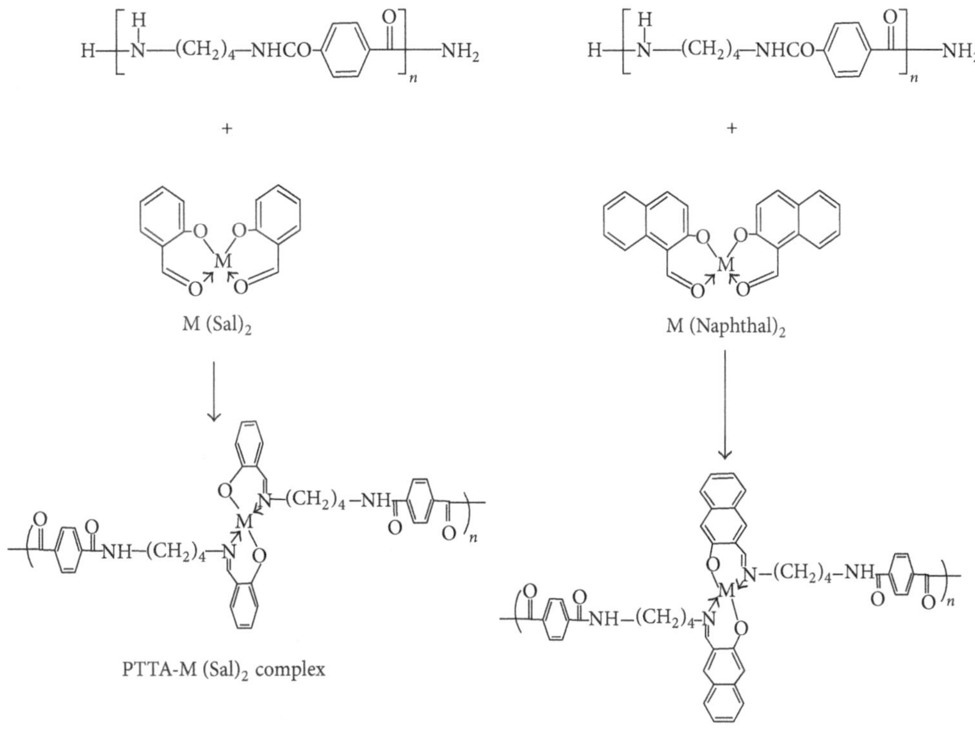

where M = Co(II), Cu(II), or Ni(II),
Sal = salicyladehyde,
and Naphthal = 2-hydroxy-1-naphthaldehyde

FIGURE 1

TABLE 3: Magnetic moment values for polyamide-metal complexes.

Compounds	Magnetic moment μ B.M.	Geometry proposed
PTTA-Co(II)-Sal′	4.59	Tetrahedral
PTTA-Co(II)-Naphthal′	4.44	Tetrahedral
PTTA-Ni(II)-Sal′	Diamagnetic	Square planar
PTTA-Ni(II)-Naphthal′	Diamagnetic	Square planar
PTTA-Cu(II)-Sal′	1.72	Square planar
PTTA-Cu(II)-Naphthal′	1.78	Square planar

slightly greater than the residual weight for the ligand. The order of stability on the basis of residual weights can be given as

$$\text{PTTA-Ni (II) -Sal}' > \text{PTTA-Co (II) -Sal}' > \text{PTTA-Cu (II) -Sal}' > \text{PTTA}. \quad (2)$$

On the basis of energy of activation calculated by Fuoss method, the following order of thermal stability was observed:

$$\text{PTTA-Cu (II) -Sal}' > \text{PTTA-Ni (II) -Sal}' > \text{PTTA-Co (II) -Sal}' > \text{PTTA}. \quad (3)$$

TABLE 4: TGA data of PTTA and its metal complexes.

Compounds	Temperature in °C							
	100	200	300	400	500	600	650	700
PTTA	1.0	5.0	10.0	50.0	80.0	98.0	—	—
PTTA-Co(II)-Sal′	0.0	0.0	0.0	19.0	72.0	92.0	93.0	94.0
PTTA-Co(II)-Naphthal′	2.0	3.0	7.0	25.0	91.0	92.0	92.0	92.0
PTTA-Ni(II)-Sal′	0.0	2.0	8.0	42.0	85.0	90.0	91.0	92.0
PTTA-Ni(II)-Naphthal′	4.0	7.0	10.0	22.0	69.0	76.5	78.0	79.0
PTTA-Cu(II)-Sal′	0.0	2.0	9.0	39.0	77.0	94.0	94.0	94.5
PTTA-Cu(II)-Naphthal′	0.0	7.0	13.2	37.5	78.0	90.0	94.0	—

The expected enhancement in the stabilities could not be obtained in the present case contrary to our previous observation [18, 19]. It may be due to the fact that sometimes the metal ion catalyses the degradation of the polymeric ligand. It is also possible that slight changes in polymeric environment influence the thermal degradation phenomenon to a large extent. The ligands which show two-step decomposition indicate the formation of intermediate products during thermal degradation process due to the rupture of polyamide chain. The degradation of amide takes place through the random scission of polymer chain into low molecular weight organic compounds and various volatile gaseous products like CO_2, NH_3, and water are evolved. Therefore it is expected that the different chain environments may be responsible for the different degradation mechanisms. Sazanov et al. [20] in their comparative thermal analysis (CTA) studies with thermally stable polyimides and model compounds have shown that the process of thermooxidation is of successive character; that is, the defective fragments are "burned out" before the main structure is oxidized.

4. Conclusion

Introduction of metal ion in polyamide chain through terminal groups generally results in increase in thermal stability of resulting polymer complexes. Further, this work can be extended to prepare similar kinds of polyamides, and, technique may also be used for increasing the thermal stability of fibres, films, sheets this and blocks of amino group containing polyamides.

Slight lowering in the initial decomposition of few complexes than the ligand PTTA may be due to the catalytic effect of metal ion on thermal degradation process.

Another application of these polyamide complexes may be as antimicrobial agents because of Schiff base moiety.

References

[1] U. D. N. Bajpai, S. Rai, and A. Bajpai, "Synthesis and characterization of metal-containing coordination polymers of poly (methylene diphenylene terephthalamide)," *Journal of Applied Polymer Science*, vol. 48, no. 7, pp. 1241–1248, 1993.

[2] U. D. N. Bajpai, A. Bajpai, and S. Rai, "Heat resistant coordination polymers based on amino group terminated oligomer of hexamethyleneadipamide," *Synthesis and Reactivity in Inorganic and Metal Organic Compounds*, vol. 24, no. 10, pp. 1719–1732, 1994.

[3] C. Selvi and D. Nartop, "Novel polymer anchored Cr(III) Schiff base complexes: synthesis, characterization and antimicrobial properties," *Spectrochimica Acta A*, vol. 95, pp. 165–171, 2012.

[4] N. Nishat, R. Rasool, S. Parveen, and S. A. Khan, "New antimicrobial agents: the synthesis of Schiff base polymers containing transition metals and their characterization and applications," *Journal of Applied Polymer Science*, vol. 122, no. 4, pp. 2756–2764, 2011.

[5] A. K. Kaura and M. Kaura, "Synthesis, spectral and comparative antimicrobial study of schiff bases," *International Journal of Chemical and Pharmaceutical Sciences*, vol. 3, no. 4, pp. 24–29, 2012.

[6] P. L. Maurya, B. V. Agarwala, and A. K. Dey, "Condensation polymerization of bis-(salicylaldehydo) metal(II) with diamines," *Polymer Bulletin*, vol. 3, no. 5, pp. 253–259, 1980.

[7] A. Prakash and D. Adhikari, "Application of Schiff bases and their metal complexes-a review," *International Journal of ChemTech Research*, vol. 3, no. 4, pp. 1891–1896, 2011.

[8] S. Kumar, D. N. Dhar, and P. N. Saxena, "Applications of metal complexes of Schiff bases-a review," *Journal of Scientific and Industrial Research*, vol. 68, no. 3, pp. 181–187, 2009.

[9] H. Aoki, D. R. Coffin, T. A. Hancock et al., "Synthesis, characterization, rheological, and fiber formation studies of p-linked aromatic polyamides," *Journal of Polymer Science Polymer Symposium*, no. 65, pp. 29–40, 1978.

[10] Y. P. Khanna, E. M. Pearce, B. D. Forman, and D. A. Bini, "Aromatic polyamides-1. Synthesis and characterization of some aromatic polyamides and their model diamides," *Journal of Polymer Science A*, vol. 19, no. 11, pp. 2799–2816, 1981.

[11] R. Puffer and J. Sebenda, "On the structure and properties of polyamides. XXVII. The mechanism of water sorption in polyamides," *Journal of Polymer Science C*, vol. 16, p. 79, 1977.

[12] J. Bellamy, *The Infrared Spectra of Complex Molecules*, John Wiley & Sons, New York, NY, USA, 1964.

[13] N. B. Colthup, L. H. Daly, and S. E. Wiberly, *Introduction To Infrared and Raman Spectroscopy*, Academic Press, New York, NY, USA, 1969.

[14] E. K. Barefield, D. H. Busch, and S. M. Nelson, "Cobalt and nickel complexes having anomalous magnetic moments," *Quarterly Reviews*, vol. 22, p. 457, 1968.

[15] H. B. Gray and C. J. Balhaueen, "A molecular orbital theory for square planar metal complexes," *Journal American Chemical Society*, vol. 85, p. 260, 1963.

[16] W. E. Hatfield, Y. Muto, H. B. Jonassen, and J. S. Paschal, "Substituted pyridine N-oxide complexes of copper(II) halides," *Inorganic Chemistry*, vol. 4, no. 1, pp. 97–99, 1965.

[17] R. M. Fuoss, I. O. Salyer, and H. S. Wilson, "Evaluation of rate constants from thermogravimetric data," *Journal of Polymer Science*, vol. A2, p. 3147, 1962.

[18] U. D. N. Bajpai, A. Bajpai, and M. Khandwe, "Synthesis and characterization of poly(thiooxamide) metal complexes," in *Metal Containing Polymeric Materials*, C. U. Pittman, Ed., Plenum Publishing Corporation, New York, NY, USA, 1996.

[19] J. H. Hodgkin and H. Heller, "Synthesis of conjugated Schiff-base polymers," *Journal of Polymer Science C*, vol. 29, pp. 37–46, 1970.

[20] Y. N. Sazanov, L. A. Shibaev, and T. A. Antonova, "Comparative thermal analysis (CTA) of thermally-stable polymers and model compounds. Polyimides and model compounds," *Journal of Thermal Analysis*, vol. 18, no. 1, pp. 65–75, 1980.

3

Rheological and Mechanical Characterization of Renewable Resource Based High Molecular Weight PLA Nanocomposites

P. J. Jandas, S. Mohanty, and S. K. Nayak

Laboratory for Advanced Research in Polymeric Materials (LARPM), Central Institute of Plastics Engineering and Technology (CIPET), Bhubaneswar, Orissa 751024, India

Correspondence should be addressed to S. K. Nayak; papers.journal@gmail.com

Academic Editor: Cornelia Vasile

The present study discusses structural aspects of nanocomposites and the ability of layered nanosilicates to alter the flow behaviour of poly(lactic acid) (PLA) melts. In addition, dynamic and static mechanical properties of PLA nanocomposites prepared from melt mixing method have been also discussed. A comparative study of nanocomposite properties has been conducted using two different nanoclays, natural montmorillonite modified with alkyl ammonium surfactant (OMMT), and commercially available organosilicate, Cloisite 30B, as reinforcements within the PLA matrix. Since OMMT has undergone better intercalation within the matrix, the corresponding nanocomposite showed superior mechanical and rheological characteristics than its C30B counterpart.

1. Introduction

Environmental impact generated by accumulating petroleum based plastic wastes is one of the most pondering topics among the researchers and environmentalists today. In this view, renewable resource based completely biodegradable aliphatic polyesters like poly(lactic acid) (PLA) and poly (hydroxyalkanoates) (PHA), carbohydrate based materials like starch and cellulose are under the scrutiny as apt replacement for petroleum based plastics. But still processability and cost are major drawbacks of these materials to clinch their niche in the market for various specific applications including "disposable materials." Researchers like Ray et al. reported about the improvements in processability and performance characteristics of PLA by reinforcing it with layered nanosilicates [1]. Also, nanocomposites of PHAs, starch, and petroleum based biodegradable materials like poly(butylene adipate-co-terephthalate) (PBAT) and poly(caprolactone) (PCL) with improved processability; mechanical and thermal characteristics also have been reported by various researchers. Among all, PLA is highly crystalline in nature with inherent biodegradability as well as enough life to maintain mechanical properties without rapid hydrolysis. As per the literature, mechanical and thermal properties of PLA show considerable improvement as a function of extent of intercalation/exfoliation of nanoclay layers within the matrix [2–5].

However, the relationship between structural features and rheological characteristics of PLA melts in its nanocomposites still demands detailed analysis through fundamental point of view to better understand the materials processability under a given set of parameters. In this view, the current study discusses the structural aspects of nanocomposites and the ability of layered nanosilicates to alter the flow behaviour of PLA melts. Further characterizations have been conducted through dynamic and static mechanical analysis. Commercially available organosilicate, Cloisite 30B (C30B), has been used as nanoreinforcement in the study. In addition, natural montmorillonite has been modified organically using hexadecyl trimethyl ammonium bromide (HTAB) surfactant and used to prepare nanocomposite with PLA.

2. Materials

Poly(lactic acid) (PLA 4042 D) with a number average molecular weight (M_w) of 120,000 g/mol was obtained from NatureWorks (USA) and used as base matrix. Melt flow index of PLA has been observed in between 4 and 8 g/10 min (190°C, 2.19 kg). Organically modified nanoclay, Cloisite 30B (C30B), and natural montmorillonite (NaMMT) were procured from Southern Clay Products (USA). Hexadecyl trimethyl ammonium bromide (HTAB) and all other laboratory reagents have been procured from Sigma Aldrich (Germany) and used without any modification.

3. Methods

3.1. Surface Modification of Natural Montmorillonite. In order to reduce the hydrophilicity of montmorillonite, cation exchange reaction was conducted using HTAB as follows. 10 g of montmorillonite was dispersed in 300 mL of distilled water using a mechanical stirrer under 2000 cycles/minute rpm for 3 hours followed by sonication for 1 hour. The well dispersed montmorillonite in distilled water was further mixed with surfactant (0.1 molar solution in distilled water) and 2 mL of concentrated HCl. The mixture was mechanically stirred again for 2 hours followed by centrifugation and filtration till the complete removal of bromide group from the filtrate. Residual bromide was detected using $AgNO_3$ test. Further, the organically modified nanoclay (OMMT) was dried at 80°C for 12 hours, cryomilled, and sieved.

3.2. Preparation of Nanocomposites. Nanocomposites have been prepared in a corotating twin-screw microcompounder (DSM Xplore 15, the Netherlands) attached with a mini-injection molder. The extruder is equipped with conical twin screw of length 150 mm with L/D ratio 18 and net capacity of 15 cc. Prior to compounding, PLA and nanoclays were predried at 80°C in a vacuum oven for 12 hrs and 4 hrs, respectively. Subsequently, the nanocomposites have been prepared using the corotating twin-screw miniextruder with various weight ratios of nanoclays. The processing temperature has been kept as 170, 175, and 180°C for the three successive zones of the extruder. Screw speed has been maintained at 40 rpm throughout the total mixing time of 4 minutes. The melt has been collected from the extruder in a preheated container and injection moulded according to the various ASTM standards as mentioned in the characterization section. Injection moulding parameters were fixed as mould temperature at 40°C, melt temperature at 170°C, injection time at 5 s, and injection pressure at 145 psi.

4. Characterization

4.1. Mechanical Testing. Tensile measurements of PLA and its nanocomposites were carried out in Universal Testing Machine (Instron, USA). Dumbbell shaped specimens of dimension 165 × 12.7 × 3 mm as per ASTM D 638 have been used for the testing. Gauge length was fixed at 50 mm with a cross head of 5 mm/min for conducting the test. Specimens of dimension 63.5 × 12.7 × 3 mm were used for impact measurements by means of an Impact tester (CEAST, Italy) as per ASTM D 256. The specimens have been notched at angle of 45° and depth of 2.54 mm using notch cutter (CEAST, Italy) prior to test.

4.2. WAXD Analysis. Wide Angle X-ray Diffraction (WAXD) analysis was performed for nanocomposites to study the extent of intercalation/exfoliation within the polymer matrix using Shimadzu X-ray diffractometer 7000L, Japan, (graphite monochromator Cu $K_{\infty 1}$ radiation with $\lambda = 0.15406$ nm) at a scanning rate of 0.5°/min within a range of 1° to 50°. The basal spacing of the silicate layer (d) was calculated using Bragg's equation, $\lambda = 2d \sin\theta$.

4.3. TEM Analysis. TEM images of blend nanocomposites have been taken in a transmission electron microscope (TEM 140, JEOL, Japan) operated at an accelerated voltage of 100 kV. The sample for imaging has been prepared by slicing it to ultrathin size of less than 100 nm using Cryo Leica EM UC6 instrument (Leica Microsystems, Switzerland) equipped with a diamond knife, and it was viewed without staining.

4.4. Dynamic Mechanical Analysis (DMA). Viscoelastic properties of PLA and its biocomposites were analysed using dynamic mechanical analyzer (DMA, Q800, TA Instruments, USA). Samples of dimension 63.5 × 12.7 × 3 mm were used for testing under a temperature range of 40 to 140°C and a fixed frequency of 1 Hz. The DMA results are expressed as storage modulus (E') corresponding to the elastic response to the deformation and damping factor ($\tan \partial$) values.

4.5. Rheological Characterization. Rheological behaviour of nanocomposites was investigated using parallel plate rheometer (Mars III, Thermo Fisher Scientific, Germany). To determine melt rheological parameters like storage modulus (G'), loss modulus (G''), and complex viscosity (η^*), dynamic frequency sweep experiment was performed within a range of 0.01 to 100 rad/s using 25 mm diameter plate. All measurements were performed at 180°C and under 1% strain.

5. Results and Discussion

5.1. Mechanical Properties of PLA Nanocomposites. Variation in mechanical properties of PLA and its nanocomposites as a function of clay loading and clay type is enumerated in Table 1. Incorporation of organoclay results consistent increase in tensile modulus with increased amount of filler content within the matrix. The tensile modulus of PLA increased to the tune of 5.43, 12.98, and 17.07% with 1, 3, and 5 wt% of nanoclay loading in case of PLA/C30B nanocomposites than that of V-PLA. This behavior reveals effective stress transfer from the C30B to the matrix as observed in case of other filled systems reported by Ray et al. [6]. Nanocomposites prepared using OMMT nanoclay have showed optimum tensile modulus with 4.08%, 17.09%, and 21.69% increment for 1, 3, and 5 wt% OMMT loading as compared with V-PLA, which is possibly due to good level of OMMT intercalation/exfoliation and effective interaction between clay layers and matrix polymer. Similarly, PLA/OMMT nanocomposites exhibited an increase in tensile strength of 18.68% as compared to that of the matrix polymer. This further confirmed that the intercalation/exfoliation expanded clay galleries of

Table 1: Mechanical properties of PLA and PLA nanocomposites.

Sample	Wt% of nanoclay	Tensile modulus (MPa)	Tensile strength (MPa)	Elongation at break	Impact strength (J/m)
V-PLA	0	3550 ± 50	38 ± 4	2.91	24.7
	1	3743 ± 60	29.3 ± 4	2.1 ± 1	25.3 ± 4
PLA/C30B	3	4011 ± 34	39.8 ± 7	2.4 ± 0.3	32.6 ± 3
	5	4156 ± 56	18.3 ± 4	0.9 ± 0.6	21.2 ± 6
	1	3695 ± 69	32.3 ± 6	2.8 ± 1	27.3 ± 4
PLA/OMMT	3	4157 ± 23	45.1 ± 4	2.3 ± 0.4	37.2 ± 3
	5	4320 ± 92	25.9 ± 6	1.5 ± 0.6	18.3 ± 5

OMMT as a result of HTAB modification. According to Yoshida and Okamoto, in case of well intercalated/exfoliated PLA nanocomposites, the matrix chains are strained by organoclay platelets due to their large interfacial area and stronger interactions [7]. Additionally, the cooperative macromolecular movement was suppressed by organoclay tethering and gallery confinement [8]. However, in all the cases, beyond 3 wt% of nanoclay loading, there was a drastic drop in the tensile strength which indicates agglomeration of the clay particles. This may induce microcracks at the interface of the nanocomposites. The elongation at break of V-PLA and its nanocomposites decreased with filler loading which is similar to the behaviour observed in case of all other filled systems.

Impact strength of PLA also showed increment with increased clay loading from 1 to 3 wt%, beyond which there was a decrease in the strength. PLA/C30B with 3 wt% of clay loading has been showed 31.98% higher values than that of V-PLA. This further confirms favourable interaction between the dihydroxyl group present in the surfactant of C30B and carbonyl groups of PLA at the interface. An optimum increase in impact strength of PLA matrix to the tune of 50.6% was observed in case of PLA/OMMT nanocomposites due to the better interaction between the silicate layers and PLA macromolecules due to the high level of intercalation/exfoliation process.

5.2. Comparative Study of Mechanical Modulus PLA Nanocomposites with Theoretical Values. Numerous micromechanical models [9] are available to predict the elastic constants of discontinuous particulate composites. These models generally depend on parameters including particle/matrix stiffness ratio E_p/E_m, particle volume fraction V_f, particle aspect ratio l/d, elasticity of the composite, and orientation. In the present study, in order to understand the reinforcing effects of organically modified nanosilicates within the PLA matrix, Halpin-Tsai (H-T) and Mori-Tanaka (M-T) micromechanical models were utilized to calculated theoretical tensile modulus values and compared with the experimental results obtained for PLA nanocomposites.

The current study mainly focuses on the prediction of longitudinal stiffness (E_c) of the nanocomposites reinforced with particulate nanofillers. Using H-T and M-T equations, exact range of aspect ratio and elasticity preserved by the nanoclay within the nanocomposite, respectively, have been calculated.

5.3. Comparative Study of Experimental and Theoretical Mechanical Modulus by H-T Model. Figures 1(a)-1(b) represent the experimentally and theoretically determined mechanical modulus of PLA nanocomposites using H-T equation as a function of different aspect ratio of the filler. As a basic assumption from the literature, aspect ratio has been assumed within a range of 10 to 100 [10–12]. Two extreme conditions of highest and lowest predicted values using parallel and series models are also depicted for reference purpose.

In case of PLA/C30B nanocomposite, the experimental value of elastic modulus shows similarities with the theoretically calculated values at lower volume fractions corresponding to an aspect ratio of 100. But at higher volume fractions of C30B, the difference between experimental and theoretical values increases further. However, the values observed corresponding to the other aspect ratios do not show much similarities with the experimental values except a cross over point noted at around $E_c/E_m = 1.07$ for the aspect ratio 75. In case of OMMT reinforced nanocomposites, the experimental values show almost an overlaid nature with values obtained from the H-T equation corresponding to an aspect ratio 75 as shown in Figure 1(b). The extent of deviation from the theoretical values is comparably lower in this case as compared with PLA/C30B nanocomposite.

The exact range of average aspect ratio of the nanoclays has been calculated through the study of variation of elastic modulus with aspect ratio under a fixed volume fraction as given in Figures 2(a)-2(b). Here, the volume fraction is kept at 0.015 for both the nanocomposites which was the optimized composition by tensile and impact properties. From the graph average aspect ratio of nanoclays was determined through the X intercepts of the experimentally determined modulus values. As per Figures 2(a) and 2(b), the average aspect ratio of C30B and OMMT was observed at around 61 and 79, respectively, corresponding to their experimentally determined nanocomposites moduli of 4011 and 4157 MPa.

Even though, the aspect ratio observed from TEM images of the nanocomposites is comparatively better than that of H-T equation suggested values for both C30B and OMMT. The values observed at around 90 and 105, respectively, for C30B and OMMT within the PLA matrix and corresponding elastic moduli values of the respective nanocomposites were observed at 4284 and 4412 MPa. The values have been determined by the extrapolation of graph towards Y intercepts of the figure. The values are around 6.8 and 6.1% higher than those of experimental values reported for PLA/C30B and PLA/OMMT, respectively. The observed variation in

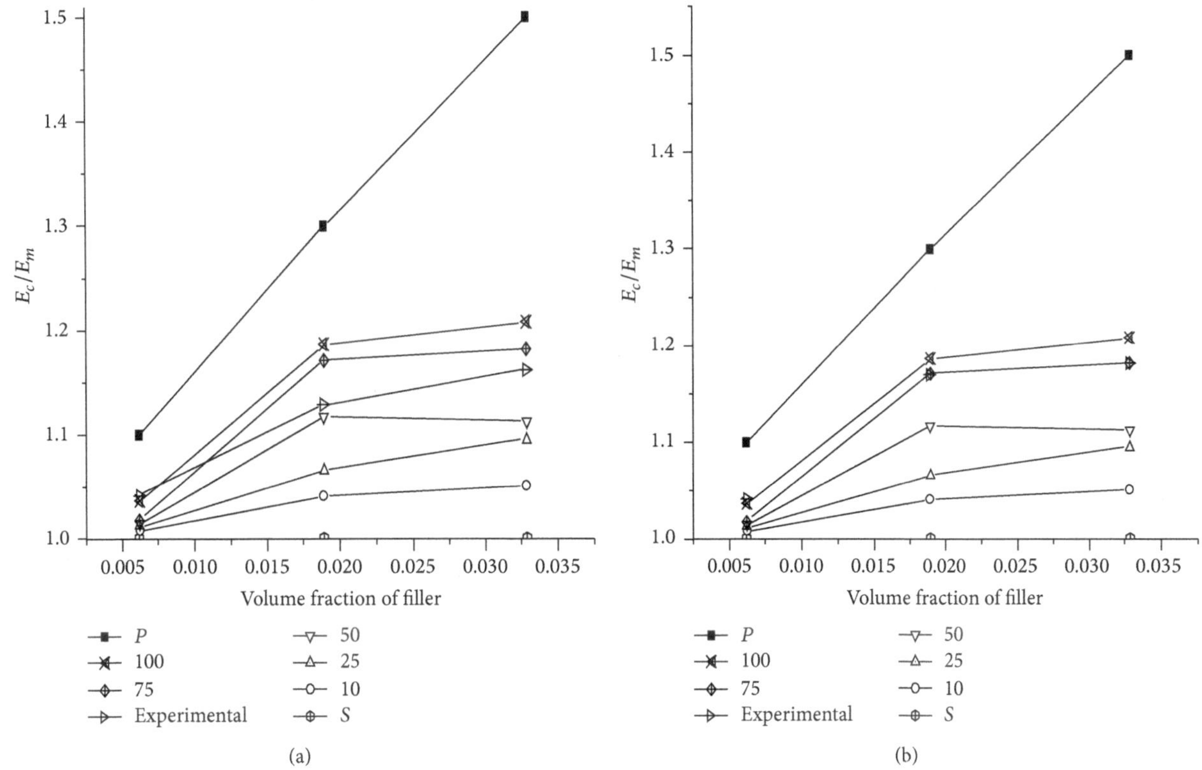

FIGURE 1: Comparative account of experimental mechanical modulus with H-T equation.

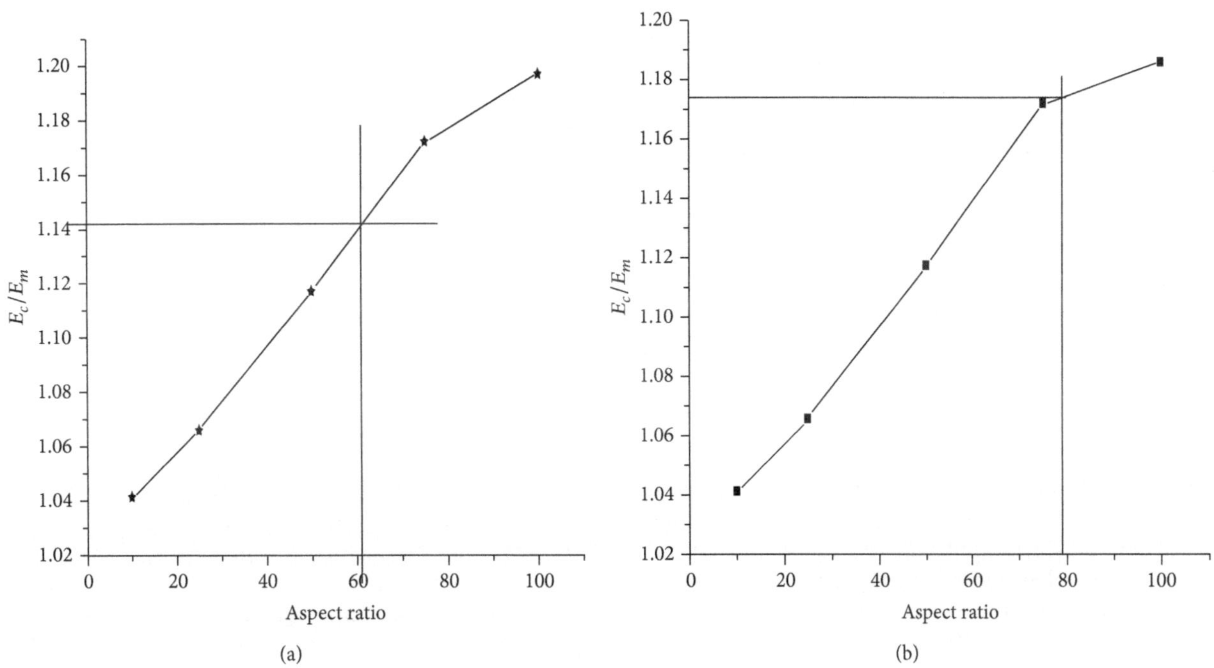

FIGURE 2: Predicted aspect ratio of nanoclays within PLA by H-T equation.

elastic modulus corresponding to the aspect ratios obtained from H-T model and TEM images can explain through the basic assumptions used for the derivation of H-T equation. There are a number of assumptions inherent in the H-T equation: (i) properties of both the matrix PLA and the nanoclays within the nanocomposites are identical to those of the pure materials; (ii) matrix and filler lamellae are linear elastic, isotropic, and perfectly bonded; (iii) the filler lamellae are perfectly aligned, asymmetric, and uniform in shape and size. The above assumptions are very much

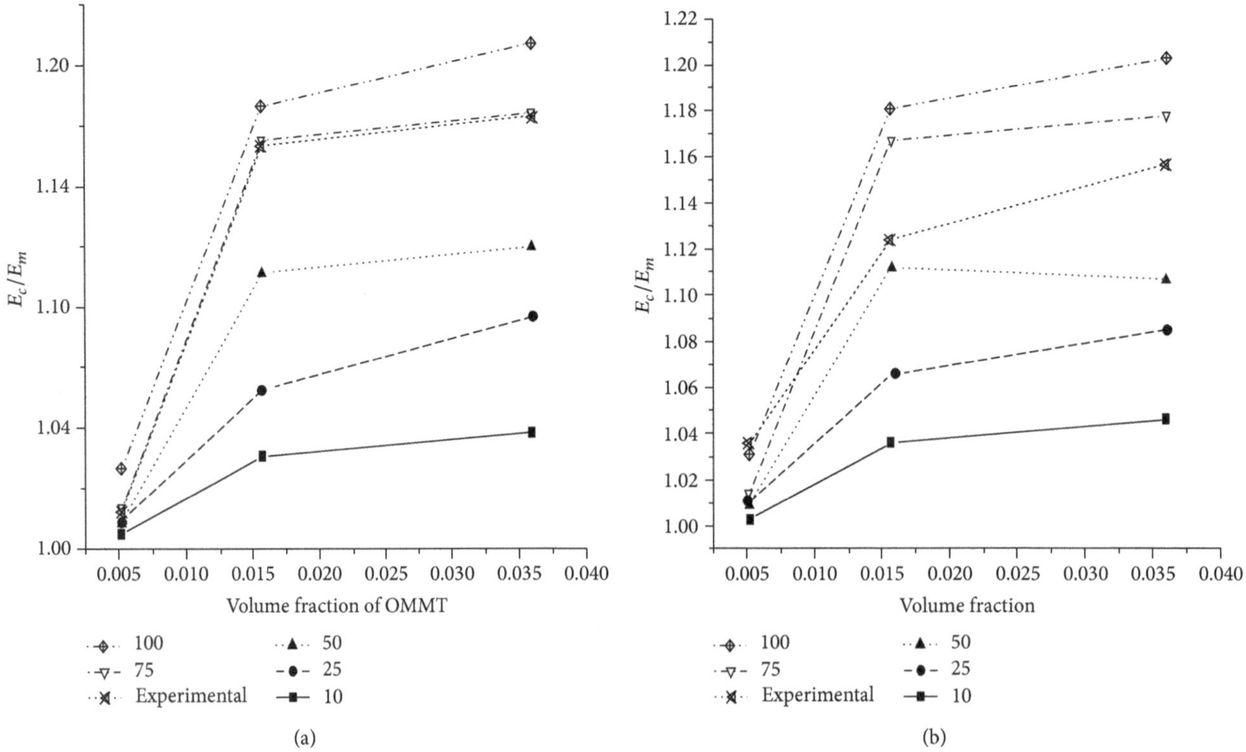

FIGURE 3: Comparative account of experimental mechanical modulus with M-T equation.

close to an ideal composite behaviour, rather than practical composite nature. These disparities in between ideal and practical situations must create the variation in predicted and real (experimental) behaviour of complex systems such as polymer nanocomposites. Mainly the second array of assumptions like linear elastic and isotropic nature of the filler and a perfect interface between matrix and filler may play vital roles in the present study. The linear elastic and isotropic nature for the filler is a hypothetical condition which is practically unachievable properties as well as the perfect matrix/filler interface also. Even though, there is an enhanced range of interaction in between PLA and layered silicate due to the organic modification of the silicate layer, it is difficult to achieve a perfect distribution of bonds in between the matrix and the intercalated/exfoliated nanoclays. As a result even at higher aspect ratios of silicate layers within the PLA matrix, the resulted nanocomposites cannot show perfect interface which results in lower elastic modulus than the expected or calculated values. However, some deviation from the experimental results has been observed, H-T model is still useful for the prediction of mechanical properties of nanocomposites. Experimental results suggesting that using exact range of aspect ratio of the filler materials, efficiency of H-T equation can improve significantly.

5.4. Comparative Study of Experimental and Theoretical Mechanical Modulus by M-T Model. Since the elasticity of nanocomposites affects significantly the predicted values of mechanical properties by H-T model, M-T model is used in the study to estimate the extent of variation in elasticity by the nanocomposite preparation. Figures 3(a), and 3(b) represent the comparative study of elastic modulus of PLA nanocomposites as a function of volume fraction and aspect ratio of the nanoclays. Comparing with the H-T model, both the models show the same expected trends for the prediction of elastic modulus. In the present study, the predicted E_c increases almost linearly with increasing V_f and aspect ratio of the C30B and OMMT nanoclays. According to Kalaprasad et al. [13], at very low values of V_f ($V_f \ll 1$) both the equations can be linearized in the form,

$$\frac{E_c}{E_m} = 1 + BV_f, \quad (1)$$

where B is a positive value (assuming $E_f/E_m > 1$), depending on filler aspect ratio and elastic properties of the nanocomposites. Study of variation in B as a function of different volume fractions and aspect ratio will give a vague idea about the dependency of both the factors on the elasticity of the nanocomposites.

Figures 4(a), and 4(b) represent the variation in B with respect to the increase in nanoclay volume fraction and aspect ratio. From the figures, it is evident that the elasticity of the nanocomposites increases with increase in volume fraction of nanoclays upto a limit. The observation is well supported by the viscoelastic study conducted in the DMA section where the storage modulus value of the nanocomposites shows considerably higher than that of V-PLA. The curve was showed an optimum value of elasticity, and further it was reduced with increasing volume fraction. Similarly, calculated elasticity also increased with increasing aspect

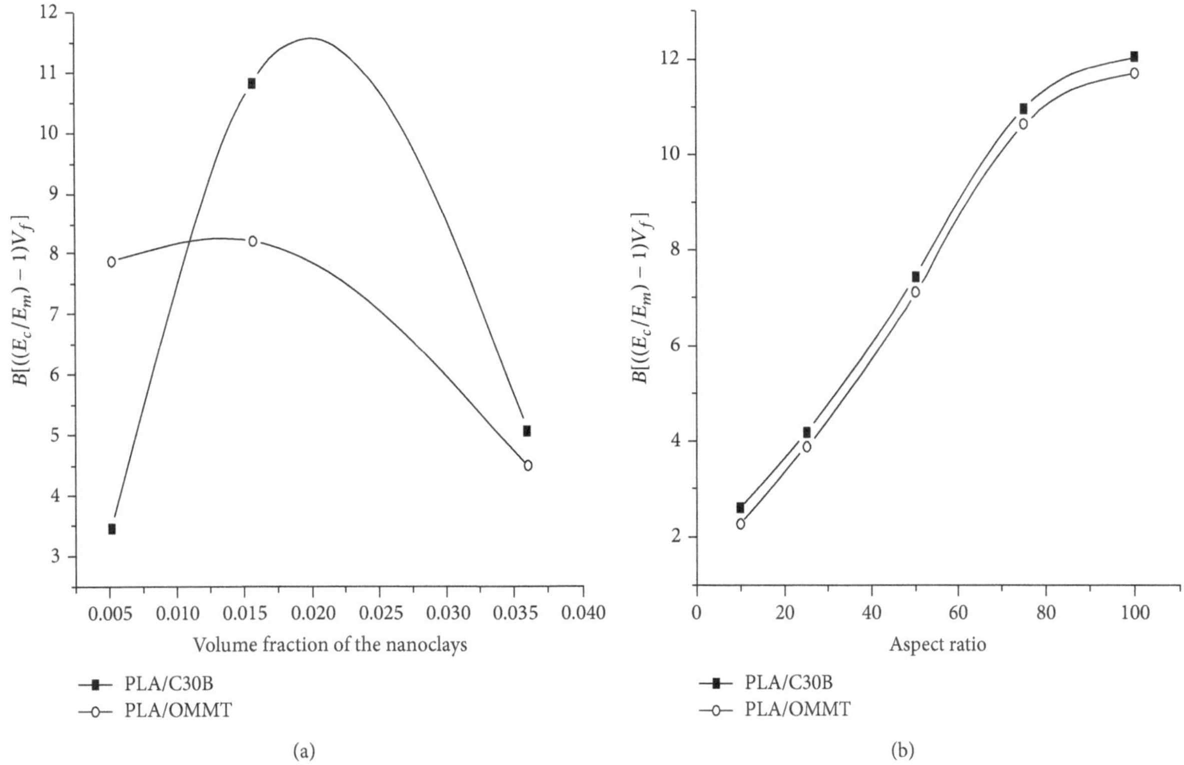

FIGURE 4: Elasticity of PLA using M-T equation.

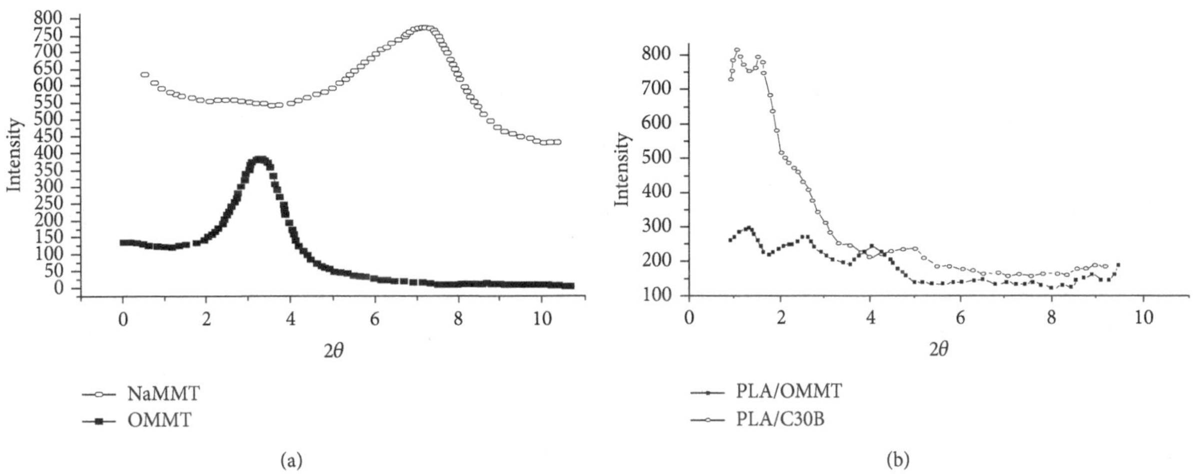

FIGURE 5: WAXD pattern for nanoclays and nanocomposites.

ratio of the nanoclay as shown in Figure 4(b). Better uniform stress transfer at higher aspect ratio of the filler may be the reason of above trend of elevation of elasticity.

5.5. Wide Angle X-Ray Diffraction Spectroscopy

5.5.1. WAXD Analysis of Natural Montmorillonite and OMMT. Interstitial spacing in the natural montmorillonite is expected to be expanded by HTAB modification. This is because the penetrated organic tail of HTAB in between the silicate layers can force the successive silicate layers to move apart. This phenomenon was confirmed through WAXD analysis according to the increase in basal spacing of montmorillonite upon HTAB modification. Figure 5(a) shows a comparative account of WAXD patterns of montmorillonite and OMMT, respectively. It is evident that OMMT shows a 2θ value of 3.3° corresponding to a d spacing of 2.66 nm as compared to natural montmorillonite which has a 2θ value of 7.2° with a d spacing of 1.33 nm. This increase in the d spacing confirms effective incorporation of HTAB chains on the natural montmorillonite by replacing Na^+ ion through cation exchange reaction.

Table 2: WAXD results of pristine nanoclays and that in nanocomposites.

Sample	2 Theta (degree)	d spacing (nm)	$2\theta^{\circ}$ in the nanocomposite	d spacing (nm) in the nanocomposite	D (nm)	$N = D/d_{001}$
NaMMT	7.2	1.33	—	—	—	—
C30B	4.78	1.85	1.2	5.3	35.5	4
OMMT	3.3	2.66	1.1	5.5	44.1	2

5.5.2. WAXD Analysis of PLA Nanocomposites. The WAXD patterns of the PLA nanocomposites of C30B and OMMT with 3 weight % loading are displayed in Figure 5(b). In case of PLA/C30B nanocomposite, multiple peaks were observed at low-angle range which confirms the intercalation of silicate layers of C30B within the PLA matrix after melt mixing. It is observable that there is a shifting of the diffraction peak of C30B to lower angle at $2\theta = 1.2^{\circ}$ (d-spacing ≈ 5.3 nm) in PLA/C30B nanocomposite from $2\theta = 4.78^{\circ}$ (d-spacing $\approx$ 1.85 nm) of pristine C30B. This indicates that the layer spacing of C30B was increased considerably, but its ordered structure was not disrupted after melt mixing. Another small peak is also observed in case of PLA/C30B due to the second registry of MMT clay. The intercalated pattern in PLA/C30B is primarily due to the hydrophilic nature of C30B clay and the formation of hydrogen-bond between the carbonyl group in the main chain of PLA molecules and the hydroxyl group in the organic modifier of C30B.

On other hand, WAXD pattern of OMMT within PLA/OMMT nanocomposite has showed a number of peaks with similar level of intensity counts, ranging from 2 to 5.4 nm. This indicates that the stacked layers of OMMT have been well intercalated/exfoliated by the polymer chains. Even though the WAXD pattern does not show a specific high intense diffraction peak, the (001) and its higher order diffractions, for example, (002) and (003), some weak crests have been identified at 1.180, 1.380, and 2.260. The corresponding increased basal spacing in the nanocomposite in comparison with the d spacing of OMMT has been recorded at 5.5 nm, 5.3 nm, and 4.41 nm, respectively.

Width of the WAXD peak, β (measured by the full-width at half maximum), is inversely proportional to the coherence length of the scattering entities and therefore reflects the coherence order of the silicate layers. This can be calculated from Scherrer equation [14] as follows:

$$D = \frac{k\lambda}{\beta \cos\theta}, \quad (2)$$

where k is a constant, generally equal to 0.9, λ is the X-ray wavelength (=0.154 nm), β is the width of WAXD peak in radian unit of full width at half maximum, and θ is the WAXD peak position. As the nanocomposite shows intercalated peaks, the coherency of the nanoclays may be significantly decreased after the melt mixing. This can be studied in more detail by calculating the number of stacked individual silicate layers in the nanocomposites, by taking the ratio of D and $d(001)$, and the data corresponding to this are displayed in Table 2. As observed from the test results reported in the table, the number of clay layers per stacks comes in a range of 35–40 and 4–6, respectively, for d and $D/d(001)$. Vaia et al. [15, 16] reported key features of the morphology of intercalated PP/clay nanocomposites using WAXD and TEM. According to them, in case of intercalated and exfoliated/intercalated systems, the particle separation is about 20–50 nm, and this is of the same order of crystal lamellae thickness. On the other hand, Lewitus demonstrated that in an exfoliated/intercalated system of PLA nanocomposites, a single organoclay comprises of ~4 layers and an average interlayer spacing of $d(001)$ ~2.4 nm [17].

5.5.3. Transmission Electron Microscopy (TEM). The internal structure of the PLA nanocomposites in the nanometre scale has been investigated from the TEM micrographs. Figures 6(a) and 6(b) show the TEM micrographs of the PLA/C30B and PLA/OMMT, respectively. The dark entities indicate cross section of intercalated or stacked clay layers, which have and an original thickness of around 1 nm, average length of 70–100 nm, and interlayer distance of 100–120 nm. The stacked silicate layers are due to clustering or agglomeration. The bright fields in the TEM micrograph represent the matrix. The nanocomposites show both intercalated and exfoliated regions. OMMT shows better exfoliation or dispersion of clay as compared to C30B within the PLA matrix. Further, few regions showing flocculated as well as intercalated clay galleries were noticed in case of PLA/C30B nanocomposites.

5.6. Dynamic Mechanical Analysis of PLA and Its Nanocomposites. Figures 7(a) and 7(b) represent the storage modulus (E') and $\tan\delta$ of V-PLA and its nanocomposites prepared from C30B and OMMT with 3 wt% loading, as a function of temperature. As evident from the figures, three different regions could be identified; the glassy state (below 60°C), the glass transition (approximately 60–80°C), and the rubbery plateau (approximately between 80–90°C). V-PLA exhibited a glass transition of 73°C, obtained from the $\tan\delta$ curve. Beyond this transition, the virgin matrix become soft which is evident from the reduction in E' by more than 1 order of magnitude. Finally, at 80–90°C in the rubbery region, the matrix polymer became extremely soft.

Further, it is observed that, the presence of organoclays had a strong influence on the viscoelastic response of the resulting nanocomposites in the entire investigated range of temperature. In the glassy region, storage modulus E' (Figure 7(a)), increased significantly with the incorporation of organically modified nanoclays which indicates the reinforcing effect imparted by the nanosilicate layers. The nanocomposite prepared using C30B nanoclay exhibited higher magnitude of E' as compared with the virgin matrix. This behaviour indicates strong inter/intramolecular hydrogen bonded network between the OH groups of methyl tallow

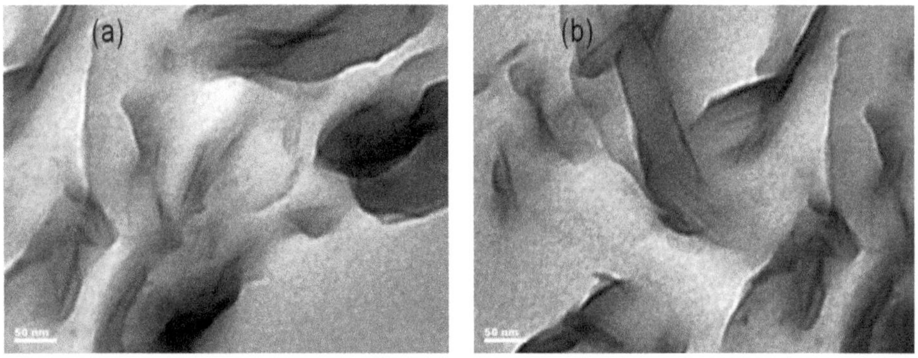

(a) (b)

FIGURE 6: TEM of nanocomposites.

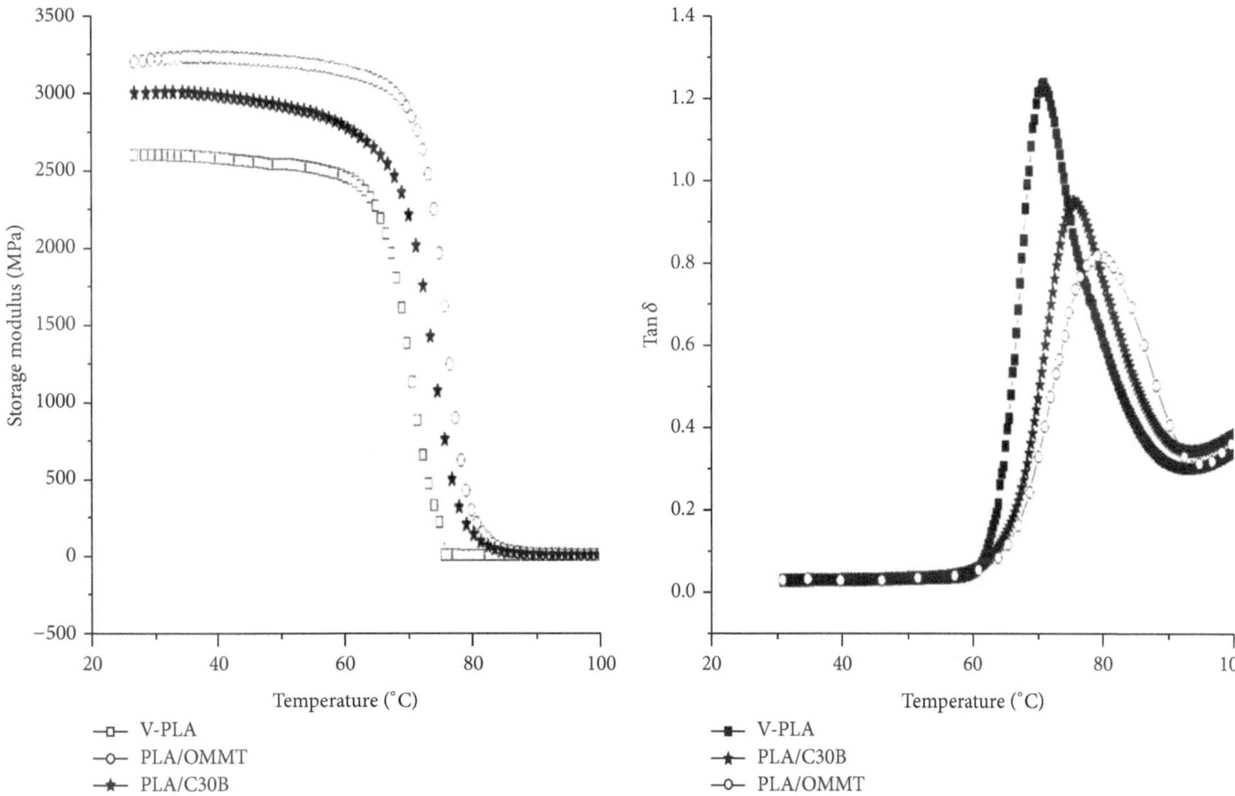

(a) (b)

FIGURE 7: DMA diagrams of PLA and its nanocomposites.

bis-hydroxyethyl intercalant of C30B and the C=O group of PLA matrix.

Near the glass transition region of PLA, there was a sudden drop in the matrix modulus. However, with the addition of nanoclays, the drop in modulus is compensated with the presence of fillers, due to the segmental immobilization of matrix chains leading to an increase in E' and decrease in $\tan\delta$ intensity (Figure 7(b)). Further, it is also noticed that the T_g of PLA increased to 81.53°C in case of PLA/C30B nanocomposite. This phenomenon is attributed to improved compatibility of the nanoclay with the bulk PLA matrix. PLA/OMMT based nanocomposite system displayed T_g at 83.21°C which is marginally less than PLA/C30B probably due to the plasticizing effect of the clay; the cay has showed better impact and elongation characteristics in the static mechanical analysis.

In the rubbery region, an important reinforcing effect was observed due to the presence of the nanoclays. As evident from the $\tan\delta$ curves, a maximum in the relaxation beyond 79°C was observed in the filled systems. V-PLA revealed a relaxation maximum around 75°C which subsequently increased in both nanocomposites. The nanocomposites prepared using OMMT and C30B nanoclay with 3 weight % filler loading showed an optimum increase in the relaxation

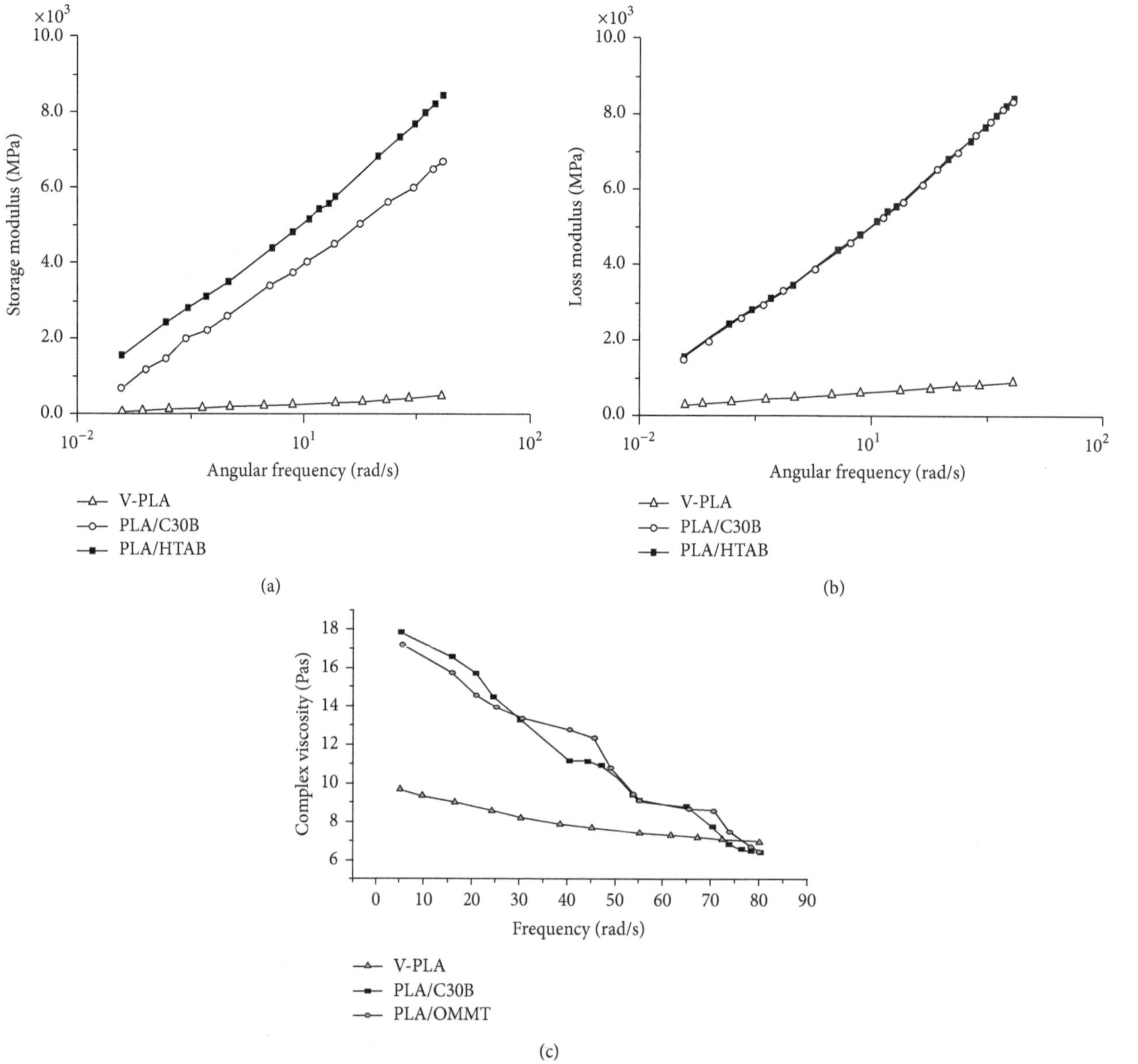

FIGURE 8: Rheological diagrams of PLA and its nanocomposites.

maximum to 80–82°C thus indicating improved dispersion characteristics of the organically modified clays due to the larger surface area between the reinforcement phase and the polymer matrix.

5.7. Rheological Properties of PLA and Its Nanocomposites. The variation of dynamic storage modulus (G'), dynamic loss modulus, and complex viscosity of virgin PLA and its optimised nanocomposites is depicted in Figures 8(a), 8(b), and 8(c), respectively. As observed from Figure 8, the nanocomposites exhibited an increase in the G' and G'' of PLA, within the experimentally investigated frequency range. This indicates slow transition from the classical liquid like characteristics to pseudosolid nature. Even though, the transition nature has observed the whole analysis region $G' < G''$, indicating the nanocomposites also does not come out of classical liquid-like behaviour of the molten V-PLA. Higher G' and G'' for both the nanocomposites compared with the virgin matrix reveals the reinforcing effect of nanoclays. In case of the matrix polymer, the G' and G'' varied from 200 to 550 Pa and 400 to 850 Pa, respectively, from low to high frequency regions. However, the nanocomposites showed higher magnitude of G' and G'' from 900 to 7000 Pa and 1500 to 8500 Pa for the nanocomposites. This behaviour is primarily due to the increase in the rigidity of network structure in the matrix with the incorporation of nanoclay.

As discussed in the earlier sections in DMA studies, the crystalline domain of the PLA has been well improved due to the nucleating effect of the intercalated/exfoliated nanoclays. Similar facts of increased stiffness have been corroborated with the increased rheological properties (G' and G'') of PLA in the nanocomposites. Ray and coworkers described similar behaviour of increment in melt state rheological properties of polyester nanocomposites. They found like other properties, rheological properties of nanocomposites are also directly related to the degree of dispersion of the silicate layers in the matrix polymer and level of interfacial interactions between the layered silicate surface and polymer chains [18].

In accordance with the time temperature superposition (TTS) concept, at high frequency region ($a_T\omega > 10$) and low frequency region ($a_T\omega < 10$) both the G' and G'' displayed similar characteristics of $G' < G''$, and even though the cross over point could not be located within the experimental range, the difference in G' and G'' considerably decreased with the increase in frequency. This indicates that the frequency independent liquid-like ($G'\alpha\omega^2$ and $G''\alpha\omega$) behavior slowly transmitted to "pseudo-solid like" behavior in case of both V-PLA and the nanocomposites. Intensity of this phenomenon was higher in case of nanocomposites than that of V-PLA. This may be due to the increased melt strength by the reinforcing effect of the nanoclays exfoliate/intercalated within the PLA matrix. Also the slopes of G' and G'' in the terminal region of the master curves of PLA matrix were 1.43 and 0.87, respectively. On other hand, the slopes of G' and G'' for PLA/C30B nanocomposites were observed at 0.62 and 0.55. This further confirms transition from "liquid like" to "pseudo-solid like" behaviour at lower to higher angular frequency for V-PLA as well as nanocomposites.

The variation of complex viscosity (η^*) as a function of angular frequency (ω) of V-PLA and its nanocomposites with 3 weight % filler loading is depicted in Figure 8(c). It is evident that both the virgin polymer as well as the nanocomposites exhibited transition to non-Newtonian pseudoplastic behaviour within the studied deformation region of 0.01 rad/sec to 100 rad/sec. As observed from the figures, the viscosity decreases consistently with the increase in angular frequency, thus revealing typical characteristics of a completely frequency dependant nature. Both PLA and the nanocomposites exhibited a shear thinning effect unlike other thermoplastic matrices [19]. But the effect is more pronounced in the case of nanocomposite as compared with that of the virgin matrix. According to Pluta et al., the real part of the complex viscosity is an energy dissipation term similar to the imaginary part of the complex modulus [20].

The magnitude of η^* in case of V-PLA matrix was observed in the range of 10 to 7 Pa s which increased considerably in the nanocomposites from 20 to 7 Pa s. This increase in the η^* over the entire experimentally investigated frequency region reflects the reinforcement effect of the nanoclay within the molten polymer matrix. Iannace and Nicolais reported such kind of improvement in η^* of PLA nanocomposite than that of the virgin matrix [21]. According to the author, the reinforcing effect results from the interaction between the components due to hydrogen bonding of carbonyl groups of the PLA matrix and hydroxylated organic "surfactant" ends of the nanoclay that opposes the applied external shear effectively.

According to Ray and Okamoto [19], the increment in complex viscosity of the PLA nanocomposite was observed in comparison with the virgin matrix. The author attributed this behaviour due to the presence of stacks of intercalated/exfoliated silicate layers that are randomly oriented in the polymer matrix, forming a 3D network structure. These intercalated/exfoliated layers have only translational motion. A large amplitude oscillatory shear is able to orient these structures thereby increasing the coefficient of viscosity. Another explanation for such kind of behaviour of improved η^* is probably due to the phenomenon of physical jamming of the dispersed silicate layers owing to their highly anisotropic nature. On the basis of intercalated/exfoliated mesoscopic pattern observed for the nanocomposites in the present investigation, the individual silicate layers are incapable of free rotation and when subjected to shear, they are prevented from a completely relaxed state. This incomplete relaxation due to the physical jamming or percolation of the nanoscopic fillers leads to the pseudosolid like behaviour for nanocomposites with intercalated/exfoliated structures.

6. Conclusion

Exfoliated/intercalated PLA nanocomposites based on two different organomodified natural montmorillonites have been successfully prepared by melt mixing method. WAXD and TEM studies confirmed exfoliated/intercalated pattern of both the nanoclays within the PLA matrix. Improved mechanical properties as tensile and impact strength point the reinforcing effect of clay layers within PLA. High level of exfoliated OMMT within the matrix tends to enhance the mechanical properties of corresponding nanocomposite, whereas polar hydroxylated modifier end of C30B has compatibilized the PLA/C30B nanocomposite. Similar with the static mechanical analysis, dynamic mechanical analysis also revealed reinforcing effect of both the nanoclays with 3 weight % loading. Formation of delayed rubbery plateau for nanocomposites (in the range of 81–83°C) in comparison with V-PLA (≈72°C) was observed during DMA analysis. On the other hand, the increased processability has been justified by the observed intense transition from "liquid like" to "pseudoplastic" nature in case of nanocomposites than V-PLA during rheological analysis. Even though both the nanocomposites showed properties improvement, PLA/OMMT showed superior characteristics than that of PLA/C30B in terms of mechanical and rheological properties.

References

[1] S. S. Ray, P. Maiti, M. Okamoto, K. Yamada, and K. Ueda, "New polylactide/layered silicate nanocomposites. 1. Preparation, characterization, and properties," *Macromolecules*, vol. 35, no. 8, pp. 3104–3110, 2002.

[2] J. Y. Nam, S. S. Ray, and M. Okamoto, "Crystallization behavior and morphology of biodegradable polylactide/layered silicate nanocomposite," *Macromolecules*, vol. 36, no. 19, pp. 7126–7131, 2003.

[3] K. Ueda, S. S. Ray, K. Yamada, and M. Okamoto, "New polylactide-layered silicate nanocomposites. 2. Concurrent improvements of material properties, biodegradability and melt rheology," *Polymer*, vol. 44, no. 3, pp. 857–866, 2002.

[4] S. S. Ray, K. Yamada, M. Okamoto, A. Ogami, and K. Ueda, "New polylactide/layered silicate nanocomposites. 3. High-performance biodegradable materials," *Chemistry of Materials*, vol. 15, no. 7, pp. 1456–1465, 2003.

[5] R. Krishnamoorti, R. A. Vaia, and E. P. Giannelis, "Structure and dynamics of polymer-layered silicate nanocomposites," *Chemistry of Materials*, vol. 8, no. 8, pp. 1728–1734, 1996.

[6] S. S. Ray, K. Yamada, M. Okamoto, and K. Ueda, "Biodegradable polylactide/montmorillonite nanocomposites," *Journal of Nanoscience and Nanotechnology*, vol. 3, no. 6, pp. 503–510, 2003.

[7] O. Yoshida and M. Okamoto, "Direct melt intercalation of polylactide chains into nano-galleries: interlayer expansion and nano-composite structure," *Macromolecular Rapid Communications*, vol. 27, no. 10, pp. 751–757, 2006.

[8] S. S. Ray, K. Yamada, M. Okamoto, A. Ogami, and K. Ueda, "New polylactide/layered silicate nanocomposites, 4. Structure, properties and biodegradability," *Composite Interfaces*, vol. 10, no. 4-5, pp. 435–450, 2003.

[9] A. Bouaziz, F. Zaïri, M. Naït-Abdelaziz, J. M. Gloaguen, and J. M. Lefebvre, "Micromechanical modelling and experimental investigation of random discontinuous glass fiber polymer-matrix composites," *Composites Science and Technology*, vol. 67, no. 15-16, pp. 3278–3285, 2007.

[10] K. C. Shive and Y. T. Guang, "Micromechanical modeling of material damping in discontinuous fiber three-phase polymer composites," *Composites Engineering*, vol. 1, no. 1, pp. 49–60, 1991.

[11] G. Kalaprasad, K. Joseph, S. Thomas, and C. Pavithran, "Theoretical modelling of tensile properties of short sisal fibre-reinforced low-density polyethylene composites," *Journal of Materials Science*, vol. 32, no. 16, pp. 4261–4267, 1997.

[12] G. Lin, X. Zhang, L. Liu, J. Zhang, Q. Chen, and L. Zhang, "Study on microstructure and mechanical properties relationship of short fibers/rubber foam composites," *European Polymer Journal*, vol. 40, no. 8, pp. 1733–1742, 2004.

[13] G. Kalaprasad, K. Joseph, S. Thomas, and C. Pavithran, "Theoretical modelling of tensile properties of short sisal fibre-reinforced low-density polyethylene composites," *Journal of Materials Science*, vol. 32, no. 16, pp. 4261–4267, 1997.

[14] J. R. Fried, *Polymer Science and Technology*, Prentice-Hall, Delhi, India, 2nd edition, 2005.

[15] R. A. Vaia, R. K. Teukolsky, and E. P. Giannelis, "Interlayer structure and molecular environment of alkylammonium layered silicates," *Chemistry of Materials*, vol. 6, no. 7, pp. 1017–1022, 1994.

[16] D. Lewitus, S. McCarthy, A. Ophir, and S. Kenig, "The effect of nanoclays on the properties of PLLA-modified polymers Part 1: mechanical and thermal properties," *Journal of Polymers and the Environment*, vol. 14, no. 2, pp. 171–177, 2006.

[17] M.-A. Paul, C. Delcourt, M. Alexandre, P. Degée, F. Monteverde, and P. Dubois, "Polylactide/montmorillonite nanocomposites: study of the hydrolytic degradation," *Polymer Degradation and Stability*, vol. 87, no. 3, pp. 535–542, 2005.

[18] S. S. Ray, K. Yamada, M. Okamoto, Y. Fujimoto, A. Ogami, and K. Ueda, "New polylactide/layered silicate nanocomposites. 5. Designing of materials with desired properties," *Polymer*, vol. 44, no. 21, pp. 6633–6646, 2003.

[19] S. S. Ray and M. Okamoto, "New polylactide/layered silicate nanocomposites, 6a melt rheology and foam processing," *Macromolecular Materials and Engineering*, vol. 288, no. 12, pp. 936–944, 2003.

[20] M. Pluta, A. Galeski, M. Alexandre, M.-A. Paul, and P. Dubois, "Polylactide/montmorillonite nanocomposites and microcomposites prepared by melt blending: structure and some physical properties," *Journal of Applied Polymer Science*, vol. 86, no. 6, pp. 1497–1506, 2002.

[21] S. Iannace and L. Nicolais, "Isothermal crystallization and chain mobility of poly(L-lactide)," *Journal of Applied Polymer Science*, vol. 64, no. 5, pp. 911–919, 1997.

4

A Preliminary Investigation of Ductility-Enhancement Mechanism through *In Situ* Nanofibrillation in Thermoplastic Matrix Composites

Bhaskar Patham[1] and M. P. Poornendu Thejaswini[2]

[1] *General Motors Technical Centre India Pvt. Ltd., Vehicle Performance Centers, Vehicle CAE Methods Group, International Tech Park Ltd., Whitefield Road, Bangalore 560 066, India*
[2] *General Motors Technical Centre India Pvt. Ltd., Vehicle Performance Centers, Materials and IMDS Group, International Tech Park Ltd., Whitefield Road, Bangalore 560 066, India*

Correspondence should be addressed to Bhaskar Patham; bhaskar.patham@gm.com

Academic Editor: Zhong-Ming Li

A preliminary investigation of interrelationships between tensile stress-strain characteristics and morphology evolution during deformation is conducted on a commercially available thermoplastic composite with a low-surface-energy nanofibrillating poly(tetrafluoroethylene) (PTFE) additive. In this class of composites, the deformation-associated nanofibrillation of the low-surface-energy additive has been hypothesized to provide an additional dissipation mechanism, thereby enhancing the ductility of the composite. This class of composites offers potential for automotive light weighting in exterior and interior body and fascia applications; it is therefore of interest to investigate processing-structure-property interrelationships in these materials. This study specifically probes the interrelationships between the plastic deformation within the matrix and the fibrillation of the low-surface-energy additive; tensile tests are carried out at two different temperatures which are chosen so as to facilitate and suppress plastic deformation within the matrix polymer. Based on these preliminary investigations, it is noted that PTFE fibrillation acts synergistically with the ductile deformation of the matrix resin resulting in higher strains to failure of the composite; the results also suggest that the mechanism of fibrillation-assisted enhancement of strains to failure may not operate in the absence of matrix plasticity.

1. Introduction

Thermoplastic matrix composites comprise an attractive class of materials holding high potential for light-weighting applications; these composites are typically designed to combine the ductility, toughness, and processability of thermoplastic matrices, and the stiffness, strength, thermo-mechanical, and other special functionalities of an additive material. Typical reinforcements that are added to polymeric matrices to enhance the thermo mechanical properties—for example, stiffness, strength, and heat-deflection temperature—include fibrous fillers (such as glass and carbon fibers), flakes and discs (such as talc), and nanofillers (such as nanoclay platelets). Typically, the addition of a stiffening reinforcement also leads to a compromise of the ductility of the composite; the resulting composite is rendered more brittle (with a lower strain to failure) compared to the unreinforced matrix. The loss of ductility relative to the unfilled polymer may be attributed to the rigid filler hindering certain degrees of freedom of the polymer to reconfigure its state upon being stressed.

While structural members fabricated from reinforced thermoplastics are typically overdesigned in terms of stiffness so as to avoid the onset of yield strains, in most of the service conditions involving semistructural or nonstructural applications of plastics, it would be more optimal to retain the ductility and toughness of the polymeric matrix while at the same time enhancing the stiffness and thermal service properties. It is highly desirable, but usually difficult, to simultaneously accomplish these opposing requirements (high stiffness, high strength, and high ductility) within the same material formulation.

To improve the strains-to-failure of the composite, it would be critical to retain—within the composite—one or more of the significant matrix deformation mechanisms which contribute to its ductility. An investigation of tensile deformation of polypropylenes carried out *in situ* with small-angle X-ray scattering measurements [1] has associated the ductile deformation of semicrystalline polymers with the dual mechanism of cavitation and stable shear of fibrillated spherulites. Clearly if analogous matrix deformation modes were to be preserved in the composite formulation, the resulting composite would display greater ductility, while improving in stiffness due to the rigid fillers.

By contrast, if compositing with rigid fillers results in the loss of these critical matrix deformation modes, then it would be appropriate to include an additive which *simulates* an analogous fibrillation behavior within the composite to increase its failure strains. One category of fibrillating inclusions that has been studied in this context is liquid crystalline polymers (LCP)—which have been shown to fibrillate during draw. In one implementation involving modification of a polycarbonate matrix with LCP [2], an increase of LCP crystallite size at higher draw ratios was seen to result in an increase of composite stiffness. Another category of additives, which has received increased attention in the context of ductility enhancement recently [3], and which is relevant to the current study, includes low-surface-energy fluoropolymers. For example, poly(tetrafluoroethylene) (PTFE) has a unique set of physical and chemical properties such as low coefficients of friction, thermal stability, chemical inertness, and high toughness; it is well researched for its low-surface-energy properties, for "nonstick" applications [4].

PTFE is known to show two low-temperature thermal transitions below its melting point—a first-order transition from triclinic to hexagonal crystalline form at around 19°C and another one at 30°C [5–7]. The melting peaks for PTFE range from about 320°C to about 350°C and are dependent on whether it is in nodular form or fibrillated form [3, 4, 8]. While PTFE powders sheared below 19°C typically show particle-flow characteristics, imposition of shear at temperatures typically above 30°C (but below the melting point of PTFE) has been shown to result in an unwinding of the packed spherulitic morphology leading to fibrillation [4, 9, 10]. More recent studies involving extrusion of PTFE pastes (e.g., refer to Ariawan et al. [9]) also revealed a similar evolution of PTFE fibrils from interconnected PTFE particles under shear.

When added to polymeric matrices during melt processing, the extent of fibrillation is seen to be a function of the strength of flow field and the residence time within the flow field. It has also been shown that PTFE nanofibrillation results in better melt processability, higher melt strength and drawability, and greater melt elasticity in composites with polypropylene or polycarbonate matrices [8, 11–14].

The addition of PTFE to other polymeric matrices with the aim of imparting higher ductility and increasing the strains to failure of the composite is a relatively recent development [3]. It has been hypothesized [3, 15] that the deformation-induced *in situ* nanofibrillation of PTFE provides an additional dissipative mechanism during deformation of the composite that results in improved ductility.

The promise of these systems is that the composites may now be formulated with the right combination of stiffness and ductility that is optimal to the component design and service conditions, rather than employing formulations which are overdesigned in terms of stiffness. The composition of such novel composites—involving a single- or multiphase polymeric matrix, a rigid structural filler, and a low surface energy PTFE additive, and the synergistic and antagonistic effects of these components in governing the composite performance—provides an interesting system for fundamental investigations of processing-structure-property interrelationships.

Objective. The overall focus of this research was to investigate the interrelationships between the global deformation behavior of this class of composites and its local microstructure evolution as qualified by the state of dispersion and distribution of the low surface energy PTFE additive. This study very specifically aimed to further the understanding on the interrelationships between the plastic deformation within the matrix and the fibrillation behavior; it sought to answer a particular question: does fibrillation (and the subsequent fibrillation-aided ductility enhancement of the composite) occur independently of matrix ductility? This investigation is essential to facilitate better design of composites with fibrillating additives for performance at extremely low temperatures, where the matrix might be rendered brittle. This understanding would also aid in better design of structural composites with high rigid-filler loading—in which the matrix may be prevented from experiencing yield (due to the overdesign for stiffness)—where the fibrillating filler may be added for impact modification. To start addressing this particular question, it was most important in this study to establish conditions which would effectively facilitate or suppress plastic deformation within the matrix. This was accomplished by carrying out tensile tests—on a commercially available thermoplastic matrix composite comprised of a polymer blend matrix reinforced with rigid talc fillers as well as PTFE additive—at two different temperatures: room temperature at which the matrix blend displays ductile behavior and −40°C at which both the components of the matrix blend are in the vicinity of their ductile-brittle transition temperatures and would be expected to display brittle behavior. Quasistatic tensile tests were conducted up to specimen failure; the structure of the fractured surfaces was then probed using scanning electron microscopy (SEM) and differential scanning calorimetry (DSC). Finally, qualitative correlations between the mechanical behavior of the composite and its structure were inferred.

2. Experimental

2.1. Material. A commercial grade of HMD-Xenoy was supplied by SABIC India Pvt. Ltd., Vadodara, India. This grade of

HMD (high modulus ductile) Xenoy is composed of a blend of polycarbonate (PC) and polybutylene terephthalate (PBT) as the matrix; it contains talc for enhancement of stiffness and strength of the matrix and PTFE for retaining and enhancing the ductility of the composite. The formulation and the exact composition of the composite, including the PTFE content, was not evaluated in this study as it is proprietary to SABIC. Further, it is important to note that this investigation did not specifically address the impact of the PTFE content on this mechanism; neither was this study aimed at evaluation of the performance of the material with respect to its specific intended use.

2.2. Tensile Specimen Preparation and Tensile Testing

2.2.1. Specimen Preparation. Tensile test specimens conforming to the ISO-527 standard (ISO 527 Type 1A for rigid sheets with a thickness specification of 4.0 ± 0.2 mm) [16, 17] were prepared by injection molding carried out at STEER Polymer Science and Diagnostic Centre, Bangalore, India, using a 100 Ton capacity L&T S Tech-100 machine. An end-gated ISO 527 Type 1A mold with an L&T LTM 200 mold temperature controller was employed. Prior to injection molding, the composite pellets were dried and dehumidified at 110°C for 5 hours in an L&T LMD 8 vacuum drier. The temperature settings in the various zones of the injection-molding machine were set as listed in Table 1. The injection-molding processing parameters employed for fabrication of the tensile specimens are listed in Table 2. Injection-molding of tensile specimens was carried out at an injection speed (measured as the linear velocity of the melt front within the mold during mold filling) of 20 mm/min, which translates to a nominal processing shear rate of 28 s^{-1}.

2.2.2. Tensile Testing. Tensile tests were carried out on an Instron Universal Testing Machine (UTM) Instron 5582, equipped with a retractable environmental chamber with a temperature range of −150°C to 600°C. Quasistatic tensile tests at a nominal initial strain rate of 10^{-5} s^{-1} were conducted at room temperature and at a low temperature of −40°C. Measurement of strains was conducted using a clip-on extensometer with an initial gage length of 1 inch mounted on the gage section of the tensile specimen (and centered with respect to the length of the specimen). Measurement of loads was conducted using a 100 kN load cell mounted on the crosshead. The load and displacement were recorded until specimen failure. The data recorded by the strain gage after the breakage of the specimens was discarded.

The tests at −40°C were carried out within the enclosure of the environmental chamber that was connected to a pressurized liquefied nitrogen tank. The temperature was maintained by automated control of the flow rate of liquefied nitrogen gas into the enclosure. Prior to the tests, the samples were placed between the test fixtures loosely and allowed to equilibrate at −40°C for about 5 min. The test fixtures were tightened, and the strain gage was mounted only after the sample had sufficiently equilibrated at this low temperature; this was done in order to avoid recording of spurious stresses

Table 1: Processing temperature of injection molding.

Zones	Temperature (°C)
Feed zone	50
MH1	250
MH2	260
MH4	270
Nozzle	265
Mold	70

Table 2: Injection-molding processing parameters.

Injection pressure (bar)	80
Injection speed (mm/sec)	20
Holding pressure (bar)	60
Holding time (sec)	5
Cooling time (sec)	5
Back pressure (Bar)	5
Refilling screw speed (%)	40% (100 rpm)

and strains due to confined shrinkage of the tensile specimen and the test fixture.

A few specimens were also subjected to high-rate tensile tests (employing nominal initial strain rates typically in the range of 10^{-1} s^{-1}) at −40°C. For these tests, however, the strain gage could not be employed due to the significant inertial effects involved; also, the stress-strain data obtained from these tests could not be attributed a high certainty. These tests were conducted mainly to subject the composite to faster deformation kinematics and thereby modify the conditions for deformation-induced microstructure evolution. The data from these tests are currently being analyzed and will be presented in a subsequent report.

An auxiliary question that this study intended to address was the impact of "as processed" microstructure of the undeformed specimens on their tensile stress-strain behavior and strains to failure. To investigate this aspect, in addition to injection molding at 20 mm/s (28 s^{-1}), tensile bar specimens were also injection molded at 50 mm/s and 80 mm/s—with the corresponding nominal processing strain rates of 70 s^{-1} and 112 s^{-1}. The premise here was that the initial state of dispersion and distribution of the low surface energy fluoropolymer additive within the undeformed specimen (which in turn would be brought about by the processing conditions employed during injection molding) would have a bearing on the fibrillation behavior during tensile deformation. This premise was based on the reports of fibrillation behavior during extensional deformation of composite melts with PTFE additives (cf. [11–14]). The injection-molding flow front offers a strong extensional flow field, and therefore, the injection rates were varied to effectively vary the extensional kinematics experienced by the melt during injection molding, which was further expected to affect how the fluoropolymer additive would fibrillate during subsequent deformation.

Unfortunately, this aspect could not be investigated in complete detail. This was due to the premature failure observed in the majority of the specimens injection molded at high strain rates; this is obvious from an inspection of the representative stress-strain curves obtained for the specimens injection molded at the three injection rates shown in Figure 1. All curves overlap in the elastic region of the response, but only the specimen injection molded at the lowest processing strain rate of 28 s^{-1} during injection molding shows significant plastic deformation before failure. The tensile specimen molded at 28 s^{-1} breaks at an engineering strain of around 20%, while those molded at 70 s^{-1} and 112 s^{-1} prematurely break at significantly lower strains of 2% and 1.5%, respectively.

These premature failures were always associated with signatures of macroscopic voids on the fracture surface. These macroscopic features were observable on both halves of the fractured surfaces of the specimens that failed prematurely, and in some cases were big enough to cover a significant portion of the failure surface. These voids were clearly detectable by the naked eye, and the typical elevation between the "hill" and "valley" regions was of the order of 1 mm or more in some instances. These macroscopic dimensions are clearly several orders of magnitude larger compared to the submicron dimensions that can be typically associated with the deformation-induced voids that are formed during spherulitic voiding of ductile polymers. Therefore, it may be concluded that such voids associated with the premature failure of specimens are processing defects that are present even in the undeformed specimens; clearly these voids are causative of the premature failure rather than the result of deformation. It is understood that such processing-induced voids may be caused due to moisture uptake by the specimen during storage or due to air entrainment during injection molding. In the current scenario, since such defects were observed only in specimens injection molded at high processing strain rates, these voids may be attributed primarily to air entrainment at high melt flow rates during injection molding.

Since the high rate processed specimens did not always fail in the expected fashion, a thorough investigation of whether the flow kinematics imposed during melt processing (and the resulting morphology of the injection-molded samples) has a significant impact on the microstructure evolution and deformation behavior of the specimens could not be carried out. A more rigorous study—with a more controlled injection-molding setup to avoid void formation at high injection rates—to further investigate these rate effects will be carried out in future.

The defects in specimens molded at 70 s^{-1} and 112 s^{-1} also resulted in similar scatter in the tensile tests conducted at the low temperature (−40°C); therefore, the low temperature results obtained with these specimens are not being reported. Nevertheless, the fracture surfaces obtained at room temperature and −40°C for the specimens injection molded at both 28 s^{-1} (defect free) and 112 s^{-1} (containing void defects) were employed for microscopy.

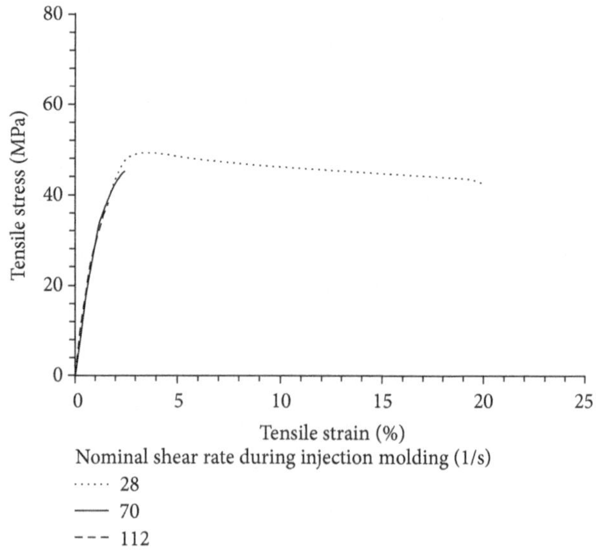

FIGURE 1: Quasistatic tensile stress-strain curves obtained at room temperature for tensile specimens injection molded at nominal shear rates of 28 s^{-1}, 70 s^{-1}, and 112 s^{-1}. The specimens processed at 70 s^{-1} and 112 s^{-1} failed prematurely owing to the presence of process-induced macroscopic void defects.

2.3. Microscopy. Microstructural analysis was conducted on the fractured surfaces of the specimens tensile tested in the quasistatic mode up to failure at room temperature and at −40°C. The intention of the scanning electron microscopy (SEM) analysis was (a) to probe the presence (or absence) of PTFE fibrils and a fibrillation network and (b) to evaluate whether these features connected with fibril dispersion and distribution may be correlated with the global stress-strain characteristics displayed by the specimen. In order to facilitate better resolution of the microstructural features, including nanoscale fibrillated PTFE (if present), a Zeiss 1540XB CrossBeam SEM was used in the field emission mode (FESEM), enabling the use of higher current compared to conventional SEM.

The fracture surfaces were subjected to SEM without any further mechanical treatments to the surface to avoid any alteration of the surface microstructure cumulatively evolved through processing (injection molding) and deformation (tensile testing). For SEM specimen preparation, the fractured half of the tensile bar was held upright using plastic clips, then placed on double sided adhesive carbon tapes for grounding the sample, and then wrapped with aluminum foils to prevent exposure of regions that were not relevant for microscopic analysis. The exposed region (the fractured surface) of the specimen, being nonconductive, was then prepared for SEM environment by depositing a conductive coating of gold-palladium (80/20) on the surface using an SC 7620 Mini Sputter Coater for 150 s.

2.4. Differential Scanning Calorimetry (DSC). DSC was carried out using a Q2000 DSC from TA Instruments. DSC thermograms were generated for the unprocessed pellets and the specimen cut from the fracture surface of the failed tensile bars. Argon gas was used for thermal equilibration within the DSC environment. The samples were characterized in ramp mode with a ramp rate of 10°C per min starting from 40°C (standby temperature) up to 400°C. This temperature range covered all major primary and secondary transitions of all the polymeric components within the composite (except the low-temperature primary peaks of PTFE [5–7]). In order to probe the first-order melting transitions of PTFE more sensitively, the DSC experiments were conducted in a "modulated" mode (MDSC); the temperature ramp was superimposed with a temperature modulation with amplitude of 1°C for 30 s imposed every 5 min. This frequency domain characterization, superimposed on the time-domain based standard ramp test, provided a convoluted heat signal, which was then deconvoluted by the TA DSC software into reversible and irreversible portions. The deconvoluted heat curves were expected to provide a significantly more sensitive probe of the PTFE first-order thermal transition compared to the conventional time-domain temperature ramp experiment. Thermograms of the fractured specimens were compared to that of an unprocessed pellet to infer any signatures of microstructure development during deformation.

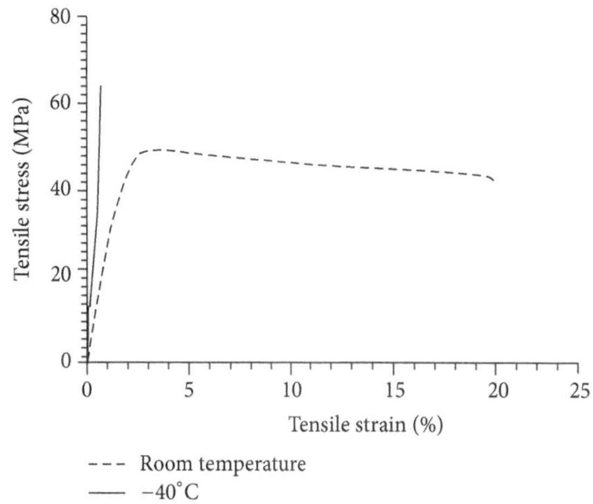

FIGURE 2: Quasistatic tensile stress-strain curve measured at room temperature compared with that measured at −40°C.

3. Results and Discussion

3.1. Tensile Stress-Strain Behavior. In Figure 2, the quasistatic tensile stress-strain curve measured at room temperature is compared with that measured at −40°C for the defect-free tensile bar specimen injection molded at a nominal shear rate of 28 s^{-1}. Both of the stress-strain curves reported are obtained up to the point of specimen failure. The specimen tested at room temperature showed significant plastic deformation before failure. The stress-strain curve obtained at −40°C displays a higher initial slope (higher modulus) as well as higher strength compared to that measured at room temperature. On the other hand, the failure strain observed for the specimen at −40°C is significantly lower compared to that observed in room temperature testing; the specimen fails before onset of yield.

"Control" tests—involving the composite system without the low surface energy fluoropolymer additive—could not be conducted. This was because the system studied was a commercially formulated composite, and an equivalent commercially available composite with exactly the same composition but without the fibrillating additive could not be identified.

However, there is data available at room temperature on other equivalent (but not exact) systems from the SABIC-IP Xenoy family which contain a polycarbonate-polyester blend as the matrix with mineral fillers. For example, for the room temperature test, the recorded tensile yield strength of about 50 MPa is comparable to the values reported in literature for commercially available PC/PBT blends containing mineral fillers and PTFE additives [15]. The secant modulus of 2.5 Gpa evaluated at 1% strain for this specimen is also comparable to (only slightly lesser than) the typical values reported in the literature for commercially available PC/PBT blends containing mineral fillers and significantly higher than the values of moduli reported for unfilled PC/PBT blends [15]. The elongation strain-to-failure of 20% for this specimen falls short of the values (up to 100%) reported in literature [15] for commercially available PC/PBT blends containing PTFE additives. The difference may be tied to the exact nature and content of mineral fillers as well as the amount of PTFE additive in the batch of the composite used in this study, compared to that within the batches for which these values were reported in [15]. Notwithstanding these differences, these strains-to-failure are still significantly larger than the values—ranging from 3% [18] to 10% [15, 19]—reported in the literature for mineral-filled PC/PBT blends without the fibrillating additives; fibrillation, if present in the fractured specimen, may therefore be unambiguously correlated with the increased strain to failure.

Unlike in case of the room temperature specimen, control data on strains to failure at low temperatures for equivalent mineral filled PC/PBT blends without fibrillating fillers was not available in the literature. In the absence of such baseline information, it would not be possible to unambiguously discern whether the response of the composite at −40°C is dominated by the brittleness of the matrix or whether fibrillation (if present in the fracture surface of the specimen) has any impact on the strain to failure. However, as discussed in the subsequent, the lack of such baseline information does not significantly alter the conclusions from this study, since the microscopy of the specimen fracture surfaces reveals stark contrasts in fibrillation behavior at room temperature and −40°C.

3.2. Morphology-Property Interrelationships. FESEM micrographs of the fracture surfaces of the tensile specimens quasistatic tensile tested at −40°C are shown in Figures 3 and 4, respectively, for specimens injection molded at nominal shear rates of 112 s^{-1} and 28 s^{-1}. In both micrographs, the failure patterns in the form of rough wavy topography are clearly visible. Another typical feature is the presence of faceted topography, with a typical wavelength of 2 microns. Flake like structures, of the order of 2 microns, appear to project out of the fracture surfaces; in some areas, these projections are adjacent to voids, suggesting a pullout of particulate inclusions, also of the order of about 2 microns. All these

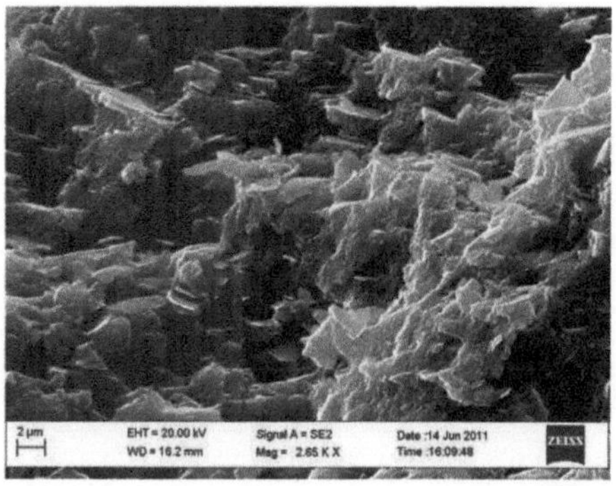

FIGURE 3: FE SEM of quasistatic tensile fractured surface obtained at −40°C for the tensile specimen injection molded at a nominal shear rate of 112 s^{-1}.

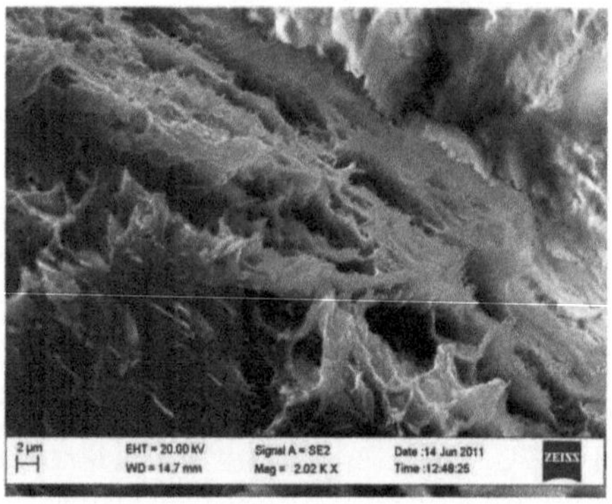

FIGURE 4: FE SEM of quasistatic tensile fractured surface obtained at −40°C for the tensile specimen injection molded at a nominal shear rate of 28 s^{-1}.

features appear to be related to talc fillers, given the flaky nature of the particulate inclusions, and the rectangular shape of the voids associated with particle pullout. The "faceted" morphology is indicative of inadequate deformation of the matrix and is consistent with the brittle failure characteristic at the low temperature. While the faceted morphology was predominant in the failure surfaces of the specimens tested at −40°C, another less prominent feature, which was more observable in the fractured surface of the tensile specimen processed at 28 s^{-1}, was the presence of intermittent "floral" patterns pulled out in the tensile direction, indicating matrix deformation (refer to Figure 4). However, PTFE fibrillation was not discernible in any of the failure surfaces generated at −40°C.

FESEM micrographs of the fracture surfaces of the tensile specimens quasistatic tensile tested at room temperature are shown in Figures 5 and 6, respectively, for the specimens injection molded at nominal shear rates of 112 s^{-1} and 28 s^{-1}. Just as in failure surfaces obtained with the specimens tested at −40°C, the features associated with flaky fillers may be clearly observed in these micrographs also. However, the "floral" topography associated with matrix deformation is much more pronounced in the failure surfaces generated in the room temperature tested specimens. More importantly, in contrast to low temperature tested samples, a fibrillated

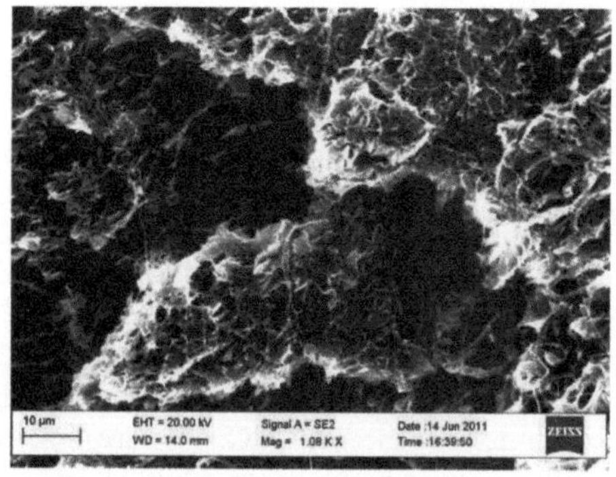

FIGURE 5: FE SEM of quasistatic tensile fractured surface obtained at room temperature (24°C) for the tensile specimen injection molded at a nominal shear rate of 112 s^{-1}.

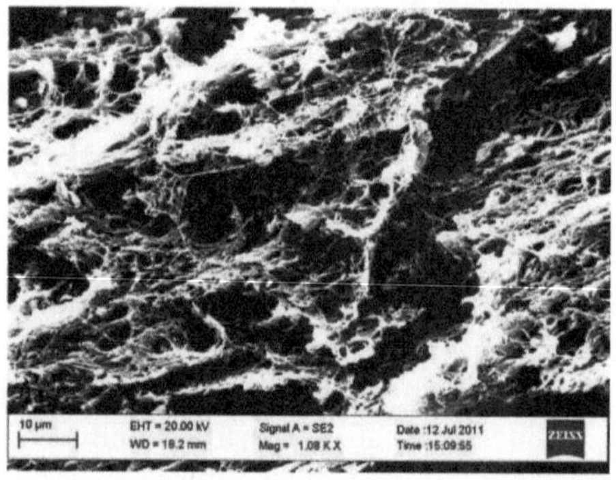

FIGURE 6: FE SEM of quasistatic tensile fractured surface obtained at room temperature (24°C) for the tensile specimen injection molded at a nominal shear rate of 28 s^{-1}.

network is clearly discernible in both Figures 5 and 6. Since no other component of the composite shows fibrillation, this feature may be unambiguously attributed to PTFE fibrillation.

Further evidence of deformation-associated fibrillation of PTFE during the ductile failure of the composite may be seen in the MDSC thermograms plotted in Figure 7. In this figure, the nonreversible heat input during MDSC of the unprocessed HMD Xenoy composite pellet is compared with those of undeformed tensile specimen and the fractured region of a room temperature quasistatic tensile tested specimen; both tensile specimens were injection molded at 28 s^{-1}. A prominent first order transition at a temperature of about 350°C is observable only in the ductile-fractured region of the room temperature tensile tested specimen, indicating the evolution of PTFE fibrillation and orientation during the deformation. The quasistatic failed specimen also shows a small peak in the vicinity of 100°C which can be attributed to the presence of moisture in the specimen; this is not surprising, given the hygroscopic nature of polycarbonate, which is one of the components of the polymer blend matrix. This detail is not of significant importance in the context of fibrillation of the low surface energy fluoropolymer additive (none of the thermal transitions of the latter are expected around 100°C).

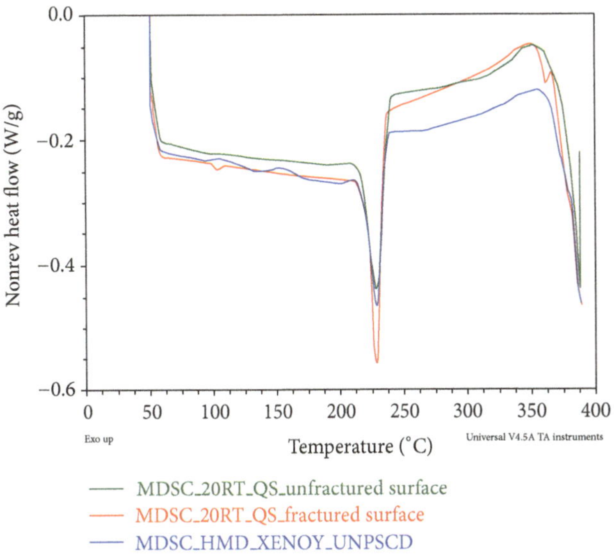

Figure 7: Nonreversible heat flow MDSC thermogram of unprocessed as-received HMD Xenoy composite pellet compared with those of undeformed tensile specimen and the fractured region of a room temperature quasistatic tensile-tested tensile specimen; both tensile specimens were injection molded at a nominal shear rate of $28\,s^{-1}$.

3.3. Discussion. In room temperature tests, the tensile specimen processed at $112\,s^{-1}$ which had a void defect discernible on the failure surface, failed at a significantly lower strain compared to that processed at $28\,s^{-1}$ which did not have any defects. While the specimen processed at $112\,s^{-1}$ did not display significant yield before failing, the specimen processed at $28\,s^{-1}$ showed significant plastic deformation before failure. Both the specimens however displayed fibrillated network on the failure surfaces. On the other hand, the significantly lower strain to failure before onset of yield observed for the specimens at −40°C is consistent with the fact that both the components of the matrix blend are expected to display brittle characteristics. None of the specimens tested at −40°C displayed fibrillation.

Based on the previous observations, two trends may be noticed.

First is the clear contrast in terms of fibrillation behavior at the two test temperatures. At room temperature, both the components of the matrix polymer blend—PC and PBT—are significantly above their brittle-ductile-transition temperature. While both PC and PBT are below their respective glass transition temperatures, PC significantly so ($T_{g,\mathrm{PC}}$ = 150°C; $T_{g,\mathrm{PBT}}$ ~ 60°C), they still display significant plastic deformation before failure at room temperature. The generally accepted brittle-ductile-transition temperatures for PC and PBT are −25°C and −40°C, respectively [20]. Therefore, at −40°C, the matrix phase is expected to be brittle and not undergo plastic deformation before failure. Comparing the strain to failure of the specimen processed at $28\,s^{-1}$ (20%) with that reported for equivalent mineral filled PC/PBT composites without fibrillating fillers (3–10%) [15, 18, 19], the increase in composite ductility may be unambiguously correlated with the presence of fibrillation. If fibrillation was observed in the failed specimens tested at −40°C, an analogous conclusion would have been difficult to arrive since information about strains to failure of equivalent composite systems without fibrillating fillers was not available at −40°C. However, no fibrillation was observed in the specimens tested at −40°C. Clearly, the response of these composites was dominated by the brittleness of the matrix. More importantly, the suppression of matrix ductility seems to have prevented the onset of fibrillation in the composite system. Clearly, there appears to be a synergistic effect between the tendency of PTFE to fibrillate and the mechanical behavior of the matrix resin in governing the resulting ductility of the composite.

Second is the observation of fibrillation in room-temperature-failed specimens irrespective of whether the composite failed after showing enhanced ductility (as in the specimen processed at $28\,s^{-1}$) or prematurely (as in the specimen processed at $112\,s^{-1}$). This observation is a little counter intuitive, because the global stress-strain behavior observed for the specimen processed at $112\,s^{-1}$ does not display pronounced plastic deformation. However, since the specimen was tested at room temperature, it is understood that the matrix blend has a tendency to deform in a plastic fashion, even though the specimen prematurely failed at a relatively low strain. Two inferences may be made from this observation: firstly, as long as matrix ductility is not suppressed, fibrillation may be observed in the specimen, irrespective of the strain to failure. And secondly, that fibrillation must be associated with localized matrix plasticity; localized plasticity in the absence of specimen-level necking may occur due to unequal load transfer across the specimen cross-section. Of course at −40°C, even these local plasticity effects would be suppressed.

4. Summary

Room temperature tensile tests reveal ductile behavior of the composite, especially the defect-free specimens processed at $28\,s^{-1}$. But the tensile tests at −40°C result in significantly lower strains-to-failure, driven by the brittleness of the PC-PBT blend matrix, both the components of which were at or below their ductile-brittle transition temperatures at −40°C. The low-temperature tested samples, in which the PC/PBT blend matrix was rendered brittle, did not show any traces of PTFE fibrillation. The room temperature failure which occurs in a ductile mode is clearly associated with PTFE nanofibrillation. The nonreversible portion of the heat flow in MDSC thermogram also contains the signature of deformation-associated PTFE fibrillation and orientation during the ductile failure of the specimen, in consistency with the micrographs.

At room temperature, when both the components of the matrix blend display the tendency for plastic deformation before failure, the increase in composite ductility (relative to equivalent composite systems without the fibrillating filler) may be unambiguously correlated with the presence of fibrillation. At or below the ductile-brittle transition temperature of the matrix resin, the mechanical response of the composite is dominated by the brittleness of the matrix. In other words, the suppression of matrix ductility seems to prevent the onset of fibrillation in the composite system. Clearly, there appears to be a synergistic effect between the tendency of PTFE to fibrillate and the mechanical behavior of the matrix resin in governing the resulting ductility of the composite.

5. Open Questions and Future Work

As discussed earlier, an auxiliary investigation on the impact of "as processed" microstructure of the undeformed specimens (brought about by variation of the strengths of the extensional flow field imposed by varying the injection rate) on their tensile stress-strain behavior and strain to failure could not be investigated in complete detail; this was due to the premature failures observed in the majority of the specimens injection molded at high strain rates (due to the presence of process-induced voids). A more rigorous study—with a more controlled injection-molding setup to avoid void formation at high injection rates—to further investigate these rate effects will be carried out in future. This study, combined with detailed quantification of the PTFE surface area on the micrographs of fractured specimens, may result in a better understanding of correlations between the failure strain and the extent of PTFE fibrillation.

In order to more fully understand the structure property relationships in this composite system, similar investigations are required at more testing temperatures, for example at temperatures where one or both of the components of the blend may be at or above their respective glass transition temperatures.

The presence of fibrillation in the prematurely failed specimen in room temperature studies (which was processed at $112\,s^{-1}$ and which had a void) is an important finding and presents the motivation to investigate further (1) the effect of localized plastic deformation within the matrix (particularly in scenarios involving uneven stress transfer across the specimen cross section) on the fibrillation and ductility-enhancement characteristics and (2) the conditions, especially in terms of a localized threshold plastic strain within the matrix, associated with the onset of fibrillation. These studies are underway and will be presented in a subsequent report.

As noted earlier, a few specimens were also subjected to high-rate tensile tests (employing strain rates typically in the range of $10^{-1}\,s^{-1}$) at $-40^{\circ}C$ so as to subject the composite to faster deformation kinematics and thereby modify the conditions for deformation-induced microstructure evolution. The data from these tests are currently being analyzed and will be presented in a subsequent report.

Conflict of Interests

The authors hereby declare that they have no conflict of interests.

Acknowledgment

The authors acknowledge several constructive remarks from the reviewers.

References

[1] X. C. Zhang, M. F. Butler, and R. E. Cameron, "The ductile-brittle transition of irradiated isotactic polypropylene studied using simultaneous small angle X-ray scattering and tensile deformation," *Polymer*, vol. 41, no. 10, pp. 3797–3807, 2000.

[2] Q. Lin, J. Jho, and A. F. Yee, "Effect of drawing on structure and properties of a liquid crystalline polymer and polycarbonate in-situ composite," *Polymer Engineering and Science*, vol. 33, no. 13, pp. 789–798, 1993.

[3] S. Bandyopadhyay, R. Kamalakaran, S. Gupta, R. Mazumdar, B. Sathapathy, and A. K. Ghosh, "Fibrillation of teflon in polymer matrices," in *Proceedings of the International Conference on Advances in Polymer Technology (APT '10)*, pp. 71–74, Kochi, India, February 2010.

[4] S. Ebnesajjad, *Fluoroplastics, Volume 1: Non-Melt Processible Fluoroplastics*, Plastics Design Library, New York, NY, USA, 2000.

[5] C. W. Burn and E. R. Howells, "Opinion," *Nature*, vol. 174, pp. 569–573, 1954.

[6] C. A. Sperati and H. W. Starkweather, "Fluorine-containing polymers. II. Polytetrafluoroethylene," *Fortschritte der Hochpolymeren-Forschung*, vol. 2, no. 4, pp. 465–495, 1961.

[7] S. Mazur, "Paste extrusion of poly (tetrafluoroethylene). Fine powders," in *Polymer Powder Technology*, M. Narkis, Ed., pp. 441–481, John Wiley, New York, NY, USA, 1995.

[8] M. A. B. M. Ali, S. Nobukawa, and M. Yamaguchi, "Morphology development of polytetrafluoroethylene in a polypropylene melt (IUPAC technical report)," *Pure and Applied Chemistry*, vol. 83, no. 10, pp. 1819–1830, 2011.

[9] A. B. Ariawan, S. Ebnesajjad, and S. G. Hatzikiriakos, "Paste extrusion of polytetrafluoroethylene (PTFE) fine powder resins," *Canadian Journal of Chemical Engineering*, vol. 80, no. 6, pp. 1153–1165, 2002.

[10] J. G. Drobny, *Technology of Flouropolymers*, CRC Press, Boca Raton, Fla, USA, 2008.

[11] A. B. M. M. Ali, K. Okamoto, M. Yamaguchi, T. Kasai, and A. Koshirai, "Rheological properties for polypropylene modified by polytetrafluoroethylene," *Journal of Polymer Science B*, vol. 47, no. 20, pp. 2008–2014, 2009.

[12] T. Murata, J. Takimoto, and K. Koyama, "Melt property and extrusion foaming processability of polypropylene blended with polytetrafluoroethylene," *Zairyo/Journal of the Society of Materials Science, Japan*, vol. 45, no. 12, pp. 1300–1305, 1996.

[13] T. Kurose, T. Takahashi, and K. Koyama, "Uniaxial elongational viscosity of FEP/ a small amount of PTFE blends," *Nihon Reoroji Gakkaishi*, vol. 31, no. 4, pp. 195–200, 2003.

[14] T. Kurose, T. Takahashi, M. Sugimoto, T. Taniguchi, and K. Koyama, "Uniaxial elongational viscosity of PC/ a small amount of PTFE blend," *Nihon Reoroji Gakkaishi*, vol. 33, no. 4, pp. 173–182, 2005.

[15] L. M. Sherman, "Plastics technology," February 2009, http://www.ptonline.com/articles/novel-pc-polyester-blends-combine-stiffness-and-ductility.

[16] ISO 527-1, "Plastics—determination of tensile properties—part 1: general principles," 2012, within ISO Standards Catalogue 83. 080. 01: Plastics in general.

[17] ISO 527-2, "Plastics—determination of tensile properties—part 2: test conditions for moulding and extrusion plastics," 2012, within ISO Standards Catalogue 83. 080. 01: Plastics in general.

[18] "Xenoy Resin 6370 Datasheet," 2013, SABIC Innovative Plastics, http://www.sabic-ip.com/gepapp/eng/weather/weatherhtml?sltRegionList=1002002000&sltPrd=1002003020&sltGrd=1002011525&sltUnit=0&sltModule=DATASHEETS&sltVersion=Internet&sltType=Online.

[19] "Xenoy Resin XD1647 Datasheet," 2013, SABIC Innovative Plastics, http://www.sabic-ip.com/gepapp/eng/weather/weather-html?sltRegionList=1002002001&sltPrd=1002003066&sltGrd=1002012923&sltUnit=0&sltModule=DATASHEETS&sltVersion=Internet&sltType=Online.

[20] Omnexus Polymer Selector, "Ductile/brittle transition temperature of plastics," http://www.datapointlabs.com/ExpCapShowDetails.asp?InstId=34.

5

Effects of Thermal and Solvent Aging on Breakdown Voltage of TPE, PBT/PET Alloy, and PBT Insulated Low Voltage Electric Wire

Eun-Soo Park

Youngchang Silicone Co., Ltd., 481-7 Gasan-Dong, Kumchun-Gu, Seoul 153-803, Republic of Korea

Correspondence should be addressed to Eun-Soo Park; t2phagel@daum.net

Academic Editor: Changwoon Nah

Tests were performed to evaluate the effects of thermal and solvent aging on the mechanical and dielectric breakdown properties of four types of polyester resins, namely, the insulation layer of poly(butylene terephthalat) (PBT)- based thermoplastic elastomer (TPE, TPE1), poly(butylene 2,6-naphthalate)-based TPE (TPE2), PBT/poly(ethylene terephthalate) alloy (Alloy), and PBT extruded onto a copper conductor of low voltage electric wire. The tensile specimens used in this series were prepared from the same extruded resins. The prepared electric wires and tensile specimens were thermally aged in air and in toluene, xylene, TCB, and NMP. When Alloy and PBT were thermally aged in toluene, xylene and TCB at 120°C for 6 h, the tensile properties were significantly decreased compared to TPE1 and TPE2 at the same condition. The reduction of elongation at break of Alloy was more discernible than that of PBT. This result indicated that Alloy is more affected by thermal and solvent ageing. Among them, TPE2 showed the highest breakdown voltage (BDV), and it has also the highest BDV after thermal and solvent aging.

1. Introduction

The service life of electric wire is determined by the condition of the insulating material. Insulations are commonly used as a flexible polymer coating on electric wire. Defects such as voids, contaminants, and scratches in the insulation structure lead to electrical failure in polymer insulted wires [1–4]. Thermoplastic polyesters, enclosing poly(ethylene terephthalate) (PET), poly(butylene terephthalate) (PBT), copolymers, and blends, are widely used for electrical and electronic applications such as switching transformers, inverters, and information and telecommunications equipment due to their good electrical insulation and dielectric properties.

Many manufacturers of transformers get electrical insulating varnishes for secondary insulation and have to apply and transform them using a chemical curing process. The insulating varnish is applied to coils of polyester insulated wire, in the process of which it fills the gaps and micropores, and after curing it forms into a continuous resin film. It is important for mechanical stability and protection of coiled wire against dust, moisture, and chemicals which might cause insulation deterioration. Thereby, leakage currents or short circuits can be avoided so that safe and reliable electrical products can be produced [5].

However, the insulating varnishes, which are provided for this purpose, are highly reactive materials. They are usually included in one or more solvents, such as toluene, xylene, isophorone, naphtha, 1,2,4-trichlorobenzene (TCB), and 1-methyl-2-pyrrolidinone (NMP). Varnishes for impregnation of electrical coils have until lately been diluted by solvents to lower the viscosity so as to permit full penetration of the windings. The varnish is generally applied by dip process. In this method, preheated winding wire is slowly dipped in the varnish, preferably with the slots in vertical direction. After certain time, the winding is removed from the varnish tank and excess varnish is allowed to drip and cured. Depending upon the type used, the winding is cured at elevated temperature (~160°C) for ~6 h in an oven. When the coils are baked, the varnish dries and the solvent is driven off. The drying leaves very small holes through which moisture can penetrate and in which corona may form.

Ribnick et al. [6] reported that a number of organic solvents (toluene, acetonitrile, dimethylformamide (DMF), trichloroethylene, perchloroethylene, and tetrachloroethane) have been causing a high level of shrinkage in oriented semicrystalline polyester. Knox and coworkers [7] observed that aqueous and solvent-treated semicrystalline polyester samples were similar in that there was an increase in structural porosity as a result of crystallization in the swollen state. Weigmann and coworkers [8] found that the formation of void-containing structures by polar solvent treatment depends on the ability of the solvent to swell the polymer and to cause recrystallization in the swollen state. Thereby, the varnish treatment process influences the deterioration of the insulation made from polyester materials and reduces the ability of the wire to withstand short circuits.

In this study, PBT-based thermoplastic elastomer (TPE, TPE1), poly(butylene 2,6-naphthalate)-*b*-poly(tetramethylene ether glycol) [PBN-*b*-PTMG (TPE2)], PBT/PET alloy (Alloy), and PBT resins extruded onto a conductor of low voltage electric wire. The effects of thermal aging of extruded wires in toluene, xylene, TCB, and NMP on dielectric breakdown voltage (BDV) change were investigated. The changes in tensile properties of dumb-bell type specimen after the aging also have been evaluated.

2. Experimental

2.1. Materials. Toluene (Sigma-Aldrich, Inc., St. Louis, MO, USA), xylene (DC chemical Co., Seoul, Republic of Korea), TCB (Sigma-Aldrich, Inc., St. Louis, MO, USA), and NMP (TEDIA company Inc., Fairfield, OH, USA) were of reagent grade and were used without further purification. PBT (TRIBIT 1500, Samyang Co., Jeonju, Republic of Korea), Alloy (KP513 (PBT/PET = 30/70 wt%), Kolon Industries, Inc., Gwacheon, Republic of Korea), polyester type TPE (TPE1, G182D, SK Chemicals, Seoul, Republic of Korea), and PBN-*b*-PTMG type TPE (TPE2, EN5000, Plastech Service Inc., Seoul, Republic of Korea) resins were used as received. All polymer resins were predried in a convection oven for at least 12 h at 80°C to remove any moisture from the pellets before processing. The characteristics of the resins used in this study were summarized in Table 1.

2.2. Instrumentation. Scanning electron microscopy (SEM) observations of the specimens were performed on a Hitachi S-4200 model (Tokyo, Japan). The fractured surfaces of the specimens were prepared by using cryogenic fracturing in liquid nitrogen followed by a coating with gold in an SPI sputter coater. The morphology was determined using an accelerating voltage of 15 kV.

Thermal properties of the polymers were determined by differential scanning calorimetry (DSC, Perkin Elmer DSC 7, Norwalk, CT, USA). Thermal history of the samples was removed by scanning to 300°C with the heating rate of 20°C/min. After cooling down the specimen at the rate of −5°C/min to room temperature, it was reheated from 20 to 300°C with the heating rate of 20°C/min, and the DSC curves were obtained.

Thermal stability of samples was determined by thermogravimetry (TG, TGA-Q50, TA instrument, USA). The TG curves were obtained under an N_2 atmosphere at a flow rate of 4 mL/min and a scanning from 20 to 800°C with the heating rate of 20°C/min.

Table 1: Characteristics of the insulation materials.

Material	Density (g/cm^3)	Hardness (°C)	T_m (°C)	Melt flow index (MI) (g/10 min)
PBT	1.31	112	221	4 (230°C/2.16 kg)
Alloy	1.40	116	251	5 (265°C/2.16 kg)
TPE1	1.27	77	217	8 (230°C/2.16 kg)
TPE2	1.24	75	226	11 (235°C/2.16 kg)

The surface appearance of the wire after solvent aging was measured by a video microscope (Inf-500TM, charge-coupled device (CCD) video microscope, Moritex Corp., Tokyo, Japan) with a PC that contained video capture software (i-Solution Lite, Image & Microscope Technology, Daejeon, Republic of Korea).

Fourier transform infrared (FTIR) absorption spectra of the samples were measured with a Perkin Elmer infrared spectrometer (Spectrum 2000, Shelton, CT, USA) in the wave number range from 4000 to 400 cm^{-1} and analyzed using commercial software.

2.3. Extrusion Weight (g/10 min) Measurements. Extrusion weight (g) of the polymer samples was measured for 10 min using an extruder (ϕ20 Extruder, Flontec Co., Ltd., Republic of Korea) equipped with resin bypass (8 mm inner diameter and 10 mm length orifice). The extruder was preheated from room temperature to 265°C, and then the pellets of polymer sample were fed into the extruder and allowed to melt. The polymer stays in the molten form inside the extruder for approximately 10 min and then the molten polymer is vented at a screw rotation speed of 200 rpm out from the exit valve.

2.4. Tensile Test Procedure. Polymer sheets (around 0.3 mm) were prepared on a hot press at 265°C for 5 min under ca. 5 atm, followed by quenching in water. The sheet thus formed was free from any distortion problems.

Dumb-bell specimens for tensile tests were prepared in accordance with IEC 60811-1-1 specification. The test specimens were preconditioned to 20% relative humidity and 20 ± 1°C in order to standardize test conditions. Tensile properties of samples were determined with a universal test machine (UTM, Model DECMC00, Dawha test machine, Republic of Korea) at a cross head speed of 250 mm/min according to IEC 60811-1-1. The mean value of at least five specimens of each sample was taken, although specimens that broke in an unusual manner were disregarded.

2.5. Extrusion of Electric Wire. Polymers were dried in a hopper equipped with a dehumidifying dryer at least 24 hr at 80°C to remove any moisture from the pellets before extruding. The electric wire was composed of conductor and 1st and 2nd insulation layers. The 1st insulation layer was applied to the annealed copper conductor by extrusion with TPE1, TPE2, Alloy, or PBT (thickness: 30 μm). The 2nd layer was applied to the 1st extruded layer by extrusion with the same material and the thickness of 1st layer. All insulation layers were extruded (temperature setting of rear zone: 235°C, second zone: 255°C, forward zone: 260°C, adapter: 265°C, and die: 265°C) at a speed of 200 rpm using an extruder (ϕ20 Extruder, Flontec Co. Ltd., Seoul, Republic of Korea) equipped with a 1 : 3 compression ratio screw. Extruded samples were quenched by passage through a water-cooling zone at a rate of 200 m/min.

2.6. Breakdown Voltage (BDV) Measurement. The measurements of BDV were carried out at 60 Hz AC voltage by using a high voltage tester (SM-40PT08, Sungmin Instruments, Republic of Korea) with a rate of 0.5 kV/s according to UL 758 standard. The test sample consists of copper conductor (diameter of 0.3 mm) and insulation (thickness: 60 μm). The voltage was applied between the conductor and aluminum foil wrapped intimate contact with the center 152 mm of the wire. Ten samples are tested and the average voltage is calculated.

2.7. Solvent Immersion of Extrude Wire. Solvent immersion test was performed to determine the ability of wire to resist degradation when exposed to toluene, xylene, TCB, and NMP according to the modified MIL-STD-2223 method 1001. Five specimens of 600 mm extruded wires were immersed to within 150 mm of each end in a 200 mm tall test tube containing 80 mL of solvent for 6 h at 20 ± 2°C (60 ± 2°C or 120 ± 2°C). After the predetermined time, specimens were removed from the glass tube, cleaned with methanol, and dried in an oven at 60°C until constant weight. The insulation was removed for a distance of 12 mm from each end of wire sample, and the specimens were subjected to the BDV measurement.

2.8. Solvent Immersion of Tensile Test Specimen. Ten dumbbell specimens for tensile tests were immersed to a 200 mm tall test tube containing 80 mL of solvent (toluene, xylene, TCB, and NMP) for 6 h at 20 ± 1°C (60 ± 1°C or 120 ± 2°C). After the predetermined time, specimens were removed from the glass tube, cleaned with methanol, and dried in an oven at 60°C until constant weight. The specimens were subjected to the tensile test in accordance with the IEC 60811-1-1 specification.

3. Results and Discussion

3.1. Extrusion Conditions of TPE1, TPE2, PBT, and Alloy. Extrusion weight of TPE1, TPE2, Alloy, and PBT was measured in an extruder equipped with 8 mm diameter orifice at 265°C for 10 min, and the results are summarized in Table 2. For the thin film extrusion process, melt viscosity is an important property since it affects the flow of the molten polymer. They have good melt flow in processing temperature range and they are adequate for thin layer extrusion (>80 g/10 min). The extruding weight of samples increased in the order of Alloy < TPE2 < PBT < TPE1 at the same conditions due to the mobility of the polymer molecules being more restricted as the melt is contained with PET and PBN unit.

Four types of polymers were extruded onto 3 mm diameter copper conductor at die temperature of 270°C with a speed of 200 rpm using a single screw extruder. The most common extrusion fault is operating at an incorrect temperature. Too low extrusion temperature will result in a nonhomogeneous mixture while too high temperature will lead to voids or micropores being formed. The presence of voids on the insulation can significantly reduce the dielectric properties [4, 9]. SEM micrograph of the cryogenically fractured surface of the extruded TPE1, TPE2, PBT, and Alloy insulations was demonstrated in Figure 1. SEM photographs showed that the TPE1 exhibited a rough structure with numerous spikes, whereas TPE2, Alloy, and PBT have a relatively smooth fracture surface without the formation of necking indicating weak resistance to crack propagation. TPE2 shows a phase separation like structure with dispersed droplets. It is shown that the extrusion process does not cause any significant voids or areas of stress concentrations, which might affect the overall mechanical and electrical properties of the material.

3.2. Thermal Stability of TPE1, TPE2, PBT, and Alloy. Thermal degradation of extruded TPE1, TPE2, PBT, and Alloy insulations is shown in Figure 2. TPE1 shows an initial degradation temperature at 320°C which is the lowest of all investigated samples. Polyester type TPE is usually a linear segmented block copolymer composed of PBT as hard segments and flexible oligomeric polyester as soft segments. Therefore the thermal stability of TPE1 was lower than that of PBT.

In sharp contrast, TPE2 shows similar thermal degradation behavior to PBT. To improve the heat resistance of TPE, a more rigid aromatic diester monomer is needed for the hard segment [10]. The chemical structure of PBN is equivalent to that of PBT except that the benzene ring is replaced by the naphthalene ring. Wholly aromatic and multiaromatic ring structures produced high thermal stability resistance due to their rigid, rod-like aromatic chain. Incorporation of this structure in a polymer dramatically improves thermal properties [11]. The TGA traces of Alloy also show a significant shift of the weight loss towards higher temperature with stabilization 20°C higher than PBT.

As listed in Table 2, it can be seen that the Alloy exhibits a higher thermal-oxidative stability in terms of $T_{-5\%}$ (a mass loss of 5% during degradation occurs) than that of TPE1, TPE2, and PBT. With elevating temperature, the carbonaceous residues are oxidized to produce volatiles. The residues at 400°C are 64.2, 69.3, 71.4, and 88.2% for TPE1, TPE2, PBT, and Alloy, respectively, which confirms that the TPE1 easier to happen thermal degradation. The residues at 500°C show the same tendency as those at 400°C. The TG curves also show that all the samples nearly crumble at about 500°C (char yield around 8%) indicating that all the polymer mass turns into gaseous product. The decomposition pathways of a polyester composed of glycol and diacid are described in the literature as a three stage process. The first stage is elimination of terephthalic acid. The second stage, around 350°C, is possibly caused by the release of styrene and a complex mixture of

TABLE 2: Thermal properties of the TPE1, TPE2, PBT, and Alloy.

Sample code	Extrusion weight (g/10 min)	TGA thermal properties			DSC thermal properties				
					1st scan		2nd scan		
		$T_{-5\%}$ (°C)	Residues at 400°C (%)	Residues at 500°C (%)	T_{m11} (°C)	T_{m12} (°C)	T_{m21} (°C)	T_{m22} (°C)	T_c (°C)
TPE1	110	343.4	64.2	6.4	—	217	—	218	193
TPE2	98	376.0	69.3	6.7	—	226	217	226	202
PBT	100	379.7	71.4	6.8		221	215	222	195
Alloy	95	389.3	88.2	9.0	216	251	—	230	148

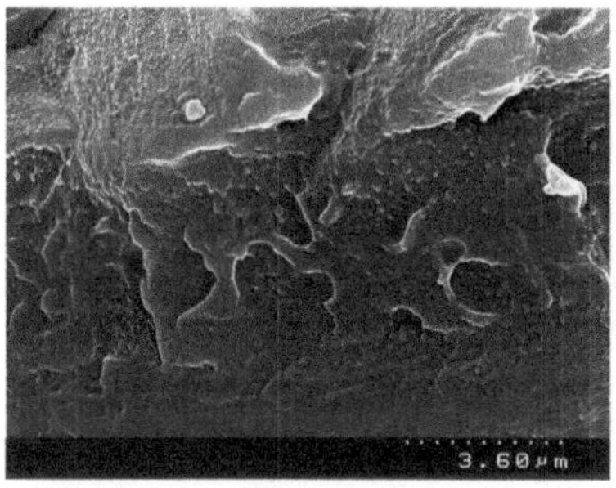

(a) TPE1

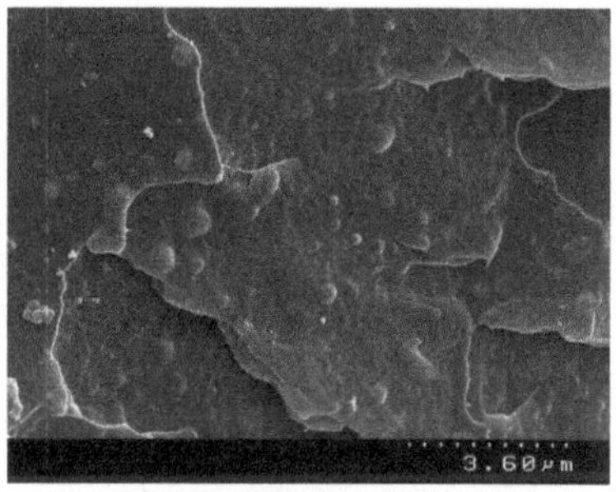

(b) TPE2

(c) Alloy

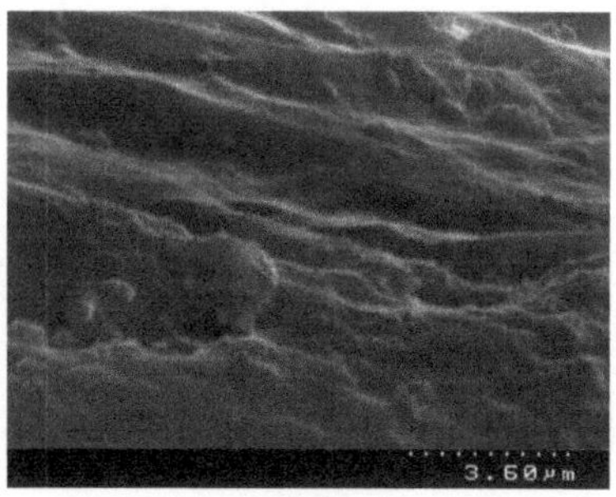

(d) PBT

FIGURE 1: SEM micrographs of the fractured surface of extruded insulation layer.

other materials, while the third stage, above 500°C, is related to the loss of high boiling tars and oxidation of the char formed [12].

3.3. DSC Thermal Properties of the Extruded TPE1, TPE2, PBT, and Alloy. The melting and the crystallization behaviors of the extruded insulations were studied by using DSC. The 1st scan DSC data of Table 2 were collected by a heating rate of 20°C/min from 20 to 300°C. When PBT melt was slowly cooled at a rate of −5°C/min and remelted, the one melting peak of 1st scan DSC thermogram was divided into two peaks (Figure 3). PBT exists in two crystalline structures, α and β forms, and the transition between two forms takes place reversibly by crystallization, mechanical deformation, and relaxation [13, 14]. The melting peak of PBN-*b*-PTMG type TPE shows similar behavior to PBT. PBN also has two crystal forms, A and B forms, and four methylene groups in the B form are more extended than those in the A form as in the case of PBT [15, 16]. In the absence of mechanical deformation, both the crystal structures of PBT and PBN favor the α form and A form, respectively. In contrast, melting temperature (T_m) of TPE1 did not show any changes.

The quenched Alloy sample has two melting peaks, which are associated with PET and PBT separate phase (Figure 3). Escala and Stein [17] found that, upon crystallization, PET and PBT did not cocrystallize. Avramova [18] also reported that, although PET and PBT formed their own crystalline phase upon crystallization of PET/PBT blend, both components could crystallize concurrently at all compositions of the blends, and the presence of one crystalline phase did not affect the crystallization rates of the other. When the Alloy melt was slowly cooled at a rate of −5°C/min and remelted, the two melting peaks of 1st scan DSC thermogram were merged into one peak and shifted to lower temperature. Alloy resin was produced with transesterification reactions occurring in the melt resulting in the formation of block copolymers with a random microstructure. These random copolymers have more irregular homopolymer sequences and have the single melting temperature when fully crystallized [9, 19]. It can be also seen that Alloy had the broad crystallization exotherm.

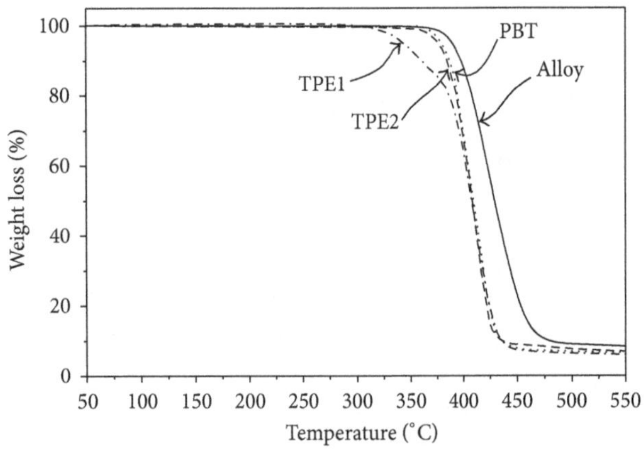

Figure 2: TGA thermograms of the TPE1, TPE2, PBT, and Alloy.

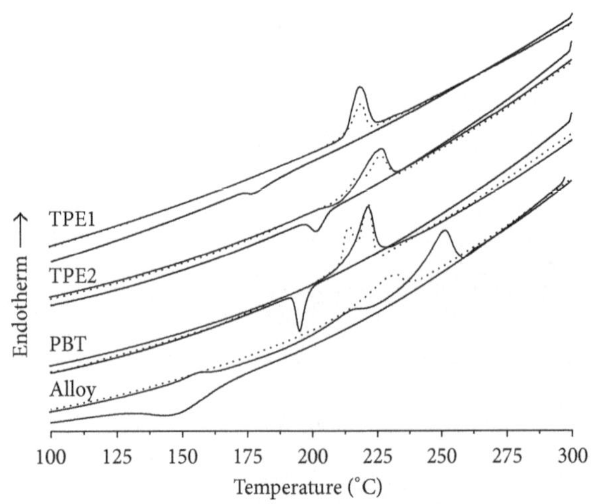

Figure 3: DSC thermograms of the TPE1, TPE2, PBT, and Alloy.

This broad peak may arise from concurrent crystallization of two different crystals at nearly the same crystallization temperature (T_c).

3.4. Tensile Properties of TPE1, TPE2, PBT, and Alloy after Thermal and Solvent Aging. Table 3 compared the effect of thermal and solvent aging on the tensile properties of TPE1, TPE2 Alloy, and PBT. The abbreviation of the sample code in Table 2, PBT-120°C-6 h, for example, means that the PBT was thermally aged at 120°C for 6 h. For TPE1, Alloy, and PBT, tensile strength and elongation at break of dumb-bell specimens were significantly decreased with increasing thermal aging temperature. In case of Alloy-180°C-6 h, an approximately 65% and 98% decrease in the tensile strength and elongation at break of 16 MPa and 6% were observed, respectively. The decrease in tensile properties after thermal aging can be related to the increase in brittleness of specimens driven by reduction of molecular weight. Thermal degradation follows the free radicals mechanism, initiated by oxygen or oxidative contamination in the material or other stresses. In the initial stages of thermal degradation of polymer, chain scission often prevails, which reduces molecular weight. Shorter polymer chains are more mobile and are thus able to crystallize more readily. In contrast, tensile properties of TPE2 were not changed compared to other samples. Incorporation of naphthalate structure in a polymer improves thermal and mechanical properties [18]. The hard PBN segment serves as a thermally reversible physical cross-linking between molecular domains, contributing thermal stability, mechanical strength, and dimensional stability to the TPE2. Thereby, TPE2 is a little more affected by the thermal aging than TPE1, Alloy, and PBT at the same temperature.

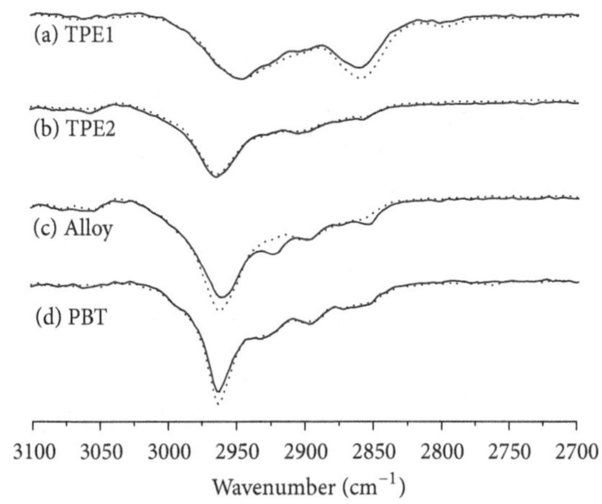

Figure 4: FTIR spectra of the insulation samples before (.....) and after (—) thermal aging at 180°C.

FTIR is an effective measurement for structure characterization. Figure 4 shows the FTIR spectra comparison of before and after thermal aging at 180°C. The peak at 2966–2855 cm^{-1} range is attributed to the vibration of –C–H bonds. The most characteristic peaks are those in the region of 1720, 1600–1500, and 1100 cm^{-1} for –C=O stretching vibration of the ester carbonyl group, stretching vibration of backbone of benzene, and –C–O– stretching vibration, respectively. For Alloy (Figure 4(c)), after thermal aging at 180°C for 6 h, new adjacent bands at 2922 and 2850 cm^{-1} can be identified, and the intensity of the band at 2964 cm^{-1} decreased significantly. In the TPE1 (Figure 4(a)), small adjacent bands at 2924 and 2905 cm^{-1} can be identified and the intensity of the band at 2985 cm^{-1} decreased. In case of PBT, the intensity of the band at 2964 cm^{-1} decreases significantly (Figure 4(d)). These could be an indication of formation of polymeric scission products and cleavage of –C–H bonds. In sharp contrast, as depicted in Figure 4(b), no chemical changes were observed by FTIR in the TPE2 specimens after thermal aging. These results are very well agreeing with tensile test.

To examine the effect of solvent aging on the tensile properties, dumb-bell specimens of TPE1, TPE2, Alloy, and PBT were immersed in toluene, xylene, TCB, and NMP at room temperature, 60°C, and 120°C for 6 h. When TPE1 and TPE2 were solvent-immersed for 6 h, the tensile properties were slightly decreased with increasing solvent immersing temperature. In sharp contrast, for Alloy and PBT, both tensile

Table 3: Tensile properties of the thermal and solvent aged polymer specimens.

Sample code	Thermally aged in air		Thermally aged in solvent: Toluene		Xylene		TCB		NMP	
	Tensile strength (MPa)	Elongation at yield (%)	Tensile strength (MPa)	Elongation at yield (%)	Tensile strength (MPa)	Elongation at yield (%)	Tensile strength (MPa)	Elongation at yield (%)	Tensile strength (MPa)	Elongation at yield (%)
TPE1	55 ± 12	754 ± 23	47 ± 6	768 ± 87	48 ± 4	727 ± 75	47 ± 6	734 ± 95	43 ± 6	714 ± 106
TPE1-60°C-6 h	55 ± 11	630 ± 77	46 ± 5	694 ± 142	48 ± 6	681 ± 152	46 ± 7	687 ± 152	45 ± 5	656 ± 69
TPE1-120°C-6 h	52 ± 12	530 ± 63	46 ± 9	568 ± 37	48 ± 11	659 ± 64	44 ± 12	659 ± 73	33 ± 10	425 ± 41
TPE1-180°C-6 h	40 ± 4	180 ± 20	—	—	—	—	—	—	—	—
TPE2	42 ± 8	725 ± 79	43 ± 3	725 ± 80	43 ± 2	695 ± 56	44 ± 3	678 ± 126	39 ± 11	659 ± 17
TPE2-60°C-6 h	52 ± 4	757 ± 79	41 ± 4	673 ± 47	40 ± 1	650 ± 58	38 ± 4	652 ± 55	38 ± 7	614 ± 141
TPE2-120°C-6 h	56 ± 12	761 ± 61	41 ± 6	646 ± 124	40 ± 9	648 ± 50	36 ± 5	636 ± 45	20 ± 10	50 ± 5
TPE2-180°C-6 h	44 ± 4	739 ± 36	—	—	—	—	—	—	—	—
PBT	68 ± 3	497 ± 51	45 ± 6	521 ± 165	46 ± 6	458 ± 135	45 ± 8	484 ± 84	46 ± 4	478 ± 198
PBT-60°C-6 h	62 ± 9	460 ± 38	42 ± 6	378 ± 72	42 ± 9	365 ± 95	39 ± 6	416 ± 170	32 ± 8	461 ± 98
PBT-120°C-6 h	57 ± 6	255 ± 77	43 ± 13	211 ± 68	40 ± 13	215 ± 71	35 ± 6	115 ± 11	28 ± 4	24 ± 5
PBT-180°C-6 h	50 ± 4	29 ± 18	—	—	—	—	—	—	—	—
Alloy	46 ± 4	335 ± 11	52 ± 8	451 ± 109	51 ± 8	463 ± 103	53 ± 6	481 ± 88	61 ± 5	499 ± 77
Alloy-60°C-6 h	46 ± 13	325 ± 44	42 ± 3	391 ± 40	41 ± 6	326 ± 26	31 ± 11	230 ± 199	27 ± 8	127 ± 22
Alloy-120°C-6 h	30 ± 17	19 ± 2	39 ± 5	42 ± 15	34 ± 1	82 ± 30	24 ± 8	29 ± 3	3 ± 0	15 ± 6
Alloy-180°C-6 h	16 ± 11	6 ± 2	—	—	—	—			—	—

strength and elongation at break were more rapidly decreased with increasing immersing temperature. This resembles the tensile behavior of thermally aged samples. After solvent-immersing at 120°C, the reduction of tensile properties of Alloy was more discernible than those of other samples. In case of Alloy-120°C-6 h-TCB, an approximately 45% and 98% decrease in the tensile strength and elongation at break of 24 MPa and 29% were observed, respectively. PBT has a faster rate of crystallization most probably due to longer and flexible chains, whereas PET has a slower rate of crystallization because of more rigid molecules in structure that hinder movement of chains. Because the tensile specimens were hot-pressed and quenched in cold water, PET region of Alloy may not be fully crystallized. As shown in Figure 3, Alloy had the broad crystallization range from 110 to 175°C and was able to crystallize more readily in thermal aging temperature range than TPE1, TPE2, and PBT. PET tends to become brittle upon thermally induced crystallization. Despite its high thermal stability, Alloy resin loses its ductility and reduces the tensile properties significantly after thermal treatment [9].

It is interesting to note that the tensile properties of Alloy were increased significantly even after solvent immersion at room temperature than neat Alloy. After immersion in NMP at room temperature for 6 h, the tensile strength of Alloy increased by 32.5%, whereas this immersion in toluene, xylene, and TCB increased it by 13.8, 11.6, and 15.6%, respectively. A minimum 34.6% overall increase in the elongation at break from 451 to 499% was observed after solvent immersion at room temperature. This improvement of the tensile properties was attributed to an increase in the degree of crystallinity and an improvement in crystal perfection without thermal effect. Ribnick et al. [6] suggested that, even at room temperature, strong interactions could be observed between some solvents and the polyester fibers. Moore and Shelden [20] reported that the polar aromatic hydrocarbons such as benzene, toluene, and benzyl alcohol enhance the crystallization of polyester. Acceleration of the crystallization in an organic solvent might be attributed to the fact that the T_g was lowered in the organic solvent.

After immersion test in NMP at 120°C for 6 h, the tensile properties of all samples were significantly decreased compared to other solvents. NMP is a noncorrosive high boiler with excellent solvent power and dissolves polymers, such as cellulose derivatives, polyamides, polyesters, and many copolymers, at higher temperature. As a result, the specimens swell and deform significantly in NMP at 120°C, and this leads to undesirable tensile properties.

3.5. BDV Measurement after Solvent and Thermal Aging. To investigate the electric breakdown properties after solvent and thermal aging of polymer samples, TPE1, TPE2, Alloy, and PBT were extruded on the conductor and quenched at a rate of 200 m/min. Table 4 summarizes dielectric BDV of extruded wire before and after thermal treatment and solvent immersion. The BDV of all samples was gradually decreased with increasing thermal aging temperature. After thermal aging at 180°C, BDV of TPE1, TPE2, Alloy, and PBT insulated wire was decreased by 50.4, 31.1, 26.3, and 34.3%, respectively, compared to corresponding unaged ones. Dielectric breakdown causes a high-temperature arc to propagate through a dielectric medium and this can lead to ignition of the medium or of surrounding objects. The thermal aging is manifested in two ways by causing formation of polymer radicals [21] and microvoids [22] in the polymer materials due to increasing chain scission and volatility of degradation products.

The morphology and microstructure changes in the insulation layer after thermal aging at 180°C for 6 h are shown in Figure 5. After aging, microvoids were formed near the fracture surfaces indicating that the cavitation was not only a surface phenomenon, but characteristic of internal damage. They usually create a weakened pathway within the material by creating permanent molecular or physical changes. Some of such microvoids can coalesce into larger voids. As soon as sufficiently large voids are formed, from them a crack can start that ultimately breaks insulation.

The dielectric breakdown causes a high-temperature arc to propagate through a dielectric medium and this can lead to ignition of the medium or of the surrounding objects. Figure 6 demonstrates the breakdown points after dielectric breakdown test. As the thermal aging temperature increased, the char formation and size of breakdown point of all wires were decreased significantly. This means that the breakdown of thermally aged wire is determined by the electrical field at the tip of the defect [23].

After the extruded wire was exposed to the solvent immersion in toluene, xylene, and TCB, the BDV of TPE2 and PBT samples is not changed significantly compared to thermally aged ones. In case of PBT, the BDV reduction was more discernible after immersion in TCB than that of toluene and xylene. In sharp contrast, TPE1 was approximately 30% of the overall decrease in the BDV from 3.0 to 6.2 kV that was observed. When the extruded wire was immersed in NMP at 120°C for 6 h, the BDV of all samples decreased significantly. An approximately over 90% overall decrease in the BDV was observed. It can be seen that the BDV of TPE2 is higher than of other samples, and it has also the highest BDV after solvent aging. The voltage required for an electrical insulation to breakdown is a function of its thickness so that the BDV increases linearly with the insulator's thickness. Weigmann and coworkers [8] found that the formation of void-containing structures by organic solvent treatment depends on the ability of the solvent to swell the polymer and to cause recrystallization in the swollen state. While crystallization could occur during thermal shrinkage at higher temperatures, the crystallization of solvated polymer chains required a dissociation of the solvate and the exclusion of the solvent molecules. Park [24] reported that chemical etching of polyester filaments has revealed a complex stress cracking behavior which varied with orientation and thermal crystallization conditions.

The images in Figure 7 demonstrate the extruded wire surface after thermal treatment and xylene and NMP immersion at 120°C for 6 h. The surface images of xylene-immersed samples were similar to those of thermally aged ones except for the weak surface crack phenomenon of TPE1 and Alloy. In sharp contrast, the surface of NMP-immersed samples had many voids and eroded much more than that of thermally

Table 4: BDV changes of the TPE1, TPE2, PBT, and Alloy in various solvents.

Sample code	Breakdown voltage (kV)				
	Thermally aged	Thermally aged in solvent			
		Toluene	Xylene	TCB	NMP
TPE1	11.7 ± 0.8	6.2 ± 0.2	4.0 ± 0.3	4.6 ± 0.5	4.5 ± 0.8
TPE1-60°C-6 h	11.9 ± 0.7	5.6 ± 0.6	3.7 ± 0.5	4.7 ± 1.5	3.1 ± 0.5
TPE1-120°C-6 h	7.8 ± 1.4	5.2 ± 0.8	3.0 ± 0.5	4.3 ± 0.9	0.5 ± 0.1
TPE1-180°C-6 h	5.8 ± 1.5	—	—	—	—
TPE2	14.8 ± 1.2	14.2 ± 1.2	13.1 ± 1.1	13.9 ± 1.2	12.1 ± 1.1
TPE2-60°C-6 h	13.6 ± 0.8	13.2 ± 1.0	12.6 ± 1.2	11.7 ± 0.9	8.5 ± 0.2
TPE2-120°C-6 h	11.0 ± 1.1	12.8 ± 1.4	11.9 ± 0.8	11.3 ± 1.2	1.2 ± 0.1
TPE2-180°C-6 h	10.2 ± 1.3	—	—	—	—
PBT	13.7 ± 1.7	13.0 ± 1.7	12.4 ± 1.5	12.4 ± 0.8	11.8 ± 1.1
PBT-60°C-6 h	12.1 ± 1.3	12.2 ± 0.8	12.2 ± 1.0	11.0 ± 2.7	8.3 ± 0.1
PBT-120°C-6 h	10.5 ± 1.4	11.8 ± 1.9	11.0 ± 1.5	10.9 ± 1.6	0.7 ± 0.1
PBT-180°C-6 h	9.0 ± 1.5	—	—	—	—
Alloy	12.9 ± 1.2	12.3 ± 2.1	11.0 ± 1.9	8.1 ± 1.2	6.1 ± 0.4
Alloy-60°C-6 h	11.9 ± 1.3	9.4 ± 1.2	8.3 ± 0.9	6.1 ± 0.7	3.7 ± 0.1
Alloy-120°C-6 h	9.1 ± 3.0	9.0 ± 1.8	8.9 ± 2.6	4.7 ± 0.6	0.5 ± 0.1
Alloy-180°C-6 h	9.5 ± 1.8	—	—	—	—

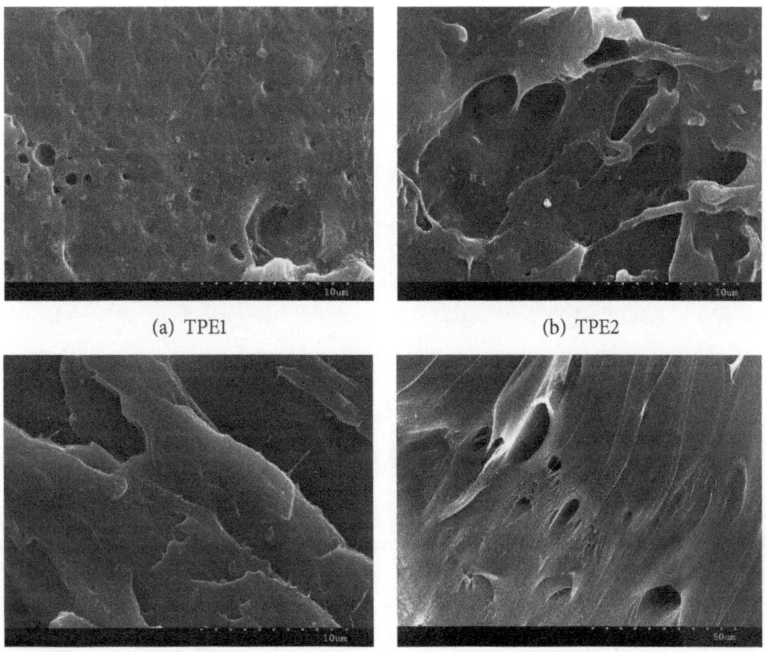

(a) TPE1 (b) TPE2

(c) Alloy (d) PBT

Figure 5: SEM micrographs of the fractured insulation layer after thermal aging at 180°C for 6 h.

and xylene-treated samples, and this agreed with the rapid BDV reduction results listed in Table 4. The presence of voids and cracking on the insulation significantly reduces the BDV [20, 22]. This is due to a combination of two factors: (1) air has lower breakdown strength than polymers and (2) presence of a void creates local electric field strength irregularities, with the lower permittivity of air causing an intensification of the electric field at the solid/void interface. Consequently, breakdown that preferentially occurs first at a solid/void interface has the consequence of disrupting the polymer structure and increasing the void size [21].

4. Conclusions

TPE1, TPE2, PBT, and Alloy were extruded as the insulation layer of low voltage electric wire, and its dielectric BDV changes after thermal and solvent aging were investigated. The presence of voids on the insulation can significantly

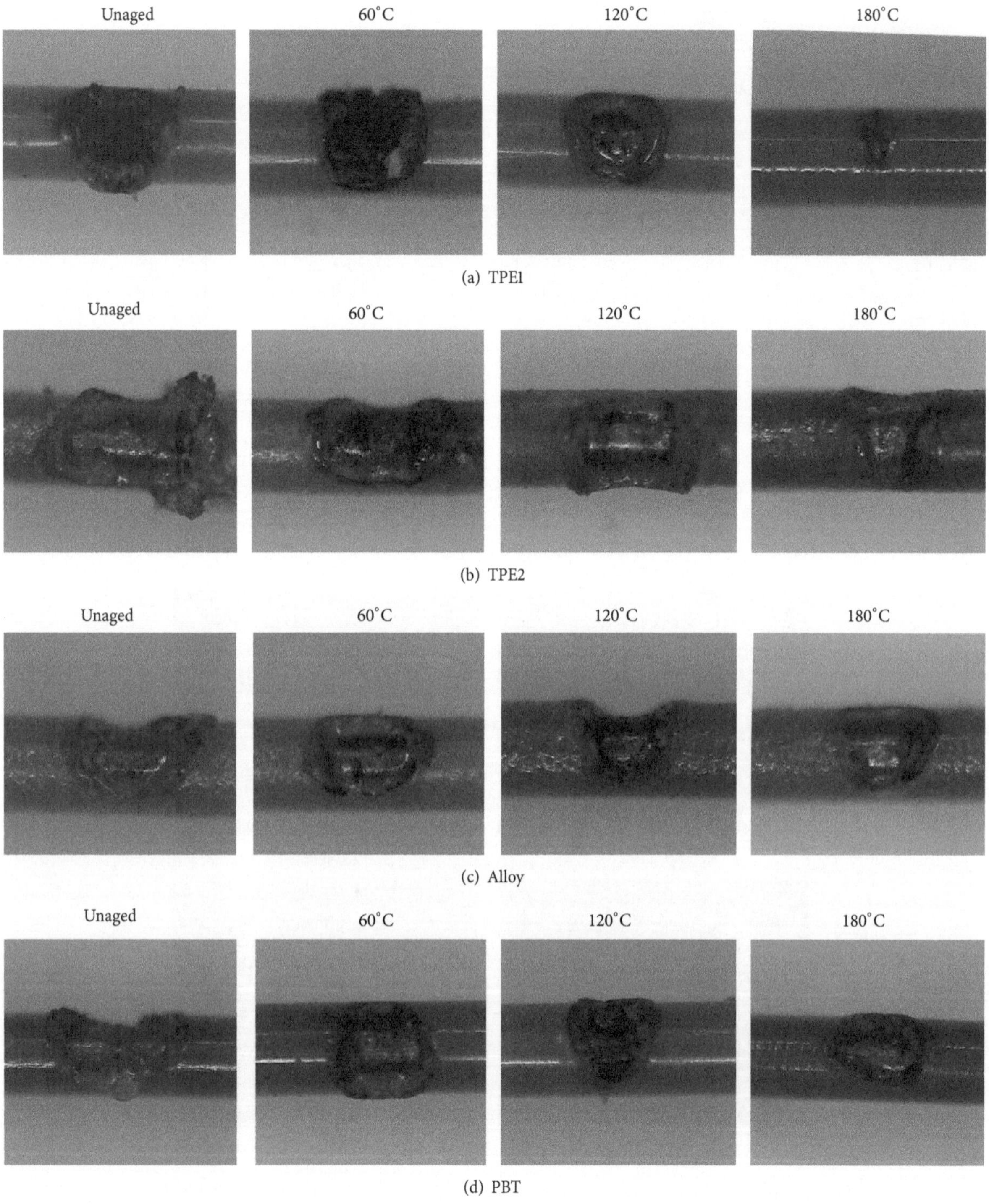

FIGURE 6: Images of the breakdown points after dielectric breakdown test (thermally aged at corresponding temperature for 6 h, 200x).

reduce the dielectric properties. SEM observation confirmed that there are significant voids or pores that were not detected in all extruded samples. In TG analysis, TPE1 shows an initial degradation temperature at 320°C which is the lowest of all investigated samples. In sharp contrast, TPE2 shows similar thermal degradation behavior to PBT because multiaromatic ring structures produced high thermal stability. Among them, Alloy showed the highest thermal stability. When Alloy and PBT were thermally aged in toluene, xylene, and TCB at 120°C for 6 h, the tensile properties were significantly decreased compared to TPE1 and TPE2. It can be also found that reduction in elongation at break of Alloy was more discernible than that of PBT. DSC traces confirmed that Alloy has the broad crystallization exotherm and it is able to

TPE1 TPE2 Alloy PBT

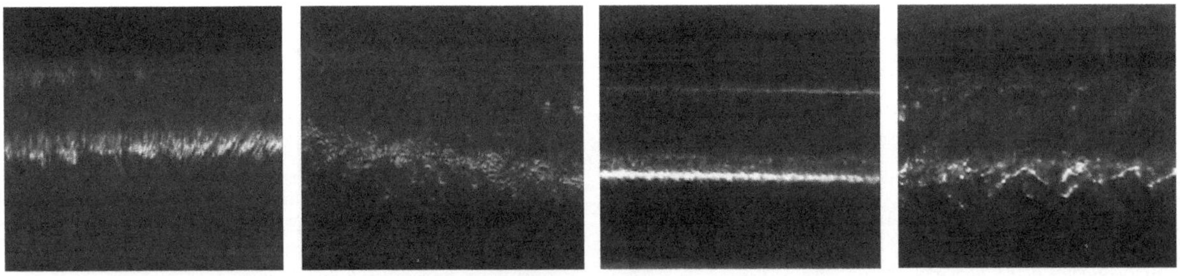

(a) At 120°C for 6 h

TPE1 TPE2 Alloy PBT

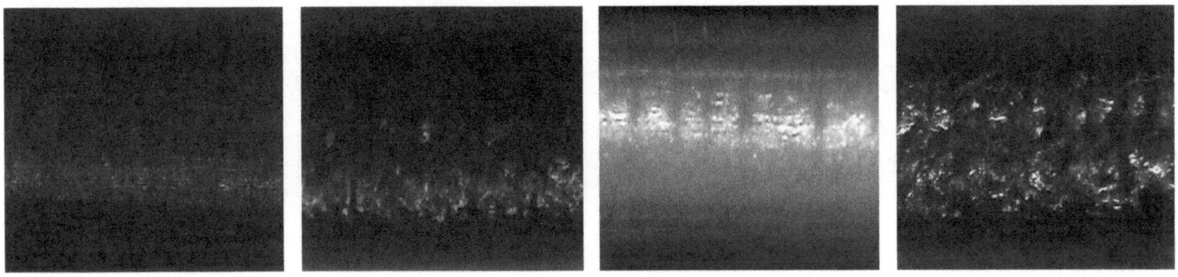

(b) At 120°C for 6 h in xylene

TPE1 TPE2 Alloy PBT

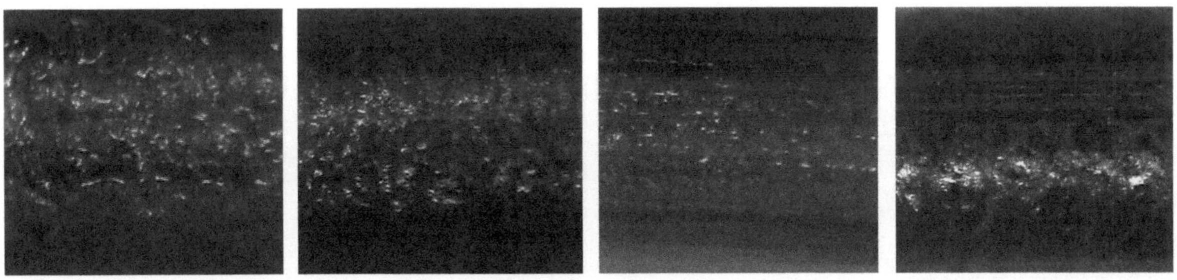

(c) At 120°C for 6 h in NMP

FIGURE 7: Microphotographs (200x) of the surface of extruded insulation layer after solvent aging at 120°C for 6 h.

crystallize more readily in thermal aging temperature range. This is because thermally degraded PET regions in Alloy become brittle upon thermally induced crystallization.

When the extruded wire was exposed the solvent immersion in toluene, xylene, and TCB, the BDV of TPE2 and PBT samples is not changed significantly compared to thermally aged ones. In sharp contrast, for the wire immersed in NMP at 120°C for 6 h, the BDV of all samples decreased significantly. Video microscope observation shows that the surface images of xylene-immersed samples were similar to those of thermally aged ones except for the weak surface crack phenomenon of TPE1 and Alloy. In sharp contrast, the surface of NMP-immersed samples had many voids and eroded much more than that of thermally and xylene-immersed samples and this agreed with the rapid BDV reduction results. Dielectric breakdown that preferentially occurs first at a solid/void interface has the consequence of disrupting the polymer structure and increasing the defects. The thermal and solvent aging process easily corrodes the extruded layer and effectively reduces its thickness causing a lowering in the dielectric properties. Deterioration of the insulating material reduces the dielectric strength and also reduces the ability of the final products to withstand short-circuit events. Thermal and solvent aging studies of the TPE1, TPE2, PBT, and Alloy showed significant reduction of BDV in accordance with aging temperature and solvent type indicating that selection of curing condition and solvents in varnish is a more important factor for determining life cycle of final products.

Conflict of Interests

This paper does not have a direct financial relation with the commercial identities mentioned in it that might lead to a conflict of interests for the author.

Acknowledgment

The author is grateful to the Small and Medium Enterprises (SMEs) Technology Innovation Program, Republic of Korea, for the financial support of this experimental work.

References

[1] M. T. Shaw and S. H. Shaw, "Water treeing in solid dielectrics," *IEEE Transactions on Electrical Insulation*, vol. 19, no. 5, pp. 419–452, 1984.

[2] R. J. Densley, "Investigation into the growth of electrical trees in xlpe cable insulation," *IEEE Transactions on Electrical Insulation*, vol. 14, no. 3, pp. 148–158, 1979.

[3] H. Y. Shin and E. S. Park, "Mechanical and dielectric breakdown properties of PBT/TPE, PBT/PBT/PET, and PBT/antioxidant blends," *Journal of Applied Polymer Science*, vol. 114, no. 5, pp. 3008–3015, 2009.

[4] E. S. Park, "Processibility and mechanical properties of micronized polytetrafluoroethylene reinforced silicone rubber composites," *Journal of Applied Polymer Science*, vol. 107, no. 1, pp. 372–381, 2008.

[5] A. M. Majdi and H. Günter, "Safety and Environmental Aspects in the Use of Impregnating Materials," in *ELANTAS Technical Presentations, Inductica Conference*, Berlin, Germany, 2008.

[6] A. S. Ribnick, H. D. Weigmann, and L. Rebenfeld, "Interactions of nonaqueous solvent with textile fibers—2. Isothermal shrinkage kinetics of a polyester yarn," *Textile Research Journal*, vol. 43, no. 3, pp. 176–183, 1973.

[7] B. H. Knox, H. D. Weigmann, M. G. Scott, and L. Rebenfeld, "Effects of an aqueous medium on the structure and physical properties of a polyester yarn," *Textile Research Journal*, vol. 51, no. 8, pp. 549–558, 1981.

[8] H. D. Weigmann, M. G. Scott, and A. S. Ribnick, "Interactions of nonaqueous solvents with textile fibers part X: effects of thermal history on polyester-solvent interactions," *Textile Research Journal*, vol. 48, no. 1, pp. 4–13, 1978.

[9] E. S. Park, "Morphology, mechanical, and dielectric breakdown properties of PBT/PET/TPE, PBT/PET/PA66, PBT/PET/LMPE, and PBT/PET/TiO_2 blends," *Polymer Composites*, vol. 29, no. 10, pp. 1111–1118, 2008.

[10] T. H. Ahn, Y. H. Park, S. H. Kim, and D. H. Baik, "Preparation and characterization of poly(ether ester) thermoplastic elastomers containing the 2,6-naphthalenedicarboxyl group," *Journal of Applied Polymer Science*, vol. 90, no. 13, pp. 3473–3480, 2003.

[11] Y. G. Jeong, W. H. Jo, and S. C. Lee, "Cocrystallization behavior of poly(butylene terephthalate-co-butylene 2,6-naphthalate) random copolymers," *Macromolecules*, vol. 33, no. 26, pp. 9705–9711, 2000.

[12] P. A. Atkinson, P. J. Maines, and G. A. Skinner, "Inorganic tin compounds as flame retardants and smoke suppressants for polyester thermosets," *Thermochimica Acta*, vol. 360, no. 1, pp. 29–40, 2000.

[13] I. H. Hall and M. G. Pass, "Chain conformation of poly(tetramethylene terephthalate) and its change with strain," *Polymer*, vol. 17, no. 9, pp. 807–816, 1976.

[14] M. Yokouchi, Y. Sakakibara, Y. Chatani, H. Tadokoro, T. Tanaka, and K. Yoda, "Structures of two crystalline forms of poly(butylene terephthalate) and reversible transition between them by mechanical deformation," *Macromolecules*, vol. 9, no. 2, pp. 266–273, 1976.

[15] H. Watanabe, "Stretching and structure of polybutylene-naphthalene-2,6-dicarboxylate films," *Kobunshi Ronbunshu*, vol. 33, no. 4, pp. 299–237, 1976.

[16] H. Koyano, Y. Yamamoto, Y. Saito, T. Yamanobe, and T. Komoto, "Crystal structure of poly(butylene-2,6-naphthalate)," *Polymer*, vol. 39, no. 18, pp. 4385–4391, 1998.

[17] A. Escala and R. S. Stein, "Crystallization studies of blends of polyethylene terephthalate and polybutylene terephthalate," *Advances in Chemistry Series*, vol. 176, pp. 455–487, 1979.

[18] N. Avramova, "Amorphous poly(ethylene terephthalate)/poly(butylene terephthalate) blends: miscibility and properties," *Polymer*, vol. 36, no. 4, pp. 801–808, 1995.

[19] G. K. Guenther and D. G. Baird, "Rheology of a textured fluid consisting of poly(ethylene terephthalate) and nylon 6,6," *Journal of Applied Polymer Science*, vol. 59, no. 5, pp. 845–859, 1996.

[20] W. R. Moore and R. P. Shelden, "The crystallization of polyethylene terephthalate by organic liquids," *Polymer*, vol. 2, pp. 315–321, 1961.

[21] L. A. Dissado and J. C. Fothergill, *Electrical Degradation and Breakdown in Polymers*, Peter Peregrinus, London, UK, 1992.

[22] R. Bartnikas and E. J. McMahon, *Engineering Dielectrics*, vol. 2B, ASTM, Philadelphia, Pa, USA, 1987.

[23] R. S. Chauhan, M. V. S. Rao, and N. E. Dweltz, "Chemical etching of poly(ethylene trephthalate) filaments," *Journal of Applied Polymer Science*, vol. 30, no. 1, pp. 19–33, 1985.

[24] E. S. Park, "Effects of electron beam irradiation on properties of ETFE insulated electric wire," *Iranian Polymer Journal (English Edition)*, vol. 20, no. 11, pp. 873–885, 2011.

6

Green Synthesis of Hydroxyethyl Cellulose-Stabilized Silver Nanoparticles

M. A. El-Sheikh, S. M. El-Rafie, E. S. Abdel-Halim, and M. H. El-Rafie

National Research Centre, Textile Research Division, El-Behooth Street, Dokki, P.O. 12311, Cairo, Egypt

Correspondence should be addressed to M. A. El-Sheikh; dr.manal.elsheikh@gmail.com

Academic Editor: Joo Hyun Kim

Green synthesis aims to minimize the use of unsafe reactants and maximize the efficiency of synthesis process. These could be achieved by using environmentally compassionate polymers and nontoxic chemicals. Hydroxyethyl cellulose (HEC), an ecofriendly polymer, was used as both reducing and stabilizing agents in the synthesis of stable silver nanoparticles, while silver nitrate was used as a precursor and water as a solvent. The formation of silver nanoparticles was assessed by monitoring UV-vis spectra of the silver colloidal solution. The size of the nanoparticles was measured using transmission electron microscope (TEM). Reaction kinetics was followed by measuring the absorbance of silver colloidal solution at different time intervals. Optimum reaction conditions revealed that the highest absorbance was obtained using HEC : $AgNO_3$ of 1.5 : 0.17 (g/100 cm^3) at 70°C for 120 min at pH 12. The Ag^0 nanoparticles colloidal solution so obtained (1000 ppm) were found stable in aqueous solution over a period of six months at room temperature (~ 25°C). The sizes of these nanoparticles were found in the range of 11–60 nm after six months of storing. FTIR spectra confirmed the interaction of both the aldehyde and OH groups in the synthesis and stabilization of silver nanoparticles.

1. Introduction

Nanotechnology is an enabling technology which deals with structures ranging from approximately 1 to 100 nm in at least one dimension. The nanosize results in specific physicochemical characteristics that may differ from those of the bulk substance or particles of larger size. This effect is mainly attributed to high surface area to volume ratio, which potentially results in high reactivity. Because of these specific characteristics, the use of substances in nanoform may have advantages over the use of bulk chemicals. Silver has been used since ancient times for jewelry, utensils, monetary and currency, dental alloy, photography, explosives, and so forth. Until the introduction of antibiotics, silver was also used for its antiseptic activity, specifically in the management of open wounds and burns. While an antibiotic can kill as many as a half-dozen different kinds of disease organisms, silver is known to kill 650 species of bacteria or viruses simply by coming into contact with them. Due to its antimicrobial properties, silver has also been used in medical textile applications [1–4] and incorporated in filters to purify drinking water and clean swimming pool water. Particle size has been the only limiting factor to silver's effectiveness [5]. Ag/Al_2O_3 catalysts are used in the catalytic reduction of NO_x [6–8].

Increasing understanding about the environment has led to focus on "green chemistry." Although metal nanoparticles exhibited properties due to quantum size effects, their synthesis protocol creates major environmental problems [9–11]. Generally, metal nanoparticles can be prepared and stabilized by physical and chemical methods; the chemical approach, such as chemical reduction, electrochemical techniques, and photochemical reduction, is the most widely used method. Chemical reduction is the most frequently applied method for the preparation of silver nanoparticles as stable, colloidal dispersions in water or organic solvents. Commonly used reductants and solvents are sodium borohydride, citrate, ascorbate, formaldehyde, N,N-dimethylformamide, hydrazine and elemental hydrogen. These chemicals are highly reactive and cause potential environmental and biological risks [12, 13]. Utilization of nontoxic chemicals, environmentally gentle solvents, and renewable materials is one of the key issues that worth important consideration in a green synthesis policy [14–16]. Biosynthesis is also an alternative green technique for synthesis of silver nanoparticles [17–21].

In earlier reports, natural polymers like starch [12, 22–28], chitosan [29], and cellulose [30–32] were reported to stabilize silver nanoparticles. Silver nanorods with average diameters of 20 nm and lengths up to 250 nm were successfully prepared in large scale using HEC at which silver nitrate was reduced by γ irradiation [31].

Silver can be present in four different oxidation states: Ag^0, Ag^+, Ag^{2+}, and Ag^{3+}. The elemental silver is Ag^0 (0 oxidation state) and free silver ion is Ag^+. Ag^0 and Ag^+ are the most abundant ones, while Ag^{2+} and Ag^{3+} are unstable in the aquatic environment. The reduction of silver ions (Ag^+) in aqueous solution generally yields colloidal silver with particle diameters of several nanometers. Initially, the reduction of various complexes with Ag^+ ions leads to the formation of silver atoms (Ag^0), which is followed by agglomeration into oligomeric clusters. These clusters eventually lead to the formation of colloidal Ag particles. When the colloidal particles are much smaller than the wavelength of visible light, the solutions have a yellow color with an intense band in the 380–400 nm range and other less intense or smaller bands at longer wavelength in the absorption spectrum. This band is attributed to collective excitation of the electron gas in the particles, with a periodic change in electron density at the surface (surface plasmon absorption). Since the polymers prevent agglomeration and precipitation of the particles, they have been frequently employed as stabilizers in chemical synthesis of metal nanoparticles [5, 33, 34].

Since the polymers prevent agglomeration and precipitation of the particles, they have been frequently employed as capping or stabilizing agents in chemical synthesis of metal nanoparticles. Among methods of green synthesis of silver nanoparticles, water is used as an environmentally benign solvent and polysaccharides as a capping agent, or in some cases, polysaccharides serve as both a reducing and a capping agent. Synthesis of starch-silver nanoparticles was carried out with starch as a capping agent and β-D-glucose as a reducing agent in a gently heated system. The starch in the solution mixture avoids use of relatively toxic organic solvents. In addition, the binding interactions between starch and silver nanoparticle are weak and can be reversible at higher temperatures, allowing separation of the synthesized particles. In a case of dual polysaccharide function, silver nanoparticles were synthesized by the reduction of Ag^+ inside of nanoscopic starch templates. The extensive network of hydrogen bonds in the templates provides surface passivation or protection against nanoparticle aggregation [23].

Cellulose and its derivatives showed unique advantages because they are the most abundant natural polysaccharide with low cost, better biodegradability, and biocompatibility. HEC is a representative derivative of cellulose with excellent water solubility and biocompatibility. HEC is manufactured by reacting alkali-treated cellulose with ethylene oxide at which a series of hydroxyethyl cellulose is produced. The manner in which ethylene oxide is added to cellulose can be described by two terms: Degree of substitution (DS) and molar substitution (MS). The degree of substitution designates the average number of hydroxyl positions on the anhydroglucose unit that has been reacted with ethylene oxide. Since each anhydroglucose unit of the cellulose molecule has three hydroxyl groups, the maximum value for D.S. is three (Scheme 1(a)). Molar substitution is defined as the average number of ethylene oxide molecules that have reacted with each anhydroglucose unit. Once a hydroxyethyl group is attached, it can further react with additional groups in an end-to-end formation. As long as ethylene oxide is available, this reaction can continue, theoretically without limit (Scheme 1(b)). The excellent properties of HEC allow it to be used in many biotechnological, biophysical, and industrial fields. Because of the existence of abundant reactive -OH groups on the HEC chains, HEC is liable to be modified to derive new materials with improved properties [35, 36].

To fulfill the growing need of environmental friendly nanoparticles synthesis, this work aims to prepare environmentally friendly silver nanoparticles using HEC (a green reagent) and simple method at mild reaction conditions. Being an organic polymer bearing abundant reactive -OH groups on its chains, HEC can act as both reducing agent for silver nitrate and stabilizing agent for silver nanoparticles formed. Silver nitrate is used as a precursor and water as a solvent. Sodium hydroxide is used to adjust the pH of the reaction to pH 12. Thus, the reaction medium only contains HEC, silver nitrate, sodium hydroxide, and distilled water. No more chemicals are used. Factors affecting the reduction and stability as well as the shape and size of the formed silver nanoparticles along with reaction mechanism are studied.

2. Experimental

2.1. Materials. Hydroxyethyl cellulose medium viscosity 1 (75–125 mPa·s) was purchased from FLUKA BioChemika. Silver nitrate and sodium hydroxide were of laboratory grade chemicals.

2.2. Method

2.2.1. Preparation of Silver Nanoparticles. Definite weight of HEC was dissolved in distilled water using heating magnetic stirrer. After complete dissolution, the pH of the solution was adjusted to 12 using dilute solution of sodium hydroxide. The temperature was raised gradually to reach to the desired temperature (60–90°C). Certain volume of diluted silver nitrate solution was then added dropwise to the mixture keeping in mind that the total volume of the reaction mixture is 100 cm^3 and silver nitrate concentration varies from 0.017 to 0.17 g/100 cm^3. The reaction mixture was kept under continuous stirring for different durations (5–180 min). After addition of silver nitrate solution, the reaction medium starts to acquire a clear yellow color which develops to brown indicating the formation of silver nanoparticles (Ag^0). After completion of the synthesis, the colloidal solution formed was checked for the presence of excess silver ions (Ag^+) using dilute NaCl solution. When no white precipitate is observed, this means the complete transformation of (Ag^+) to silver nanoparticles (Ag^0). The development of the reaction was evaluated by withdrawing aliquots from resultant Ag^0 colloidal solution at given time intervals.

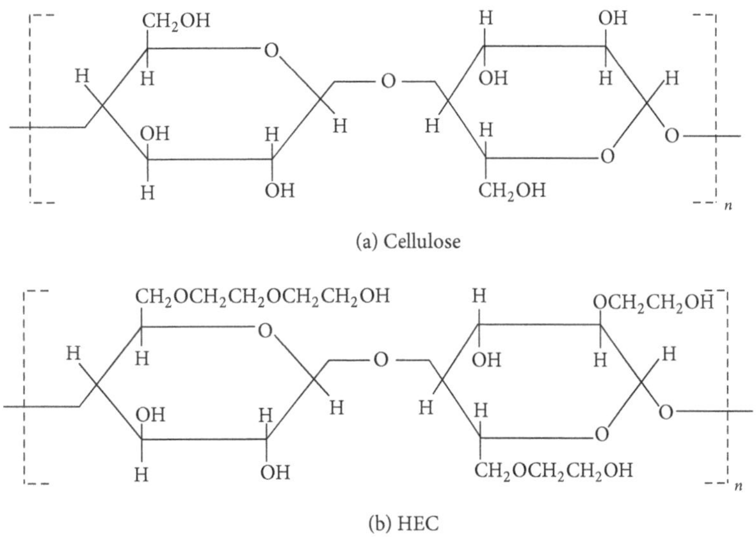

SCHEME 1: Structure of (a) cellulose and (b) HEC.

To obtain a solid substance of HEC-silver nanoparticles for the Fourier transform infrared (FTIR) measurements, the HEC-silver nanoparticles colloidal solution was precipitated using a mixture of absolute acetone and ethanol (1 : 1). After precipitation, the solution was allowed to settle down; the solvents were separated by decantation then filtration. The solid substance was air oven dried at 30°C.

2.3. Testing and Analysis. The UV-vis spectra of silver nanoparticles were recorded using UV-2401, UV-vis spectrophotometer, Shimadzu, Japan, at wavelength range from 300 to 550 nm. Formation of silver nanoparticles is expressed as absorbance of the colloidal solution of the samples under test. The absorbance, the broadening, and the wavelength of the band measure the intensity of the colloidal solution, that is, the conversion of silver ions to silver nanoparticles. A solution containing HEC alone was used as blank. Very concentrated silver nanoparticles samples did not show one smooth band (either sharp or broad) but showed a number of crowded sharp bands. This behavior leads to false readings. So, concentrated samples which showed this behavior were diluted "x" times, and the obtained absorbance value was then multiplied by "x" to obtain the actual absorbance value. The accuracy of this dilution technique was tested by comparing the absorbance readings of certain samples with moderate concentration before and after dilution. The readings were found approximately the same after multiplying by "x" times of dilutions.

Transmission electron microscope was used to characterize the shape and size of the synthesized silver nanoparticles. This was done by means of a JEOL-JEM-1200 transmission electron microscope. The samples were prepared by placing a drop of the colloidal solution on a 400 mesh copper grid coated by an amorphous carbon film and evaporating the solvent in air at room temperature. The average diameter of the silver nanoparticles was determined from the diameter of 100 nanoparticles found in several chosen areas in enlarged microphotographs.

Stability of the silver nanoparticles was tested by storing samples in sealed glass bottles in dark place at room temperature (~ 25°C) for six months then recording the spectra of the solutions using UV-vis spectrophotometer as well as the shape and size using TEM.

FTIR spectra of both HEC and HEC-silver nanoparticles were obtained using Perkin-Elmer spectrum 1000 spectrophotometer. The solid samples were mixed with KBr forming round disk for the FTIR measurements. The FTIR spectra were scanned over a range of wave numbers from 4000 to 400 (cm^{-1}).

3. Results and Discussion

3.1. Mechanism of Silver Nanoparticles Formation and Stabilization Using HEC. The reduction of transition metal salts in solution is the most widely practiced method of generating colloidal suspensions of the metals. The so-called "alcohol reduction process" is a very general process for the production of metal nanoparticles, often stabilized by organic polymers. In general, the alcohols which were useful reducing agents contained α-hydrogen and were oxidized to the corresponding carbonyl compounds [37].

Hydroxyethyl cellulose macromolecule consists of chemically modified linear polymeric cellulose. In addition to the original alcoholic groups, new alcoholic groups are introduced as a result of the hydroxyethylation process (Scheme 1(b)). Also, in addition to original reducing aldehydic end groups in the HEC, new aldehydic end groups are generated as a result of the oxidative degradation of HEC molecule during the synthesis of silver nanoparticles under the effect of sodium hydroxide (pH 12) at relatively high temperature (60–90°C) in the synthesis medium. While the reduction keeps on, silver nanoparticles grow gradually, and HEC will further form a stable protection layer on the silver nanoparticles surface. All these components with their anionic and reducing properties support the utilization of HEC as a reducing and stabilizing agent for the synthesis of silver nanoparticles.

Factors affecting the reduction and stability as well as the shape and size of the formed silver nanoparticles along with reaction mechanism are given in what follows.

3.2. Effect of HEC Concentration and Reaction Time on the Formation of Silver Nanoparticles. Figure 1 shows the effect of increasing HEC concentration (HEC) along with increasing reaction time on the formation of silver nanoparticles. Figures 1(a)–1(d) represent the spectra of silver nanoparticles colloidal solution after different lengths of time (5–120 min) at HEC of 0.15, 0.3, 0.6, and 0.9 g/100 cm^3, respectively. At HEC of 0.15 g/100 cm^3, up to 30 min, the bands are very broadened with little absorbance values indicating little conversion of Ag^+ to Ag^0. After 45 min, the broadening of the band decreases and the absorbance value increases indicating better conversion to Ag^0. From 60 to 120 min, the bands are sharp and a "bell shape" is represented with higher absorbance values. It is also clear that absorbance values at 60, 90, and 120 min are nearly the same showing a state of leveling off. Leveling off at 60–120 min means that no more conversion to silver nanoparticles will take place at the set of reaction conditions. At HEC of 0.3 g/100 cm^3, up to 15 min, the bands are broad with little absorbance, and starting from 20 min, broadening of the bands decreases while absorbance increases. A state of leveling off is reached at 30–60 min where sharp bands with bell shapes and higher absorbance values are reached. At HEC of 0.6 g/100 cm^3, the band at 15 min is broad and is not representing the formation of silver nanoparticles. At 30–120, the bands are sharp, bell shaped with high absorbance which clearly expresses the formation of silver nanoparticles. At HEC of 0.9 (w/v), at 5 min the band is broad with low absorbance, while after 10 min, the band broadening is less indicating partial conversion of silver ions to silver nanoparticles. At 15–120, the bands are sharp, bell shaped, and the absorbance is higher than after 10 min. The absorbance at 15–120 is nearly the same meaning that a state of leveling off that started after 15 min took place. It is clear from the previous results that by increasing the HEC concentration, the formation of silver nanoparticles increases at even lower synthesis time. To conclude, bell-shaped bands with high absorbance values appeared after 60, 30, 30, and 15 min for HEC 0.15, 0.3, 0.6, and 0.9, respectively. The conversion of Ag^+ to Ag^0 as a result of increasing the concentration of HEC could be attributed to the increase in the number of reducing groups and to the enhancement in the stabilization efficiency of the formed silver nanoparticles. It should also be mentioned that the least amount of HEC in the reaction medium (0.3 g/100 cm^3) is enough for full reduction of the Ag^+ to Ag^0 nanoparticles. To focus on the effect of HEC on the formation of silver nanoparticles alone while keeping the synthesis time fixed at 60 min, a small figure was introduced (Figure 1(d)(D)). Figure 1(d)(D) shows four sharp bands with the "bell shape" at wavelength 390–420 which is identical to the formation of silver nanoparticles [23, 38]. Although the increase in HEC favors the rate of formation of silver nanoparticles in its initial stage, and after 60 min, the behavior is quite different. Figure 1(d)(D) shows that the increase in HEC from 0.15 to 0.3 is accompanied by an increase in the absorbance. Further increase in HEC from 0.6 to 0.9 is accompanied by marginal decrement in the absorption intensity to reach its minimal value at 0.9 which could be attributed to some aggregation of the formed silver nanoparticles. However, the differences in the absorbance values among the four bands are not significant.

3.3. Effect of Temperature and Reaction Time on the Formation of Silver Nanoparticles. Figures 2(a)–2(d) show the effect of increasing the reaction temperature from 60 to 90°C after different lengths of time (5–120 min) on the synthesis of silver nanoparticles. As shown in Figure 2, performing the reaction at 60°C is not sufficient to synthesize silver nanoparticles, regardless of the increase in reaction time up to 120 min. This is clear from the very broad bands with no bell shape and little absorbance. However, at 70°C, up to 15 min, the bands are broad with little absorbance, and starting from 20 min, broadening of the bands decreases, while absorbance increases. A state of leveling off is reached at 30–60 min where sharp bands with bell shapes and higher absorbance values are reached. At 80°, at 5 min, the band is broad with little absorbance, and at 10 min, broadening of the band decreases, while absorbance increases. A state of leveling off is reached at 15–60 min where sharp bands with bell shapes and higher absorbance values are reached. At 90°, at 5–10 min, the bands are broad with little absorbance, and at 15 min, broadening of the band sharply decreased, while absorbance increased. A state of leveling off is reached at 20–90 min where sharp bands with bell shapes and higher absorbance values are reached. To summarize, bell-shaped bands with high absorbance values did not appear at 60°C even after up to 120 min but appeared at 30, 15, and 20 min when reaction was performed at 70, 80, and 90°C, respectively. This could be attributed to the favorable effect of temperature on the formation of new reducing groups from HEC by oxidative degradation of HEC in presence of sodium hydroxide. On the other hand, prolonging the reaction time (60 min) at a temperature higher than 70°C (80–90°C) is accompanied by decrease in the absorbance values as seen in Figure 2(d)(D). The decrease in the absorbance value as a result of increasing the reaction temperature could be explained in terms of the weak binding interactions between HEC and silver nanoparticles which can be reversible at higher temperatures, allowing separation of the synthesized particles [23].

3.4. Effect of Silver Nitrate Concentration and Reaction Time on the Formation of Silver Nanoparticles. Figures 3(a)–3(d) show the effect of increasing the concentration of $AgNO_3$ from 0.017 to 0.0.051 g/100 cm^3 after different durations (5–180 min) on the synthesis of silver nanoparticles. As shown in Figure 3, performing the reaction using $AgNO_3$ of 0.017 g/100 cm^3, up to 15 min, the bands are broad with little absorbance, and starting from 20 min, broadening of the bands decreases, while absorbance increases. A state

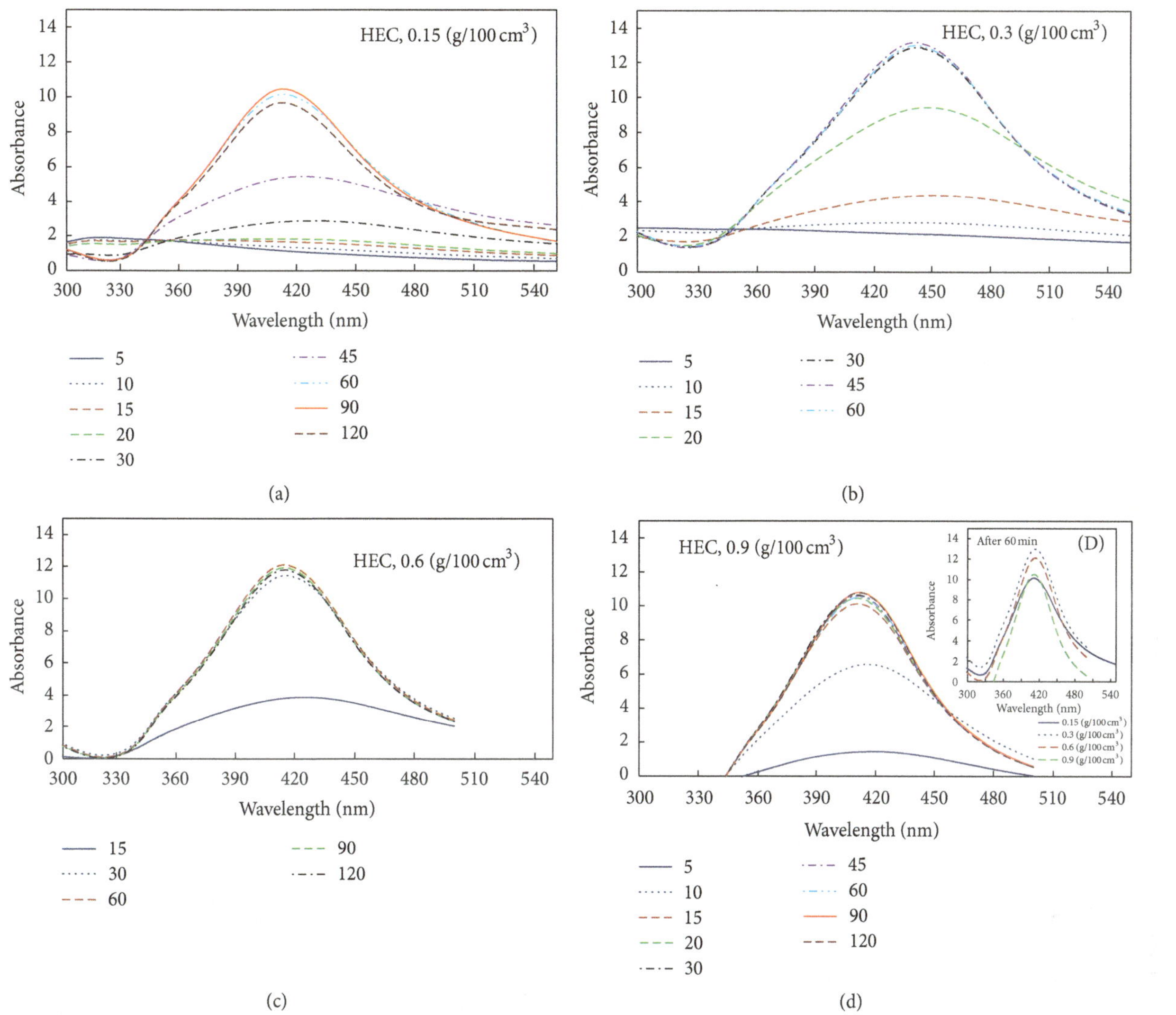

FIGURE 1: Effect of HEC concentration and reaction time on the formation of silver nanoparticles HEC, 0.15–0.9 g/100 cm^3; pH, 12; temperature, 70°C; $AgNO_3$, 0.017 g/100 cm^3; time, 5–120 min.

of leveling off is reached at 30–60 min where bands with bell shapes and higher absorbance values are reached. At $AgNO_3$ of 0.034 g/100 cm^3, up to 45 min, the bands are broad with little absorbance. At 60 min, broadening of the band decreases, while absorbance increases. The broadening continued to decrease and the absorbance to increase until it reaches its maximum at 120 min where bands with bell shapes and higher absorbance values are reached. At $AgNO_3$ of 0.051 g/100 cm^3, up to 60 min, the bands are broad with little absorbance. From 90 to 120 min, broadening of the band slightly decreased, while absorbance increased. The broadening continued to decrease and absorbance to increase until it reaches its maximum at 180 min where a band with bell shape and higher absorbance values is reached. To conclude, bell-shaped bands with high absorbance values appeared at 30, 60, and 90 min when reaction was performed using $AgNO_3$ of 0.017, 0.034, and 0.051 g/100 cm^3, respectively. Figure 3(d) which shows the effect of increasing $AgNO_3$ on the absorbance of the colloidal solution of silver nanoparticles at 60 min is in agreement with these findings. It seems from the previous results that the amount of reducing groups in HEC is only sufficient to bring silver nanoparticles when silver nitrate concentration is in the range of 0.017 g/100 cm^3. More reducing groups are required to guarantee the conversion of Ag^+ to Ag^0 when $AgNO_3$ is to be increased to more than 0.017 g/100 cm^3. This could be achieved by increasing both HEC and $AgNO_3$ together in a parallel way.

3.5. Effect of Increasing Both HEC and $AgNO_3$ Concentrations Together and Reaction Time on the Formation of Silver Nanoparticles. Figures 4(a)–4(d) show the effect of increasing both HEC and $AgNO_3$ concentrations together (0.6 : 0.068, 0.9 : 0.102, and 1.5 : 0.17 (g/100 cm^3)) and reaction time (5 : 120 min) on the formation of silver nanoparticles. As shown in Figure 4, performing the reaction using HEC : $AgNO_3$ of 0.6 : 0.068 brings bell-shaped bands with high absorbance regardless of the reaction time. Figure 4

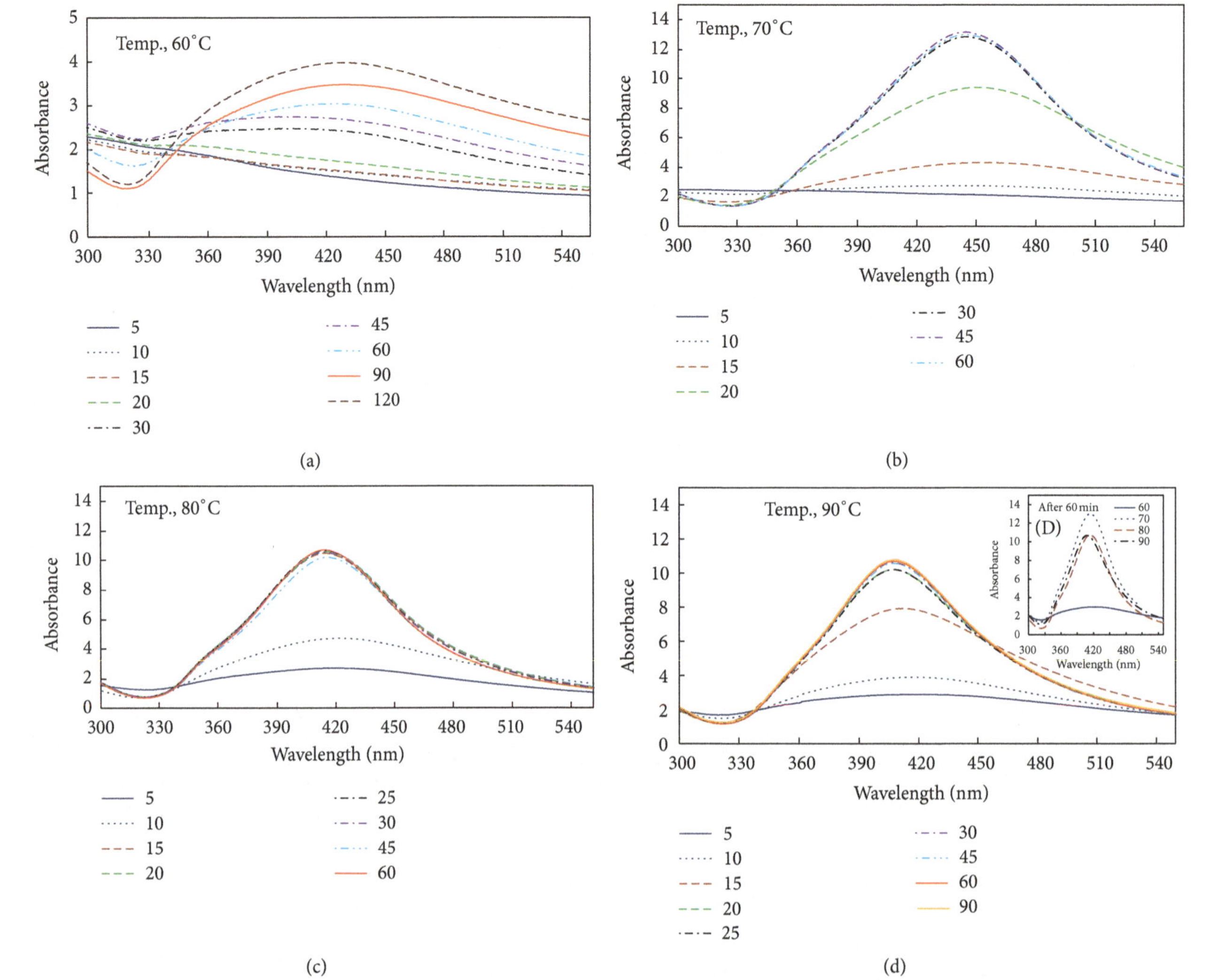

FIGURE 2: Effect of temperature and reaction time on the formation of silver nanoparticles HEC, 0.3 g/100 cm^3; pH, 12; temperature, 60–90°C; $AgNO_3$, 0.017 g/100 cm^3; time, 5–120 min.

also shows that the absorbance increases by increasing the reaction time from 15 to 30 min after which it remains constant, and a state of leveling off is reached from 30 to 120 min. By using HEC : $AgNO_3$ of 0.9 : 0.102 at 15 min the band is broad with little absorbance. From 90 to 120 min, broadening of the band sharply decreased, while the absorbance sharply increased with appearance of clear bell-shaped band. Using HEC : $AgNO_3$ of 1.5 : 0.17 (g/100 cm^3), at 15–30 min, the bands are broad with little absorbance. From 60 to 120 min, broadening of the band sharply decreased while the absorbance sharply increased with appearance of clear bell-shaped bands. The absorbance reaches its highest value at 120 min. To conclude, bell-shaped bands with high absorbance values appeared at 15, 30, and 60 min when reaction was performed using HEC : $AgNO_3$ of 0.6 : 0.068, 0.9 : 0.102, and 1.5 : 0.17 (g/100 cm^3), respectively. By making reaction time constant (60 min), as shown in Figure 4(d), the increase in both HEC : $AgNO_3$ is accompanied by an increase in the absorbance. Results of Figure 3 and results of Figure 4 assure that there must be parallel increase in both HEC and $AgNO_3$ in order to obtain high yield of silver nanoparticles where the amount of reducing groups in HEC must be sufficient to reduce all Ag^+ in the solution to Ag^0. More concentrations of $AgNO_3$ must be accompanied by parallel increase in HEC concentration.

3.6. Stability of Silver Nanoparticles after Six Months. Stability of silver nanoparticles was tested by measuring the absorbance of the colloidal solution after synthesis (1) and after six months of storing (2). Figure 5 shows absorbance of the colloidal solution of samples with different concentrations of both HEC and silver nitrate (HEC : $AgNO_3$, 0.6 : 0.068, 0.9 : 0.102, and 1.5 : 0.17) (g/100 cm^3) at identical reaction conditions. As seen in Figure 5, the absorbance curve and the absorbance maxima of each pair of samples, namely, 0.6 : 0.068 (1) and 0.6 : 0.068 (2); 0.9 : 0.102 (1) and 0.9 : 0.102 (2), are identical showing that the silver nanoparticles are stable at the storing conditions. The pair of samples 1.5 : 0.17 (1) and 1.5 : 0.17 (2) showed even higher absorbance after storing for six months. The increase of the absorbance of the colloidal solution of 1.5 : 0.17 (2) after such long period

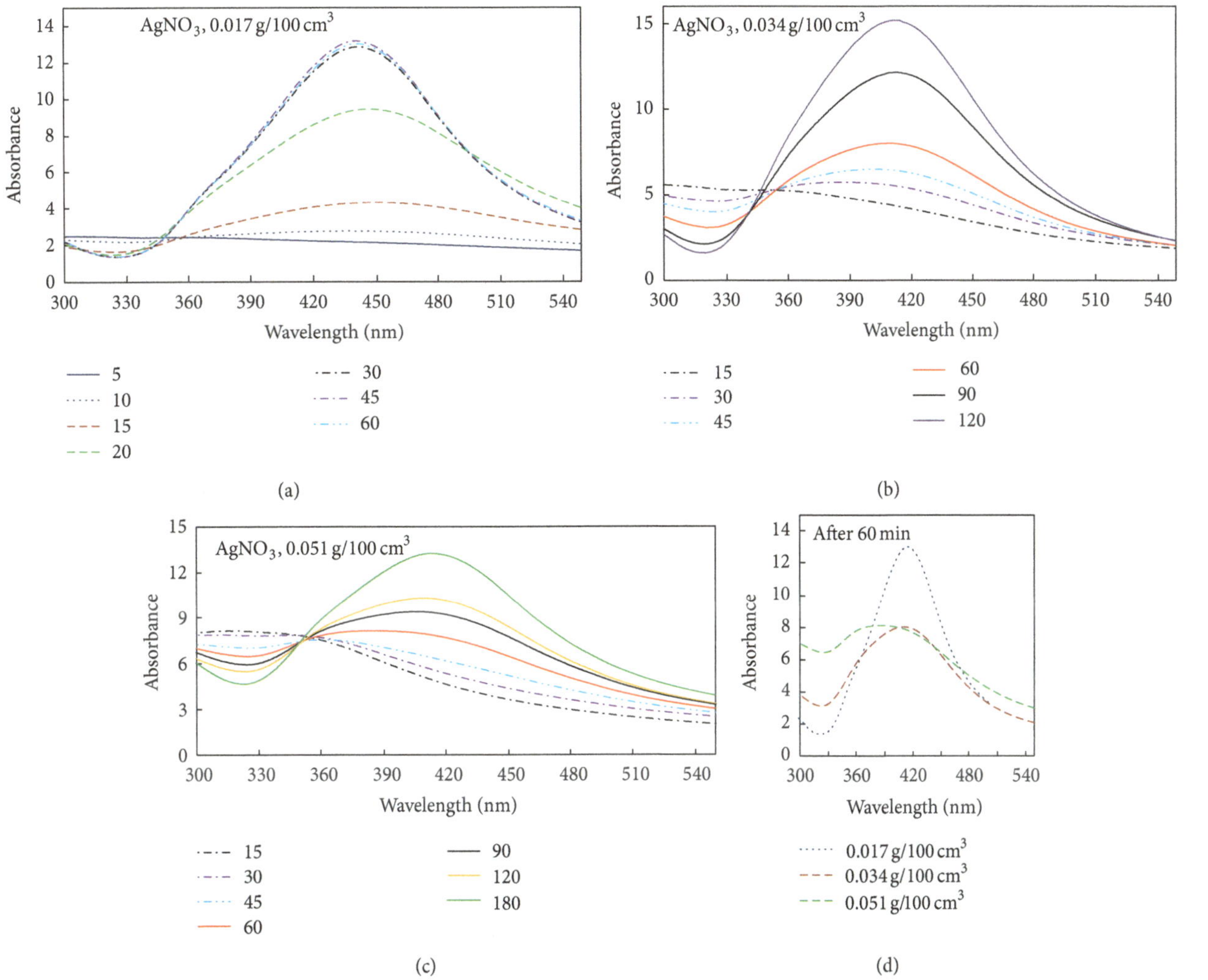

FIGURE 3: Effect of $AgNO_3$ concentration and reaction time on the formation of silver nanoparticles HEC, 0.3 g/100 cm^3; pH, 12; temperature, 70°C; $AgNO_3$, 0.017–0.051 g/100 cm^3; time, 5–180 min.

of storing means that the reaction was continued at room temperature until all silver irons are converted to silver nanoparticles followed by stability of the colloidal solution.

3.7. TEM Micrograph and Particle Size Distribution Histogram of Silver Nanoparticles. Figure 6 shows TEM micrograph and particle size distribution histogram of silver nanoparticles synthesized using different HEC : $AgNO_3$ (g/100 cm^3): (a) 0.6 : 0.068, (b) 0.9 : 0.102, and (c) 1.5 : 0.17 after six months of storing. The three micrographs show that silver nanoparticles have spherical shape morphology, fine dispersion, and mean particle sizes between 11 and 55, 6 and 55, and 11 and 60 nm for HEC : $AgNO_3$ 0.6 : 0.068, 0.9 : 0.102, and 1.5 : 0.17 (g/100 cm^3) respectively. Highest count% between 26 and 30, 31 and 35% for HEC : $AgNO_3$ 0.6 : 0.068, 0.9 : 0.102 (g/100 cm^3), respectively and 11–20 and 31–35% for HEC : $AgNO_3$ 1.5 : 0.17 (g/100 cm^3). It is clear that the increase in the HEC : $AgNO_3$ has no effect on the particle size and shape. It is worth to mention that even after six months of storing, the particle size is generally in the range of 6–60 nm.

3.8. FTIR Spectra of HEC and HEC-Silver Nanoparticles. HEC is an organic polymer bearing abundant reactive –OH groups and some aldehyde groups on its chains. High reaction temperature (70°C) favors the formation of new reducing aldehyde groups by oxidative degradation of HEC in presence of sodium hydroxide (pH 12). The interaction of both the aldehyde and hydroxyl groups in the reduction of silver ions and stabilization of silver nanoparticles obtained was confirmed by FTIR spectra of HEC-silver nanoparticles compared with HEC.

Figure 7 shows the FTIR spectra of both HEC and HEC-silver nanoparticles. The FTIR spectra of HEC show bands at 3410, 2928, 1644, and 1060 cm^{-1} characteristic for hydrogen stretching of bonded O–H, hydrogen stretching of C–H, double bond stretching of carbonyl, carboxyl and carboxyl salt, and hydrogen bending of O–H, respectively, with corresponding intensities of 57, 59, 67, and 56% *T*, respectively. Same characteristic bands for HEC-silver nanoparticles appeared at 3426, 2928, 1647, and 1060 cm^{-1} at either the same frequencies (2928 and 1060 cm^{-1}) or with

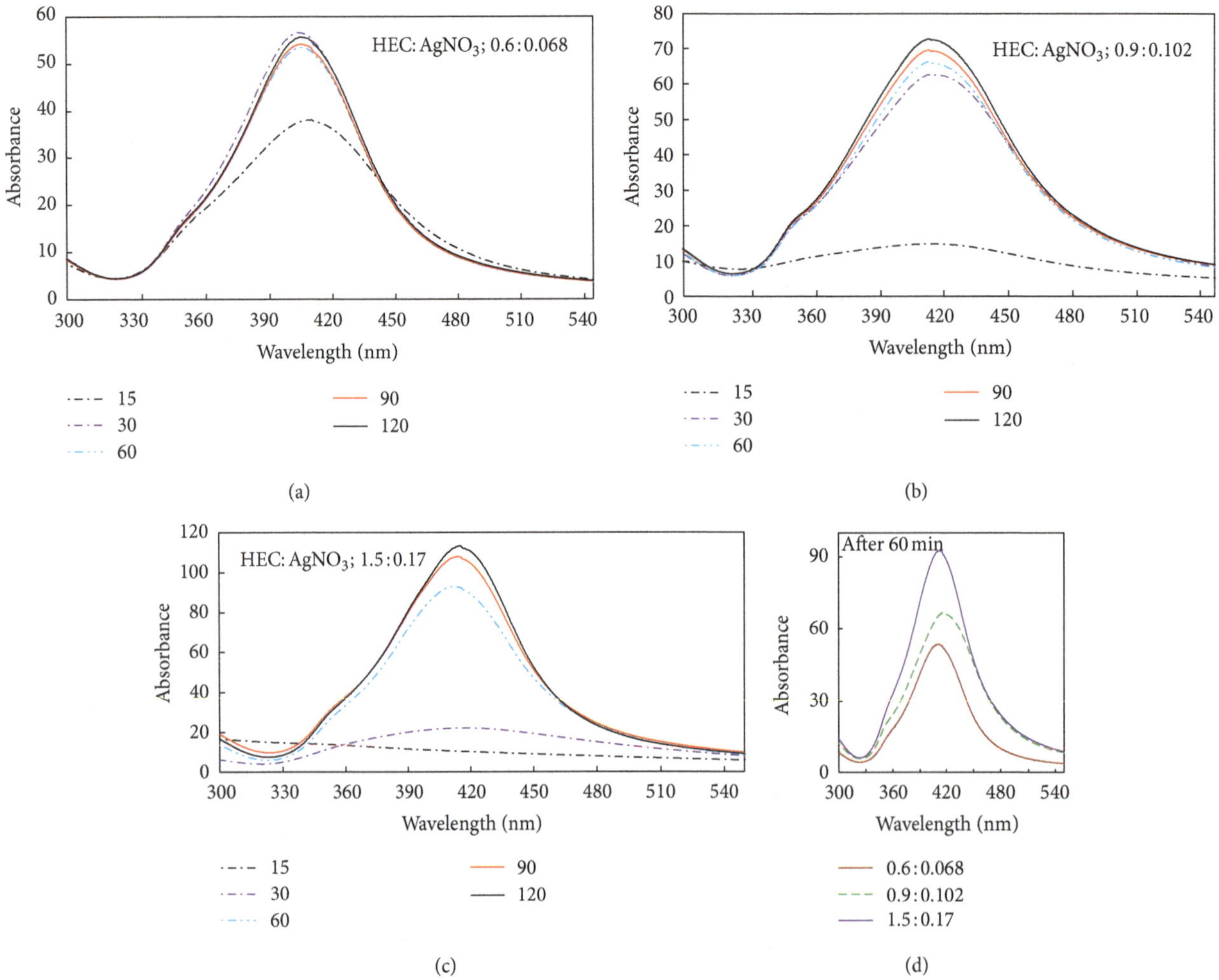

FIGURE 4: Effect of the ratio between HEC and $AgNO_3$ concentration and reaction time on the formation of silver nanoparticles HEC : $AgNO_3$, 0.6–1.5: 0.068 : 0.17 (g/100 cm^3); pH, 12; temperature, 70°C; time, 15–120 min.

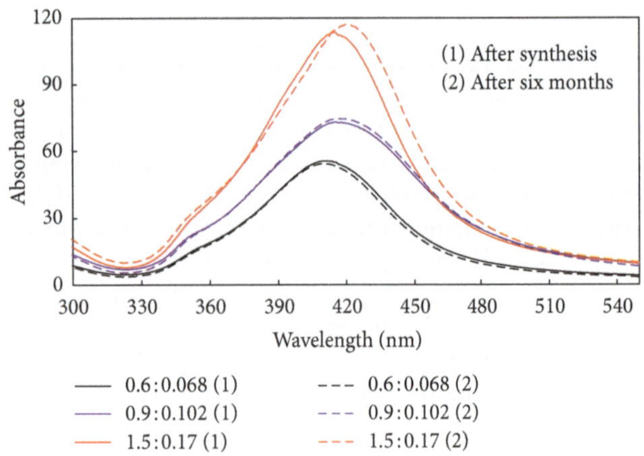

FIGURE 5: Stability of silver nanoparticles after six months. HEC : $AgNO_3$, 0.6–1.5: 0.068 : 0.17 (g/100 cm^3); pH, 12; temperature, 70°C; time, 60 min, after six months of synthesis.

little shift to higher frequencies (3426 and 1647 cm^{-1}). The strengths of the bands of HEC-silver nanoparticles (86, 90, 87, and 84% T, resp.) are completely different from HEC. These observations confirm the interaction of both the aldehyde and the OH groups in the synthesis and stabilization of silver nanoparticles.

4. Conclusions

Green synthesis of silver nanoparticles was performed using HEC as both reducing and stabilizing agent, silver nitrate as precursor, water as solvent, and sodium hydroxide to bring alkaline reaction medium of pH 12. At low reaction times, the increase in [HEC] favors the rate of formation of silver nanoparticles, while longer durations are not favorable. This observation was repeated in studying other parameters; that is, it is not recommended to extend reaction for more duration after synthesis is completed because aggregations of silver nanoparticles can take place. Low reaction temperature (60°C) and high reaction temperatures (80–90°C) have adverse effect on the synthesis of silver nanoparticles, while 70°C is optimal. Increasing silver nitrate concentration should be accompanied by increasing HEC concentration in a parallel way to guarantee complete conversion of Ag^+ to

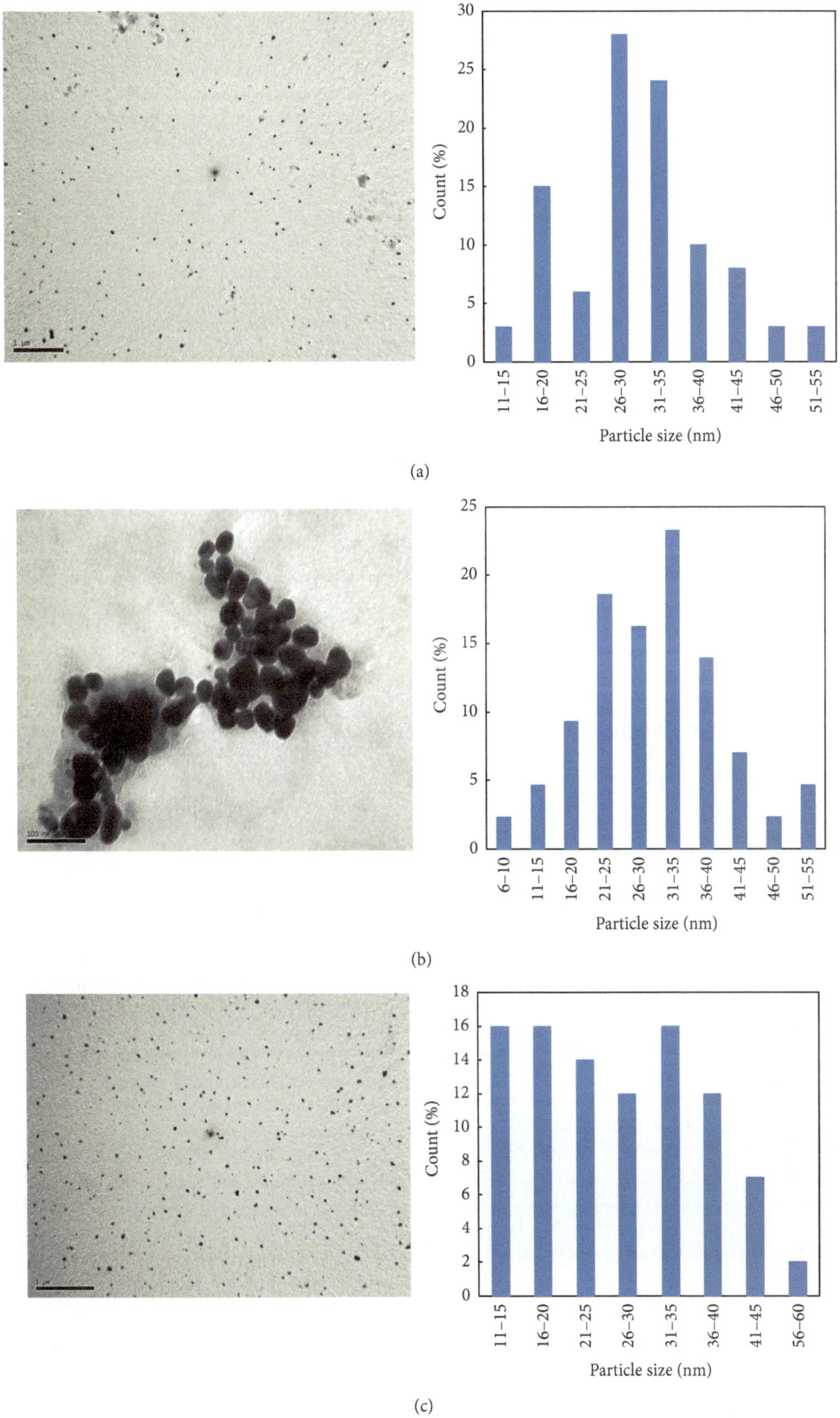

FIGURE 6: TEM micrograph and particle size distribution histogram of silver nanoparticles synthesized using different HEC : $AgNO_3$ (g/100 cm^3): (a) 0.6 : 0.068, (b) 0.9 : 0.102, and (c) 1.5 : 0.17 pH, 12; temperature, 70°C; time, six months.

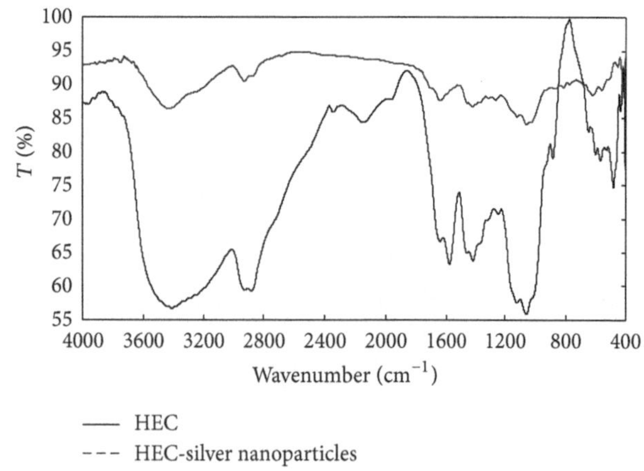

FIGURE 7: FTIR of HEC and HEC-silver nanoparticles HEC : $AgNO_3$, 0.6 : 0.068 (g/100 cm^3); pH, 12; temperature, 70°C; time, 60 min. The FTIR spectra of HEC-silver nanoparticles were obtained using the solid form of the substance.

Ag^0 if we are after high yield of silver nanoparticles. Optimal reaction conditions that yield the highest absorbance values were HEC : $AgNO_3$ of 1.5 : 0.17 (g/100 cm^3), 70°C for 120 min at pH 12. The nanoparticles so obtained were found stable in aqueous solution over a period of six months at room temperature (~25°C). The size of these nanoparticles was found in the range of 11–60 nm after six months of storing. FTIR spectra confirm the interaction of both the aldehyde and OH groups in the synthesis and stabilization of silver nanoparticles.

References

[1] A. I. Wasif and S. K. Laga, "Use of nano silver as an antimicrobial agent for cotton," *Autex Research Journal*, vol. 9, no. 1, pp. 5–13, 2009.

[2] D. Tessier, I. Radu, and M. Filteau, "Antimicrobial fabrics coated with nano-sized silver salt crystals," in *Proceedings of the NSTI Nanotechnology Conference and Trade Show—NSTI Nanotech*, M. Laudon and B. Romanowicz, Eds., pp. 762–764, Anaheim, Calif, USA, May 2005.

[3] M. Gouda, "Nano-zirconium oxide and nano-silver oxide/cotton gauze fabrics for antimicrobial and wound healing acceleration," *Journal of Industrial Textiles*, vol. 41, no. 3, pp. 222–240, 2012.

[4] M. H. El-Rafie, A. A. Mohamed, T. I. Shaheen, and A. Hebeish, "Antimicrobial effect of silver nanoparticles produced by fungal process on cotton fabrics," *Carbohydrate Polymers*, vol. 80, no. 3, pp. 779–782, 2010.

[5] S. W. P. Wijnhoven, W. J. G. M. Peijnenburg, C. A. Herberts et al., "Nano-silver—a review of available data and knowledge gaps in human and environmental risk assessment," *Nanotoxicology*, vol. 3, no. 2, pp. 109–138, 2009.

[6] J. P. Breen, R. Burch, C. Hardacre, and C. J. Hill, "Structural investigation of the promotional effect of hydrogen during the selective catalytic reduction of NO_x with hydrocarbons over Ag/Al_2O_3 catalysts," *Journal of Physical Chemistry B*, vol. 109, no. 11, pp. 4805–4807, 2005.

[7] K. Arve, F. Klingstedt, K. Eränen et al., "Analysis of the state and size of silver on alumina in effective removal of NO_x from oxygen rieh exhaust gas," *Journal of Nanoscience and Nanotechnology*, vol. 6, no. 4, pp. 1076–1083, 2006.

[8] A. Musi, P. Massiani, D. Brouri, J.-M. Trichard, and P. Da Costa, "On the characterisation of silver species for SCR of NO_x with ethanol," *Catalysis Letters*, vol. 128, no. 1-2, pp. 25–30, 2009.

[9] A. Henglein, "Small-particle research: physicochemical properties of extremely small colloidal metal and semiconductor particles," *Chemical Reviews*, vol. 89, no. 8, pp. 1861–1873, 1989.

[10] L. N. Lewis, "Chemical catalysis by colloids and clusters," *Chemical Reviews*, vol. 93, no. 8, pp. 2693–2730, 1993.

[11] A. P. Alivisatos, "Semiconductor clusters, nanocrystals, and quantum dots," *Science*, vol. 271, no. 5251, pp. 933–937, 1996.

[12] N. Vigneshwaran, R. P. Nachane, R. H. Balasubramanya, and P. V. Varadarajan, "A novel one-pot "green" synthesis of stable silver nanoparticles using soluble starch," *Carbohydrate Research*, vol. 341, no. 12, pp. 2012–2018, 2006.

[13] P. Dallas, V. K. Sharma, and R. Zboril, "Silver polymeric nanocomposites as advanced antimicrobial agents: classification, synthetic paths, applications, and perspectives," *Advances in Colloid and Interface Science*, vol. 166, no. 1-2, pp. 119–135, 2011.

[14] P. Raveendran, J. Fu, and S. L. Wallen, "Completely "Green" synthesis and stabilization of metal nanoparticles," *Journal of the American Chemical Society*, vol. 125, no. 46, pp. 13940–13941, 2003.

[15] J. M. DeSimone, "Practical approaches to green solvents," *Science*, vol. 297, no. 5582, pp. 799–803, 2002.

[16] M. Poliakoff and P. Anastas, "A principled stance," *Nature*, vol. 413, no. 6853, p. 257, 2001.

[17] P. Vankar and D. Shukla, "Biosynthesis of silver nanoparticles using lemon leaves extract and its application for antimicrobial finish on fabric," *Applied Nanoscience*, vol. 2, no. 2, pp. 163–168, 2012.

[18] Z. Sadowski, "Biosynthesis and application of silver and gold nanoparticles," in *Silver Nanoparticles*, D. P. Perez, Ed., pp. 257–276, In Tech, 2010.

[19] S. Kaviya, J. Santhanalakshmi, and B. Viswanathan, "Biosynthesis of silver nano-flakes by Crossandra infundibuliformis leaf extract," *Materials Letters*, vol. 67, no. 1, pp. 64–66, 2012.

[20] A. Saxena, R. M. Tripathi, F. Zafar, and P. Singh, "Green synthesis of silver nanoparticles using aqueous solution of Ficus benghalensis leaf extract and characterization of their antibacterial activity," *Materials Letters*, vol. 67, no. 1, pp. 91–94, 2012.

[21] M. H. El-Rafie, T. I. Shaheen, A. A. Mohamed, and A. Hebeish, "Bio-synthesis and applications of silver nanoparticles onto cotton fabrics," *Carbohydrate Polymers*, vol. 90, no. 2, pp. 915–920, 2012.

[22] M. Z. Kassaee, A. Akhavan, N. Sheikh, and R. Beteshobabrud, "γ-Ray synthesis of starch-stabilized silver nanoparticles with antibacterial activities," *Radiation Physics and Chemistry*, vol. 77, no. 9, pp. 1074–1078, 2008.

[23] V. K. Sharma, R. A. Yngard, and Y. Lin, "Silver nanoparticles: green synthesis and their antimicrobial activities," *Advances in Colloid and Interface Science*, vol. 145, no. 1-2, pp. 83–96, 2009.

[24] M. Valodkar, A. Bhadoria, J. Pohnerkar, M. Mohan, and S. Thakore, "Morphology and antibacterial activity of carbohydrate-stabilized silver nanoparticles," *Carbohydrate Research*, vol. 345, no. 12, pp. 1767–1773, 2010.

[25] A. Hebeish, M. E. El-Naggar, M. M. G. Fouda, M. A. Ramadan, S. S. Al-Deyab, and M. H. El-Rafie, "Highly effective antibacterial textiles containing green synthesized silver nanoparticles," *Carbohydrate Polymers*, vol. 86, no. 2, pp. 936–940, 2011.

[26] M. H. El-Rafie, M. E. El-Naggar, M. A. Ramadan, M. M. G. Fouda, S. S. Al-Deyab, and A. Hebeish, "Environmental synthesis of silver nanoparticles using hydroxypropyl starch and their characterization," *Carbohydrate Polymers*, vol. 86, no. 2, pp. 630–635, 2011.

[27] V. Djoković, R. Krsmanović, D. K. Božanić et al., "Adsorption of sulfur onto a surface of silver nanoparticles stabilized with sago starch biopolymer," *Colloids and Surfaces B*, vol. 73, no. 1, pp. 30–35, 2009.

[28] P. Raveendran, J. Fu, and S. L. Wallen, "A simple and "green" method for the synthesis of Au, Ag, and Au-Ag alloy nanoparticles," *Green Chemistry*, vol. 8, no. 1, pp. 34–38, 2006.

[29] H. Huang, Q. Yuan, and X. Yang, "Preparation and characterization of metal-chitosan nanocomposites," *Colloids and Surfaces B*, vol. 39, no. 1-2, pp. 31–37, 2004.

[30] J. Chen, J. Wang, X. Zhang, and Y. Jin, "Microwave-assisted green synthesis of silver nanoparticles by carboxymethyl cellulose sodium and silver nitrate," *Materials Chemistry and Physics*, vol. 108, no. 2-3, pp. 421–424, 2008.

[31] W. Liu, Z. Zhang, H. Liu, W. He, X. Ge, and M. Wang, "Silver nanorods using HEC as a template by γ-irradiation technique and absorption dose that changed their nanosize and morphology," *Materials Letters*, vol. 61, no. 8-9, pp. 1801–1804, 2007.

[32] A. A. Hebeish, M. H. El-Rafie, F. A. Abdel-Mohdy, E. S. Abdel-Halim, and H. E. Emam, "Carboxymethyl cellulose for green synthesis and stabilization of silver nanoparticles," *Carbohydrate Polymers*, vol. 82, no. 3, pp. 933–941, 2010.

[33] N. Singh and P. K. Khanna, "In situ synthesis of silver nanoparticles in polymethylmethacrylate," *Materials Chemistry and Physics*, vol. 104, no. 2-3, pp. 367–372, 2007.

[34] D. V. Goia, "Preparation and formation mechanisms of uniform metallic particles in homogeneous solutions," *Journal of Materials Chemistry*, vol. 14, no. 4, pp. 451–458, 2004.

[35] W. Wang, J. Wang, Y. Kang, and A. Wang, "Synthesis, swelling and responsive properties of a new composite hydrogel based on hydroxyethyl cellulose and medicinal stone," *Composites B*, vol. 42, no. 4, pp. 809–818, 2011.

[36] J. Reuben, "Description and analysis of hydroxyethyl cellulose," *Macromolecules*, vol. 17, no. 2, pp. 156–161, 1984.

[37] J. S. Bradley, G. Schmid, D. V. Talapin, E. V. Shevchenko, and H. Weller, "Syntheses and characterizations: 3. 2 synthesis of metal nanoparticles," in *Nanoparticles From Theory to Application*, pp. 185–238, Wiley-VCH Verlag GmbH & Co. KGaA, 2005.

[38] E. S. Abdel-Halim, M. H. El-Rafie, and S. S. Al-Deyab, "Polyacrylamide/guar gum graft copolymer for preparation of silver nanoparticles," *Carbohydrate Polymers*, vol. 85, no. 3, pp. 692–697, 2011.

7

Hydrogen Sorption onto Hypercrosslinked Polymer Decorated with Metal-Organic Framework

Renuka R. Gonte,[1] P. C. Deb,[1] and K. Balasubramanian[2]

[1] *Department of Applied Chemistry, Defence Institute of Advanced Technology, Girinagar, Pune 411 025, India*
[2] *Department of Materials Engineering, Defence Institute of Advanced Technology, Girinagar, Pune 411 025, India*

Correspondence should be addressed to K. Balasubramanian; balask@diat.ac.in

Academic Editor: Marc Visseaux

Metal-organic framework decorated styrene-maleic acid (SMA) co-polymer beads were synthesized by slow diffusion of TEA. The obtained materials were characterized by FTIR, scanning electron microscopy, and thermogravimetric analysis. The different dicarboxylic acid linkers influenced the crystal morphologies and hydrogen-storage behavior of the polymer beads. Hydrogen-storage measurements revealed slow micropore opening mechanism involved in adsorption of hydrogen.

1. Introduction

Hydrogen is considered as an interesting energy vector particularly for vehicular applications due to its attractive properties such as nontoxicity, clean combustion, and high energy density [1]. Hydrogen binds to surfaces by weak dispersive interactions (physisorption) or through stronger chemical associations (chemisorption). Physisorption correlates with surface area, high surface area favours higher gas uptake. Thus, materials with large surface areas and low densities are attractive candidates for hydrogen-storage applications [2]. The physisorption method of hydrogen storage required sorbent with a large surface area such as activated carbons and carbon nanostructures [3], zeolites [4], porous polymers [5], and metal-organic frameworks (MOFs) [6]. Due to weak sorbent-sorbate interaction, physisorption-based hydrogen-storage systems show fast kinetics with a charging time of minutes. However, the same weak interaction results in less than 2 wt% gravimetric hydrogen uptake at ambient temperature and applicable pressure [7]. However, a number of significant technological hurdles need to be overcome for realizing these applications which include safe, compact, and high capacity hydrogen-storage systems [8]. Low-density microporous solids have garnered considerable scientific attention due to their potential practical applications including emerging energy challenges, catalysis, hydrogen and methane storage, and chemical separations [9]. Hypercrosslinked polymers (HCPs) represent a class of predominantly microporous organic materials that exhibit high surface areas, "Davankov-type" resins being the most studied hypercrosslinked materials [10]. Due to light elemental compositions, additional functionalities can be introduced to the porous organic polymers through large number of synthesis routes; chemical and thermal stability and scalable technology possess potential advantage for gas sorption [11]. Porous metal-organic frameworks (MOFs) have been emerging as one type of the most promising materials for hydrogen storage because of their amenability to design and extraordinary permanent porosity [12]. Potential reaction sites on the organic ligands in MOFs render postsynthetic modification possible, leading to the introduction of additional active sites for stronger hydrogen binding. All these characteristics of MOFs make them one of the most promising sorbent materials for hydrogen storage [13].

The structural versatility of MOFs has given new insights for rational design and assembly of materials possessing novel topologies and exceptional host-guest properties. Synthesis of polymer materials and exploring specific properties of polymer confinement are achievable via nanochannels of MOFs. Polymer encapsulation is also achievable through tailored pore size optimization which has opened new dimensions for material based MOF-polymer hybridization [14]. Rodrigues et al. [15] have revealed the association of metallic core with organic linkers to yield a variety of arrangements and have demonstrated that change in the organic linker aids in control of bonded water to the central metal at molecular level in absence of any hydroxo or aquo ligands which acts as a precatalyst for polymerization of isoprene.

In the present work, hypercrosslinked polymers decorated with metal-organic frameworks were developed as materials for hydrogen storage. The effect of varying organic ligands in MOF formation was studied. Hypercrosslinked polymer was chosen for this study due to its high surface area, microporous network structure, and easy synthesis method. The MOFs were synthesized by slow diffusion of TEA. The formation of metal-organic frameworks with carboxylic acid units of polymer and $Zn(NO_3)_2{\cdot}6H_2O$ was also studied to examine the effect of formation of additional network in the polymer. These materials were characterized by scanning electron microscopy (SEM), thermogravimetric analysis (TG), and Fourier Transform Infrared (FTIR) spectroscopy. The hydrogen-storage capacity was measured at 300 K by gas sorption analyzer. The preliminary results indicate that such MOF decorated hypercrosslinked polymers can be considered for hydrogen storage.

2. Materials and Methods

2.1. Materials. Styrene (S, Merck India) and divinylbenzene (DVB, Fluka, India) were purified by washing with 1N NaOH. Maleic anhydride (MAn, Merck, India) and AIBN were recrystallised from chloroform prior to use. NaOH, methylisobutylketone (MIBK), NaCl, $Zn(NO_3)_2{\cdot}6H_2O$, 1,4-benzenedicarboxylic acid (H_2BDC), 1,6-naphthalenedicarboxylic acid (H_2NDC), biphenyldicarboxylic acid (H_2BPDC), N,N dimethylformamide (DMF), and TEA were of analytical reagent grade from commercial sources and were used without further purification.

2.2. Synthesis

2.2.1. Synthesis of Hypercrosslinked SMA Beads. Hypercrosslinked styrene-maleic acid (SMA) copolymer beads were synthesized by suspension polymerization technique using free radical initiator as reported in our previous communication [16]. Briefly, homogeneous solution of monomers (styrene and MAn), crosslinker (DVB) in MIBK was added to aqueous salt solution with in situ formed magnesium hydroxide. The mixture was stirred at 800 rpm to obtain uniform sized (200 μm–400 μm diameter) spherical beads of crosslinked styrene-maleic anhydride which were further hydrolyzed to produce crosslinked styrene-maleic acid (SMA) copolymer beads. The beads were then water soxhlated and finally acetone soxhlated and dried under vacuum at 50°C to constant weight.

2.2.2. Synthesis of Metal-Organic Framework Decorated SMA Beads. Metal-organic frameworks were synthesized by slight modification from the literature following the method of slow diffusion of TEA [17]. Solution containing $Zn(NO_3)_2{\cdot}6H_2O$ and dicarboxylic acid in DMF was diluted with toluene. Requisite amounts of SMA beads were allowed to swell in this solution. The mixture was placed under reduced pressure for 5 hr to ensure complete filling up of pores with solutions. The supernatant liquid was decanted and slow vapor diffusion of triethylamine (TEA) in toluene was allowed. The system was sealed and kept undisturbed for 2 days at room temperature. These SMA beads were then washed with DMF and dried under vacuum at 50°C to constant weight.

2.3. Characterization. The SMA beads and SMA beads decorated with MOFs were characterized using Fourier Transform Infrared (FTIR) spectroscopy (Model FTIR-8400, Shimadzu, Japan). Each sample was grounded thoroughly with KBr to form pellet which was then analyzed in the wavelength region 4000 to 400 cm^{-1}. The morphologies of the materials were observed on a JEOL ASM 6360A scanning electron microscope. TG analysis was carried out in nitrogen with a heating rate of 10 degree (°)/min.

2.4. Hydrogen Adsorption. Hydrogen gas adsorption studies were carried out in a Sorptomatic 1990 instrument of Thermo Finnigan, Italy that records the fall of pressure and automatically converts the same into amount of gas adsorbed as per ASTM standards. The instrument was standardized with hydrogen and nitrogen to determine the dead volume. As per the recommended standards, 1.0 gm of beads was evacuated (10^{-6} torr) at 70°C for 2 hr in a previously evacuated glass sample holder to ensure cleanliness of surface and availability of clean pores for adsorption. A typical measurement for sample was continued for about 24 hr at room temperature (300 K).

3. Results and Discussion

Hypercrosslinked SMA beads with rigid porous network structures were synthesized under optimized reaction conditions with S: MAnh mole ratio of 1 : 1.6 and 30 mol% DVB content at 80°C under continuous agitation at 800 rpm [16]. Figure 1(a) shows the schematic representation of the formation of SMA copolymer beads. The number average molecular weight between the crosslink (Mc) and the effective crosslink mole fraction (Fc) in the polymer was calculated using Flory-Rehner equation [18] and was found to be 438.179 and 0.114%, respectively,

$$\ln(1-\nu_2)+\nu_2+\chi\nu_2^2+\left(\frac{\mathrm{V1}_{\rho_2}}{\mathrm{Mc}}\right) \times\left(\nu_2^{1/3}\,\nu_{2r}^{2/3}-\frac{\nu_2}{2}\right)=0, \quad (1)$$

$$\mathrm{Fc}=\frac{1}{2\mathrm{Mc}},$$

where χ is the Flory polymer solvent interaction parameter taken as 0.5, ρ_2 is the polymer density, ν_{2r} is polymer volume fraction in reference state, and ν_2 is polymer volume fraction in swollen state.

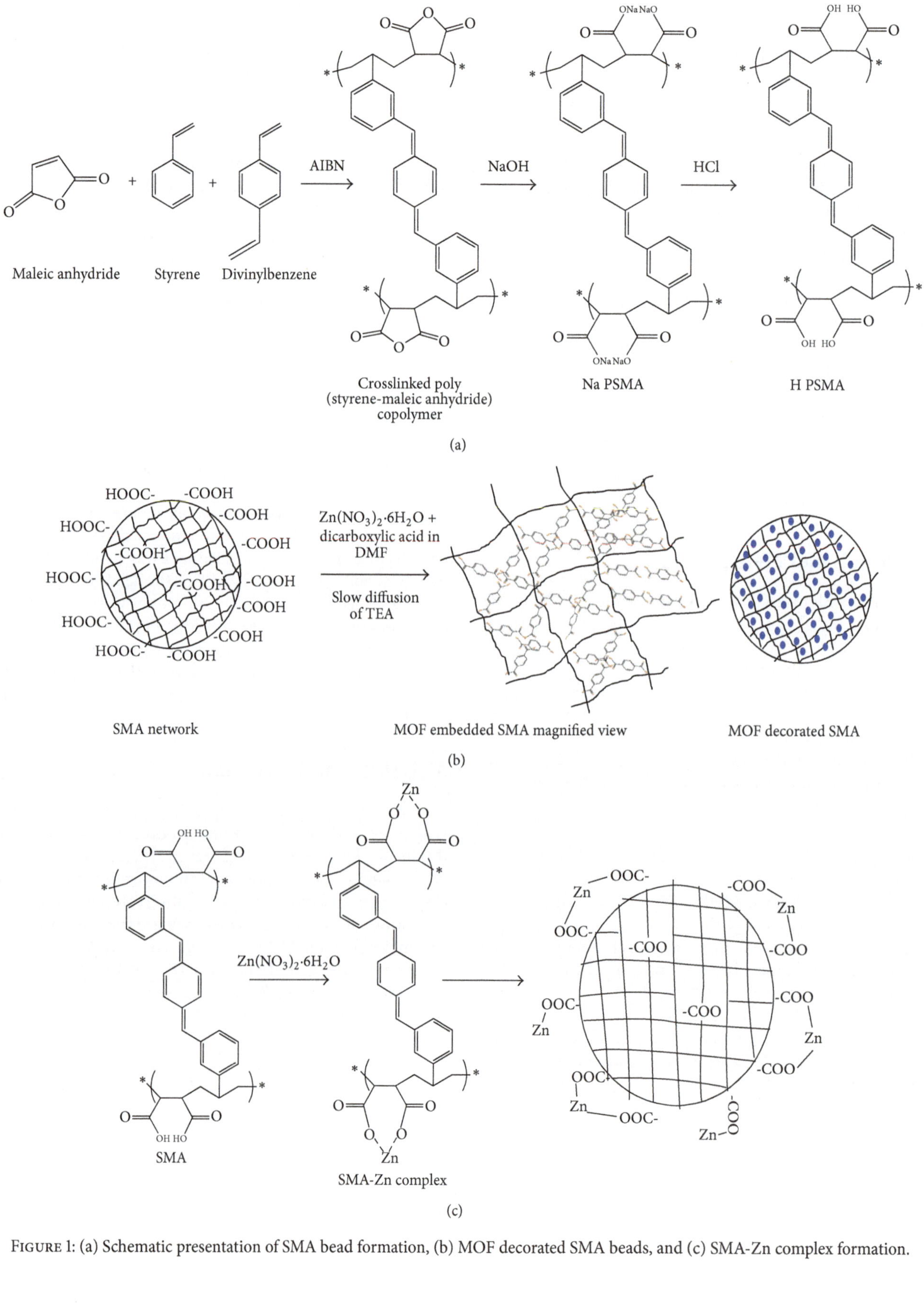

FIGURE 1: (a) Schematic presentation of SMA bead formation, (b) MOF decorated SMA beads, and (c) SMA-Zn complex formation.

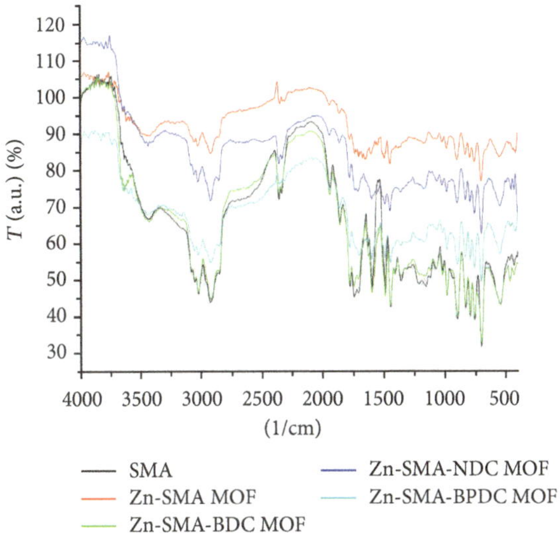

(a)

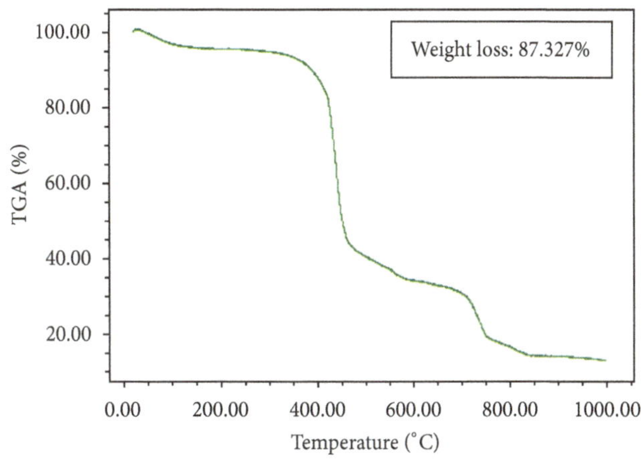

(b)

FIGURE 2: (a) FTIR and (b) TGA.

The SMA decorated with MOFs was also synthesized using under optimized reaction conditions at 300 K with $Zn(NO_3)_2{\cdot}6H_2O$: dicarboxylic acid ratio 1:2 wt%. It was observed that highly oriented crystal structures were formed in the hypercrosslinked SMA beads with reaction time of 48 hr indicating that the slow diffusion of TEA was essential for formation of MOF. No significant change in the density of the polymer was noted on formation of MOF in polymer matrix. The schematic presentation of MOF decorated SMA beads is shown in Figure 1(b). The formation polymer metal complex was observed without addition of dicarboxylic acid [19] which is schematically shown in Figure 1(c). The formations of such complexes are known to induce additional network in the polymer matrix.

The FTIR spectra of hydrolyzed SMA beads, beads containing MOF and SMA-Zn complex are shown in Figure 2(a). The adsorption peaks around 3500 cm^{-1} and 1710 cm^{-1} (shoulder peak) correspond to the free acid formed after hydrolysis of anhydride groups of copolymer. Broad band peak around 3000 cm^{-1} indicates the presence of high concentration of –COOH groups, indicating the presence of high maleic acid content. Characteristic >C=C< aromatic peak also appears at 795 cm^{-1} and 850 cm^{-1}. The peak at 1600 cm^{-1} and 1410 cm^{-1} corresponds to the carboxylate anion. The IR spectrum of in situ formed MOF in porous SMA polymer shows no specific change in the peak position, confirming the formation of MOF with only slight decrease in the intensity of the 1710 cm^{-1} peak revealing the formation of polymer metal complex which is predominantly observed in the SMA-Zn beads.

Thermogravimetric analysis revealed that the decomposition of the polymer starts at 400°C (Figure 2(b)). The initial short dip in the curve presents the decomposition of the maleic acid to maleic anhydride and the sharp decomposition curve of styrene at 400°C. The residual weight was found to be 12.7% which increased on formation of Zn-polymer complex and MOF embedded in polymer which is due to Zn content in the material.

The presence of Zn in the Zn-SMA complex beads and MOF decorated SMA beads was confirmed from the EDX spectrum. The magnified surface SEM image of hypercrosslinked SMA beads revealed porous dense network structure with presence of micropores, voids, fissure, and well-developed flow line indicating capillary action being induced in the polymer matrix (Figure 3(a)). The SEM image of polymer-Zn complex formation shows uniform distribution of the Zn complex clusters in the size range 0.2–2 μm (Figure 3(b)). The result indicates that complex formation tends to occur at region with carboxyl content. The dense polymer network aids in close proximity of carboxylic groups of the polymer which results in complex formation with the metal ion being attached to the functional group of both chains [19]. The SEM images of SMA-(Zn-BDC) indicated the formation of well defined flower like crystal growth of the Zn-BDC MOFs in the polymer matrix (Figure 3(c)). The average size of these crystals was found to be around 5 μm. Such flower shaped crystal of Zn-BDC MOFs known as MOF-5(γ) [20, 21] or MOF 69 [22, 23] is in well agreement with the literature. Changing the dicarboxylic acid showed a drastic change in the morphology of the MOF formed. It was observed that, with 1,6-naphthalene dicarboxylic acid linkers, the fine pellet shaped MOFs (Figure 3(d)) with average diameter of 2–3 μm were developed while globular structures (Figure 3(e)) of varying diameters were formed with 1,8-biphenyldicarboxylic acid. Such varied morphologies in the development of metal-organic frameworks were observed only with change in the dicarboxylic acid under similar reaction condition. With 1,4-benzenedicarboxylic acid the crystal growth was well defined with definite orientation of individual layer to form flower-like arrangement. Increasing

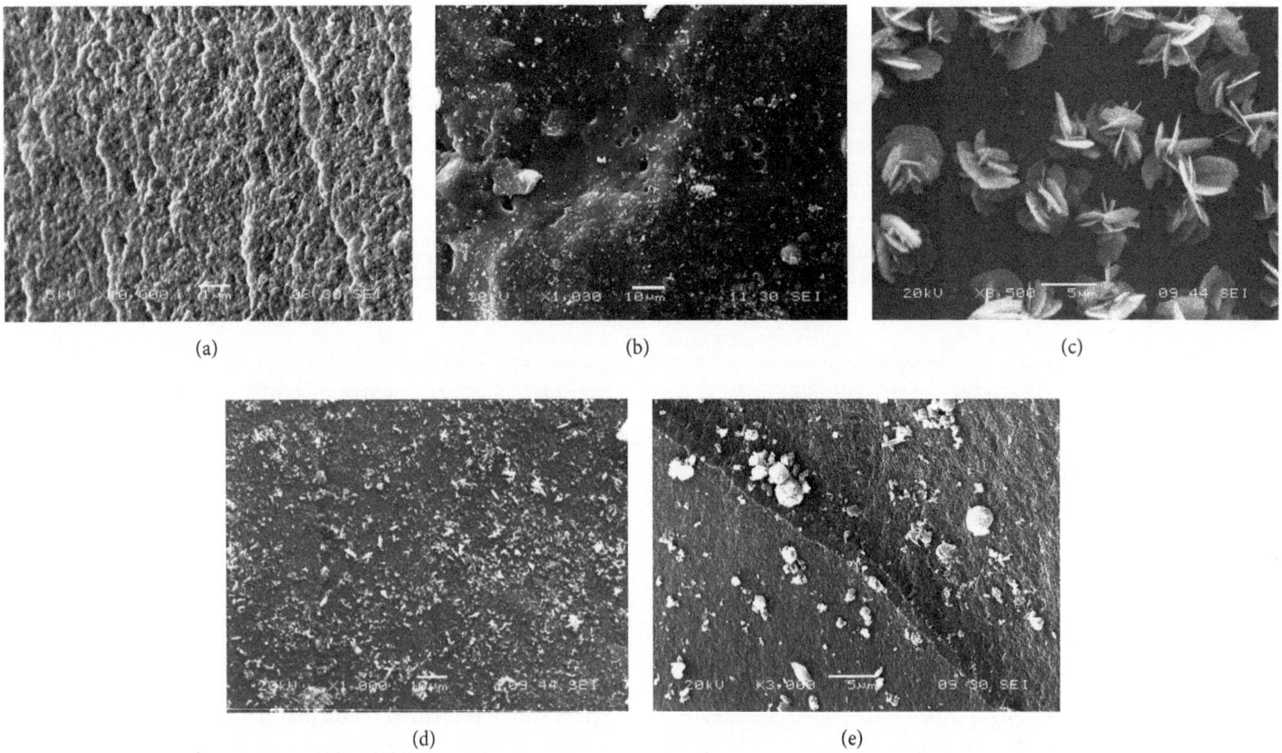

Figure 3: Magnified SEM images of (a) SMA surface, (b) SMA-Zn complex, (c) SMA-Zn-BDC, (d) SMA-Zn-NDC, and (e) SMA-Zn-BPDC.

the size of the dicarboxylic acid by using 1,6-NDC introduces steric strain in the structure which results in the formation of individual pellets structures and the pore diameter of the polymer restricts/hinders the further growth of the crystals. The globular ball shaped crystal structures originating due to the use of 1,8-BPDC are also a result of restricted pore diameter of the polymer and the increased size of dicarboxylic acid which introduces steric strain. Thus, the MOF morphology is found to be highly influenced by the steric interference between neighbouring ligands on the surface and orientation of the ligands in 3-dimensional network in the confined region/space of the pore.

These MOFs were found to be adhering to the inner pore surface with minimum contact, which can be clearly observed from the SEM micrographs. This finding supports the possibility of Zn forming complex with the polymer carboxylic group at specific sites in the pores which acts as nucleus to aid in further development of MOF crystals. Such MOFs can also be thought of having independent existence in the pores. The MOFs seem to be dispersed in the pores and are clustered on the pore surface at random spots. This is expected as the reacting constituents are mixed together and allowed to diffuse into the beads.

Porous polymers are well known for low adsorption of hydrogen at ambient temperatures and high pressures (~0.5 wt%) whereas MOFs can adsorb up to 7.5 wt% at 77 K [24] and about 1 wt% at room temperatures and 20 atmosphere [25]. Recently, 0.06 wt% hydrogen sorption at 32°C and 60 bar has been demonstrated on functionalized PEEK containing chlorosulphonated groups [26]. The H_2 storage capacity of microporous organic polymer HCP-PTMP-PVBC- 2 wt% Pt via spillover (0.21 wt%) was found to be enhanced by a factor of 1.75 as compared to 1 wt% Pt at 298.15 K and 19 bar [27]. MOF-5 synthesized by slow diffusion of TEA exhibited 2.63 wt% hydrogen adsorption at 1.74 MPa and 77 K [28].

Hydrogen adsorption in the hypercrosslinked SMA beads, SMA-Zn complex beads, and SMA beads decorated with MOFs was studied by allowing the beads to be in contact with the gas in discreet pressure ranges within 0–1 bar. The analysis was carried out by maintaining the equilibrium time of 1 min and equilibrium pressure deviation of 0.5 torr. All the measurements were carried out at room temperatures (300 K). The SMA polymers are known to adsorb hydrogen without attaining saturation limits [19, 29]. The slow micropore opening mechanism being predicted for the nonattainment of saturation. The hydrogen gets filled into large pores present on the surface; the pressure created in the pores allows the opening of micropores connected to these bigger pores through capillary action. Hydrogen thus diffuses into these pores. The same phenomenon was observed even with the SMA-Zn complex beads and MOF decorated SMA beads (Figure 4(a)). Instantaneous hydrogen adsorption on these beads for 25 hr revealed that hydrogen-storage capacity of the hypercrosslinked polymer beads (0.67 wt%) decreases slightly due to formation of metal complex and entrapment of MOFs in the pores. These findings can be attributed to the blockage of pores by the complex formation and MOF decoration. The hydrogen-storage capacities are presented in

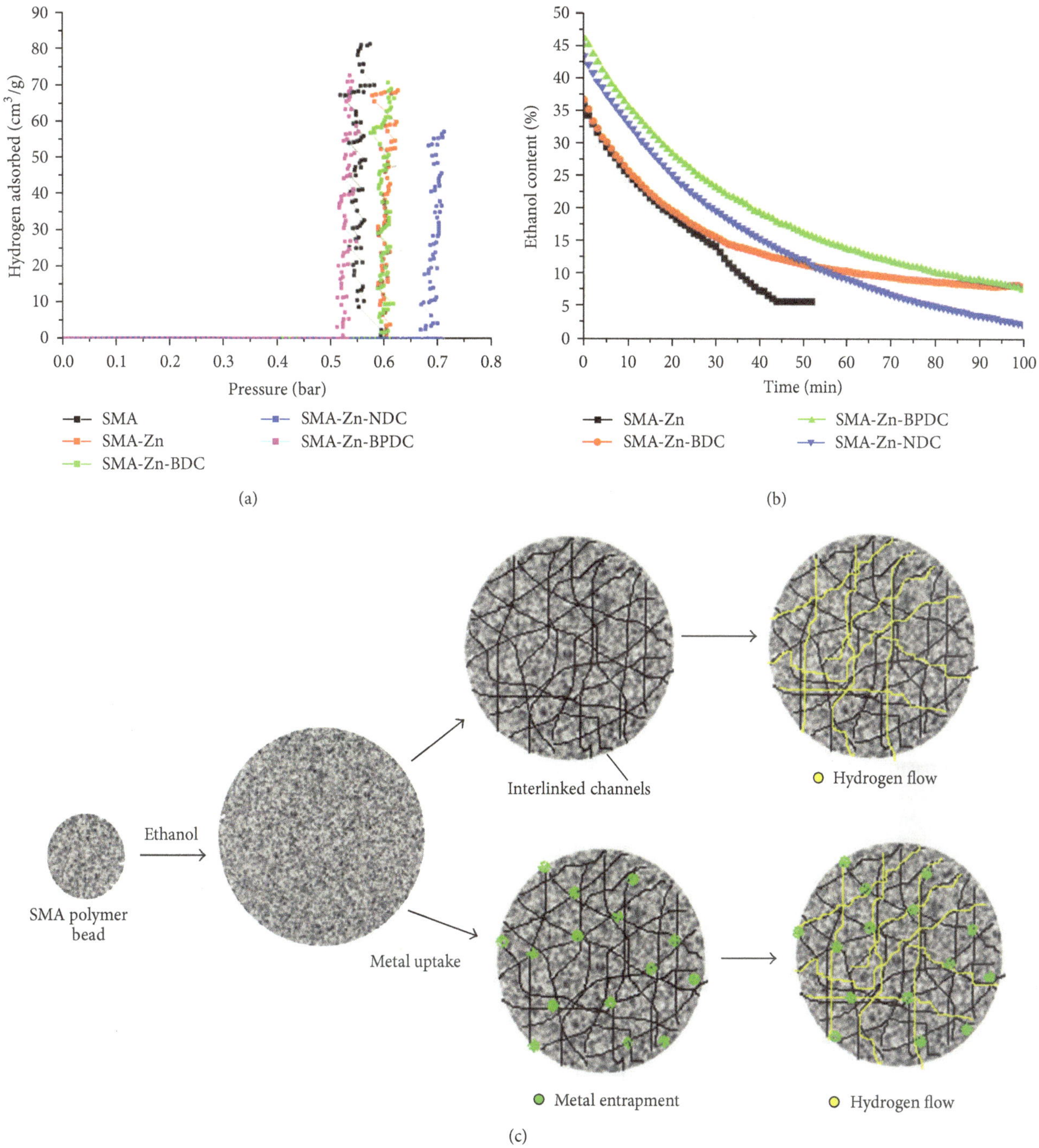

FIGURE 4: (a) Hydrogen adsorption, (b) ethanol retention plots, and (c) schematic presentation of hydrogen flow.

Table 1. Though the metal complex clusters and MOFs acts as sink for hydrogen storage, they serve as barriers during the process of micropore opening. This effect was found to diminish with long exposure time of 3 days where similar hydrogen adsorption capacity (~0.8 wt%) was noted for all types of beads.

To test the liquid hydrogen-storage capacity, the degree of swelling and solvent content in the MOF decorated SMA beads were noted with ethanol at 300 K and atmospheric pressure (Table 2). It was observed that these beads swell rapidly up to 3 times their original volume. The maximum swelling and solvent retention was observed with SMA beads decorated with Zn-BPDC MOFs, with solvent retention capacity of 46.37% (Figure 4(b)). The residual solvent could be easily removed by exposing the beads to 50°C for 5 hr. It was also observed that these SMA beads were able to adsorb Zn(II) ions from aqueous solution and exhibited maximum adsorption of 143 mg/g.

TABLE 1: Hydrogen-storage capacity of the SMA decorated beads.

Material	Pressure (bar)	Hydrogen-storage capacity (wt%)
SMA beads	0.55	0.67
SMA-Zn complex beads	0.61	0.57
SMA-Zn-BDC beads	0.60	0.59
SMA-Zn-NDC beads	0.69	0.61
SMA-Zn-BPDC beads	0.53	0.47

TABLE 2: Degree of swelling and solvent content in SMA decorated beads.

	Ethanol	
	Sw (%)	W_Q
SMA-Zn	66	35.95
SMA-Zn-BDC	58	36.82
SMA-Zn-BPDC	86	46.37
SMA-Zn-NDC	77	43.66

4. Conclusions

Hypercrosslinked SMA beads decorated with MOFs were synthesized by the slow diffusion of TEA. The dicarboxylic acid significantly influenced the morphology of the MOFs formed inside the polymer matrix and hydrogen-storage capacity. The optimized reaction conditions yielded crystalline MOF. SMA decorated MOFs samples exhibit hydrogen-storage capacities: 0.59 wt%, 0.61 wt%, and 0.47 wt% at 300 K and 0.60 bar, 0.69 bar, and 0.53 bar for SMA-MOF-5, SMA-Zn-NDC, and SMA-Zn-BPDC, respectively, in 24 hr. The adsorption of Zn(II) ions from aqueous solutions by SMA beads suggests the recovery of Zn(II) from industrial effluents. Such Zn-SMA complexes can also be explored for hydrogen storage which exhibits hydrogen-storage capacity of 0.57 wt% at 300 K and 0.61 bar. The ethanol adsorption and retention in these decorated beads can be explored for storage of liquid hydrogen in fuel cells. The use of these polymer beads in fuel cell can render advantages of low cost, light weight, embedded catalyst activity, effective solvent and simultaneous hydrogen storage, and operation at higher temperatures, thus providing a solution to most of the shortcomings faced by the present fuel cells.

Acknowledgments

The authors would like to thank Dr. Prahlada, the Vice Chancellor of DIAT (DU), for his encouragement in this research activity. Thanks are also due to the Director of Naval Materials Research Laboratory, Ambarnath, for financial support under CAR Program.

References

[1] R. Ramachandran and R. K. Menon, "An overview of industrial uses of hydrogen," *International Journal of Hydrogen Energy*, vol. 23, no. 7, pp. 593–598, 1998.

[2] L. J. Murray, M. Dinc, and J. R. Long, "Hydrogen storage in metal-organic frameworks," *Chemical Society Reviews*, vol. 38, no. 5, pp. 1294–1314, 2009.

[3] R. Ströbel, J. Garche, P. T. Moseley, L. Jörissen, and G. Wolf, "Hydrogen storage by carbon materials," *Journal of Power Sources*, vol. 159, no. 2, pp. 781–801, 2006.

[4] Y. Li and R. T. Yang, "Hydrogen storage in low silica type X zeolites," *Journal of Physical Chemistry B*, vol. 110, no. 34, pp. 17175–17181, 2006.

[5] J. Y. Lee, C. D. Wood, D. Bradshaw, M. J. Rosseinsky, and A. I. Cooper, "Hydrogen adsorption in microporous hypercross-linked polymers," *Chemical Communications*, no. 25, pp. 2670–2672, 2006.

[6] J. L. C. Rowsell and O. M. Yaghi, "Metal-organic frameworks: a new class of porous materials," *Microporous and Mesoporous Materials*, vol. 73, no. 1-2, pp. 3–14, 2004.

[7] D. Zhao, D. Yuan, and H. Zhou, "The current status of hydrogen storage in metal-organic frameworks," *Energy and Environmental Science*, vol. 1, no. 2, pp. 222–235, 2008.

[8] A. Zuttel, "Hydrogen storage and distribution systems," *Mitigation and Adaptation Strategies for Global Change*, vol. 12, no. 3, pp. 343–365, 2007.

[9] O. K. Farha, A. M. Spokoyny, B. G. Hauser et al., "Synthesis, properties, and gas separation studies of a robust diimide-based microporous organic polymer," *Chemistry of Materials*, vol. 21, no. 14, pp. 3033–3035, 2009.

[10] V. A. Davankov, S. V. Rogozhin, and M. P. Tsyurupa, "Macronet polystyrene structures for ionites and method of producing same," US, 3729457, 1971.

[11] C. D. Wood, B. Tan, A. Trewin et al., "Microporous organic polymers for methane storage," *Advanced Materials*, vol. 20, no. 10, pp. 1916–1921, 2008.

[12] O. M. Yaghi, M. O'Keeffe, N. W. Ockwig, H. K. Chae, M. Eddaoudi, and J. Kim, "Reticular synthesis and the design of new materials," *Nature*, vol. 423, no. 6941, pp. 705–714, 2003.

[13] H. K. Chae, D. Y. Siberio-Pérez, J. Kim et al., "A route to high surface area, porosity and inclusion of large molecules in crystals," *Nature*, vol. 427, no. 6974, pp. 523–527, 2004.

[14] T. Uemura, N. Yanai, and S. Kitagawa, "Polymerization reactions in porous coordination polymers," *Chemical Society Reviews*, vol. 38, no. 5, pp. 1228–1236, 2009.

[15] I. Rodrigues, I. Mihalcea, C. Volkringer, T. Loiseau, and M. Visseaux, "Water-free neodymium 2,6-naphthalenedicarboxylates coordination complexes and their application as catalysts for isoprene polymerization," *Inorganic Chemistry*, vol. 51, no. 1, pp. 483–490, 2012.

[16] R. Gonte, K. Balasubramanian, P. C. Deb, and P. Singh, "Synthesis and characterization of mesoporous hypercrosslinked poly (styrene co-maleic anhydride) microspheres," *International Journal of Polymeric Materials*, vol. 61, no. 12, pp. 919–930, 2012.

[17] H. Li, M. Eddaoudi, T. L. Groy, and O. M. Yaghi, "Establishing microporosity in open metal-organic frameworks: gas sorption isotherms for Zn(BDC) (BDC = 1,4-benzenedicarboxylate)," *Journal of the American Chemical Society*, vol. 120, no. 33, pp. 8571–8572, 1998.

[18] P. J. Flory, *Principles of Polymer Chemistry*, Cornell University Press, Ithaca, NY, USA, 1953.

[19] R. Gonte, K. Balasubramanian, and P. C. Deb, "Adsorption properties of crosslinked mesoporous (styrene-maleic acid)-Zn (II) complexes," *International Journal of Industrial Engineering and Technology*, vol. 3, no. 3, pp. 279–288, 2011.

[20] D. Saha, S. Deng, and Z. Yang, "Hydrogen adsorption on metal-organic framework (MOF-5) synthesized by DMF approach," *Journal of Porous Materials*, vol. 16, no. 2, pp. 141–149, 2009.

[21] B. Panella and M. Hirscher, "Hydrogen physisorption in metal-organic porous crystals," *Advanced Materials*, vol. 17, no. 5, pp. 538–541, 2005.

[22] N. L. Rosi, J. Kim, M. Eddaoudi, B. Chen, M. O'Keeffe, and O. M. Yaghi, "Rod packings and metal-organic frameworks constructed from rod-shaped secondary building units," *Journal of the American Chemical Society*, vol. 127, no. 5, pp. 1504–1518, 2005.

[23] T. Loiseau, H. Muguerra, G. Férey, M. Haouas, and F. Taulelle, "Synthesis and structural characterization of a new open-framework zinc terephthalate $Zn_3(OH)_2(bdc)_2{\cdot}2DEF$, with infinite Zn–($\mu_3$-OH)–Zn chains," *Journal of Solid State Chemistry*, vol. 178, no. 3, pp. 621–628, 2005.

[24] K. M. Thomas, "Hydrogen adsorption and storage on porous materials," *Catalysis Today*, vol. 120, no. 3-4, pp. 389–398, 2007.

[25] N. L. Rosi, J. Eckert, M. Eddaoudi et al., "Hydrogen storage in microporous metal-organic frameworks," *Science*, vol. 300, no. 5622, pp. 1127–1129, 2003.

[26] R. Pedicini, A. Sacc, A. Carbone, and E. Passalacqua, "Hydrogen storage based on polymeric material," *International Journal of Hydrogen Energy*, vol. 36, no. 15, pp. 9062–9068, 2011.

[27] B. Li, X. Huang, R. Gong et al., "Catalyzed hydrogen spillover for hydrogen storage on microporous organic polymers," *International Journal of Hydrogen Energy*, vol. 37, no. 17, pp. 12813–12820, 2012.

[28] J. Lia, S. Chengb, Q. Zhaoa, P. Longa, and J. Dong, "Synthesis and hydrogen-storage behavior of metal-organic framework MOF-5," *International Journal of Hydrogen Energy*, vol. 34, pp. 1377–1382, 2009.

[29] P. C. Deb, L. D. Rajput, and P. K. Singh, "Polyaniline encapsulated microporous polymer beads and their vapor and gas adsorption behavior," *Journal of Applied Polymer Science*, vol. 104, no. 1, pp. 297–303, 2007.

8

Analysis of Incomplete Filling Defect for Injection-Molded Air Cleaner Cover Using Moldflow Simulation

Hyeyoung Shin[1] and Eun-Soo Park[2]

[1] *Kumho Petrochemical R&D Center, Hwaam-Dong 57-1, Yuseong, Daejeon 305-348, Republic of Korea*
[2] *Youngchang Silicone Co., Ltd., 481-7 Gasan-Dong, Kumchun-Gu, Seoul 153-803, Republic of Korea*

Correspondence should be addressed to Eun-Soo Park; t2phagel@daum.net

Academic Editor: Pasquale Longo

A large-sized cover part for air cleaner was injection molded with ABS resin, and its incomplete filling defect was analyzed using commercial Moldflow software. To investigate the effect of processing temperature on incomplete filling defect, tensile properties, weight loss, and phase separation behavior of ABS resin were evaluated. The tensile properties of dumbbell samples were not changed up to 250°C and decreased significantly thereafter. SEM micrographs indicated no significant changes in the status of polybutadiene rubber phase below 250°C. These different test results indicated that ABS resin little affected the thermal decomposition in processing temperature range. The Moldflow simulation was performed using measured thickness of molded cover and actual mold design with the defects. As expected, the cover part showed unbalanced filling and incomplete sections. To improve these defects, two possible cases of hot runner system have been simulated. When applying modified 5-gate system, the maximum injection pressure was decreased approximately 5.5% more than that of actual gate system. In case of 6-gate system, the maximum injection pressure reduced by 23%, and the injection pressure required to fill is well within the range of the molding equipment. The maximum clamping force of 6-gate system was also significantly reduced than that of actual and 5-gate system.

1. Introduction

Injection molding is one of the most commonly used manufacturing processes for the fabrication of plastic parts in net shape with excellent dimensional tolerance [1]. A wide variety of products are manufactured using injection molding, which vary greatly in their size and complexity [2–5]. Examples of common, everyday products that require plastic injection molding include automobile bumpers, mobile phone housings, television cabinets, compact discs, toys, and lunch boxes are all examples of injection-molded parts. This process requires the use of an injection-molding machine, raw plastic material, and a mold. A plastic material is melted in the injection-molding machine and then injected into the mold, where it cools and solidifies into the final part. The most commonly used thermoplastics are polystyrene, polypropylene, polyvinyl chloride, and acrylonitrile-butadiene-styrene (ABS) [6–9]. ABS resins are among the most versatile thermoplastics in the styrenic polymers. The primary features and benefits of ABS are derived from the tree building blocks. Thermal stability and chemical resistance are derived from acrylonitrile, while butadiene provides impact resistance and toughness. Styrene imparts rigidity and processability to ABS. Therefore, they have practical toughness, high modulus, dimension stability, and good aesthetics for broad range of applications [9–11].

Injection-molding process is a complex technology with possible production problems. They can either be caused by defects in the molds or more often by part processing. Many processing factors are involved in this process and have a great influence on the quality of final products. These factors can be usually classified into four categories: materials, injection machine, model design, and process conditions [10]. Although the process and stages of custom injection molding can be as quick as just a few seconds, precision and accuracy are critical when it comes to such condition factors as time, pressure, temperature, and force. Possible defects that can occur in a molded plastic product include flash, which occurs when the clamp force is too low and molten material seeps out of the molding area and warping and unfilled sections. Each factor, accurately chosen and controlled, requires a multidisciplinary knowledge in order to improve and optimize final product performances.Simulation of injection molding is widely used to help mold designers to select appropriate injection technology and investigate product performance and processability [2–6]. It is the most successful example of simulation for any plastic forming process. Benefits that have been identified from experienced users of analysis programmed are better part quality, reduced tooling lead times, optimum cycle times, and reduce rework and scrapping of materials [3]. This may be contrasted with other polymer forming processes where tooling costs are much lower or where problems may be overcome by varying process conditions. Therefore, the simulation is usually conducted during the product design stage and the early stages of the tool design.

Many simulation studies were performed to determined potential problems and their most economical solutions. Shen et al. [2] have done numerical analysis on a thin walled part which has a fiber reinforced thermoplastic material. Moldflow software was used for the analysis. Different injection parameters (injection pressure, melt temperature, mold temperature, injection time), different fibers, different fiber ratios, and different thickness values were used. The outer shell of a notebook computer was used as an analysis model. Dairanieh et al. [4] examine the sensitivity of Moldflow's weld-line prediction algorithm to variations in material properties and processing conditions. Qualitatively, the algorithm correctly predicted the effects of changes in viscosity, density, and pressure-viscosity-temperature (PVT) relationship on weld-line strength of a poly(methyl methacrylate). Huang and Tai [5] have examined the influential effects over warpage that is seen in thin walled parts produced by injection molding. C-Mold software was used for the analysis. They have used the Taguchi experimental method to decide the injection parameters. As a result, packing pressure was found as the most important factor that affects warpage. It is seen that gate location and filling time have little effect over warpage. Shin and Park [6] have examined the residual stresses in a part which was produced by plastic injection method using Moldflow software. For analysis, screw of ice water filter machine was used. They found that the majority of the crack comes from residual stress in the final part that originated from higher injection pressure during injection-molding process.

In this study, a large-sized cover for air cleaner was injection molded with ABS resin and its incomplete filling defect was analyzed. During the manufacturing of cover, many undesired products with unfilled sections and burrs were formed repeatedly. The incomplete filling is one type of the defect that normally will appear either due to a lack of fluidity in the molten polymer or cross-section of mold that is too narrow. To determine the effect of processing temperature on these defects, the weight loss, phase separation behavior, and tensile properties of ABS resin which was used in manufacturing were investigated. At the same time, Moldflow simulation was conducted using actual mold used in manufacturing process. For the injection-molding simulation analysis, polymer rheological properties are one of the vital data that must be taken into consideration. Accordingly, the viscosity and PVT measurement of ABS resin were also performed on a capillary rheometer and on a high-pressure mercury dilatometer, respectively. Analyzing the defect leads to identifying the root cause of the defect and to finding the solution to overcome the defect in further development processes. Finally, two case studies are demonstrated to show the feasibility of the proposed optimization methods.

Table 1: ABS specifications.

Properties	Unit	Values
Density	g/cm^3	1.04
Rockwell hardness	*R* scale	108
	g/10 min	7 (200°C, 5.0 kg)
Melt index	g/10 min	50 (220°C, 10.0 kg)
	g/10 min	75 (200°C, 21.6 kg)
Heat distortion temperature	°C	85 (18.6 kg/cm^2)
Vicat softening point	°C	85

2. Experimental

2.1. Materials. ABS resin (ABS 780, Kumho Petrochemical Co., Seoul, Korea) was used as received. ABS was pre-dried in a convection oven for at least 12 h at 80°C to remove any moisture from the pellets before processing. The characteristics of the resin used in this study are summarized in Table 1.

2.2. Instrumentation. Scanning electron microscopy (SEM) observations of the samples were performed on a Hitachi S-4300 model (Tokyo, Japan). The fractured surfaces of the composites were prepared using cryogenic fracturing in liquid nitrogen followed by a coating with platinum in an SPI sputter coater. The morphology was determined using an accelerating voltage of 15 kV.

The surface appearance of samples was measured with a video microscope (Inf-500TM, charge-coupled device (CCD) video microscope, Moritex Corp., Tokyo, Japan) with a PC that contained video capture software (i-Solution Lite, Image & Microscope Technology, Daejeon, Korea).

2.3. Tensile Test Procedure. ABS sheets (ABS 780 pellets) were prepared on a hot press at 220, 230, 240, 250, and 260°C for 20 min under ca. 5 atm, followed by quenching in water. The sheet (around 1.5 mm) thus formed was free from any distortion problems. Dumbbell specimens for mechanical tests were prepared in accordance with IEC 60811-1-1 specification. The test specimens were preconditioned to 20% relative humidity and 20 ± 1°C in order to standardize test conditions. Tensile properties of samples were determined with a universal test machine (UTM, Model DECMC00, Dawha test machine, Korea) at a cross-head speed of 50 mm/min. The mean value of at least five specimens of each sample was taken, although specimens that broke in an unusual manner were disregarded.

2.4. Weight Loss Test. ABS pellets were thermal aged in dry oven maintained at 230–260°C. At predetermined intervals, specimens were removed from the dry oven and cooled to room temperature until constant weight. The cooled specimens were weighed to calculate the weight loss. Five measurements were taken on average.

Sample surface after the weight loss test were measured by a video microscope (Inf-500TM CCD Video Microscope, Moritex Corp., Japan) with a PC that contained video capture software (i-Solution Lite, Image and Microscope Technology, Korea).

2.5. Injection Molding of Cover. Cover part for air cleaner was injection molded at a speed of 100 rpm using a 2,500-ton injection-molding machine (ND2500 (screw diameter: ϕ140, maximum injection pressure: 1,640 kg/cm^2, maximum injection rate: 1,370 cm^3/s), Woojin Selex Co., Ltd., Anyang, Korea) equipped with a 1 : 5 compression ratio screw using a temperature setting of rear zone: 205°C, second zone: 225°C, and forward zone: 230–260°C. The processing parameters used in manufacturing are summarized in Table 2.

2.6. Shear Viscosity. Viscosity measurement of pristine ABS was performed on capillary rheometer (RH10, Malvern Instruments Ltd., Worcestershire, UK) with the length-to-inner diameter (L/D) of the circular die was 20/1. Studies were done at four test temperatures of 200, 220, 240, and 260°C, and the range of shear rates varied from 1 to 8000 s^{-1}.

2.7. Pressure-Volume-Temperature (PVT) Measurement. The PVT behavior was investigated on a high-pressure mercury dilatometer GNOMIX (Gnomix, Inc., USA) in pressure range 50–200 MPa at temperatures 25–260°C under isobaric conditions. The heating and cooling rates of 1°C/min were selected for experimental runs.

2.8. Moldflow Analysis. Moldflow (Moldflow 6.2, Moldflow Corporation, MA, USA) simulation was conducted using a single cavity mold with hot runner gating system. The 3D-CAD drawing of cover mold used in manufacturing was exported to the FE software using STEP file format. Therefore, mold cavity has exactly the same geometries as the manufacturing. Several iterations were carried out to establish optimum location of the sprue gate in order to achieve the most balanced flow into the cavity along with uniform pressure drop and flow front temperature. All simulations were separately carried out for each component by using configurations with five or six gates, and the process parameters are reported in Table 2. The hot runner diameter was 12 mm, and the circular sprue diameter was tapered from 12 to 6 mm. The sprue length and gate diameter were 50 and 10 mm, respectively.

TABLE 2: Processing parameters used in the manufacturing and Moldflow simulation.

Processing parameters	Unit	Manufacturing values	Simulation values
Flow rate	cm^3/s	1,000–1200	1,000
Melt temperature	°C	230–260	230
Mold temperature	°C	40	
Filling/packing	—	Switch over at 98 vol%	
Packing profile	—	80% of maximum pressure	
Holding time	s	10	
Cooling times	s	20	

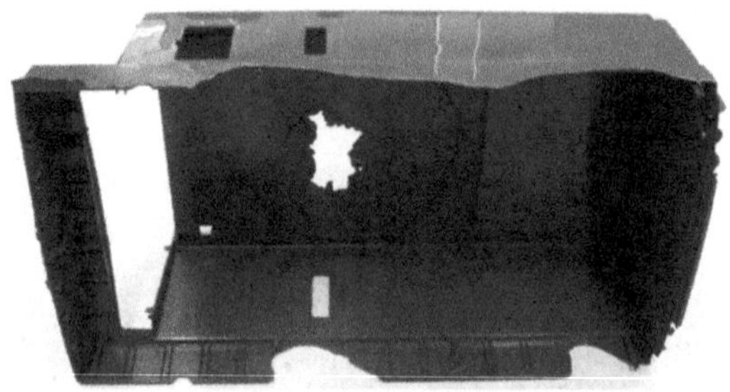

FIGURE 1: Image of the inferior cover part.

3. Results and Discussion

3.1. Problem Statements. The cover part for air cleaner was injection molded with ABS resin using a 2,500-ton injection-molding machine. The main dimension of cover was 1,070 × 580 × 520 mm^3 and the average thickness of component was 3.0 mm. During manufacturing process, many undesired products were formed repeatedly, and a representative image of inferior product was demonstrated in Figure 1. It can be seen that the cover has many unfilled sections and burrs.

Injection molding is a cyclic process. A typical injection-molding process consists of four stages: (i) filling of the molten polymer into the mold; (ii) packing of more material into the mold under high pressure to compensate for volumetric shrinkage of the material as it cools; (iii) cooling during which the material solidifies while in the mold until it is sufficiently solid; (iv) ejection of the solidified product from the mold [12].

The incomplete filling defects are caused by a wrong injection-molding temperature, resin burnout, insufficient shot volume, or low flow rate of molten plastics. During the filling stage, the hot molten polymer must fill the cold mold completely before it solidifies. It can be assumed that the processing temperature and the flow rate of molten polymer are maintained too high to eliminate unfilled sections and these lead to additional burr defect.

3.2. Injection-Molding Conditions. Computer analysis of the filling and packing phases of injection-molding process needs data on the physical properties, mechanical properties, and rheological properties of polymer is constructed and the material from which the mold is also constructed [13].

Table 3 summarized the effect of processing temperature on the tensile properties of ABS resin. The abbreviation of the sample code in Table 3, ABS-220°C, for example, means that the ABS was press-molded at 220°C for 20 min. The tensile properties of dumbbell samples were not changed up to 250°C and decrease significantly thereafter. In case of ABS-260°C, an approximately 56 and 40% decrease in the tensile strength and elongation at break of 190 kg_f/cm^2 and 3% were observed compared to ABS-220°C, respectively.

The significant drop in tensile properties at higher processing temperature can be related to the increase in brittleness of ABS specimens driven by chemical changes in the polybutadiene (PB) microstructure, which are attributed to chain scission and cross-linking [13]. Thermal degradation follows the free radicals mechanism, initiated by oxygen or oxidative contamination in the material or other stresses. In the initial stages of thermal degradation of polymer, chain scission often prevails, which reduces molecular weight. Furthermore, degradation of the PB phase at the surface forms hydroperoxide species [14–16]. Degradation of PB phase by cross-linking would also reduce polymer chain mobility and decrease free volume [14].

Figure 2 shows the influence of thermal aging at various temperatures on the weight loss of ABS 780 pellets used in manufacturing for cover. The weight loss gradually increased up to 40 min with the aging time and almost maintained thereafter. It is noted that the weight loss of ABS was drastically increased during thermal aging at 260°C. Approximately, a minimum of 20% overall increase in the weight loss was observed comparing to 230°C. The weight decrease in pellet after thermal aging is attributed to the decrease in low molecular weight components, such as volatile residues, additives, oligomers, and thermal degradation products. There was a good correspondence between the changes in tensile properties and the observed weight loss variations. This is consistent with the changes outlined above, namely, the degradation of the PB phase. Finally, they usually create a weakened pathway within the material by creating permanent molecular or physical changes. This leads to formation of voids (gas porosity) or defects of injection-molding products and lowering the quality of final products.

Figure 3 shows the phase separation behavior of the ABS sheet after the thermal aging at 230, 240, 250, and 260°C for 20 min. All the samples after thermal aging appeared to have a noticeable increase in white pattern. The white pattern was likely a product of chemical oxidation-induced phase separation by surface directed thermal decomposition and release of internal stress that had accumulated during oxidation. ABS is composed of a two-phase system having a styrene-acrylonitrile copolymer (SAN) as the continuous phase with a dispersed phase of PB rubber. The PB phase is compatibilized with the rigid SAN component through grafting. Phase separation is induced when a sample is transferred from one-phase region into miscibility gap. Usually this is accomplished by a change in temperature, upward or downward depending on the system under study [17]. However, the shape of white

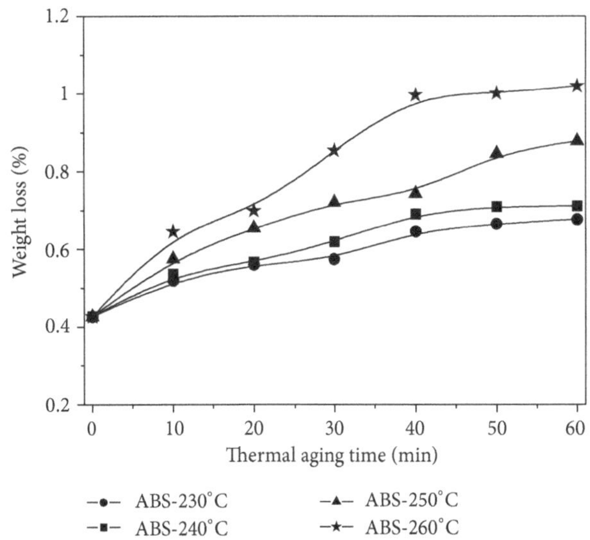

FIGURE 2: Weight loss test results of the ABS resin (t_0: after drying at 12 h at 80°C).

TABLE 3: Tensile properties of the ABS resin.

Sample code	Tensile properties	
	Tensile strength (kg_f/cm^2)	Elongation at break (%)
ABS-220°C	442 ± 18	5.0 ± 0.8
ABS-230°C	445 ± 31	4.6 ± 0.7
ABS-240°C	446 ± 14	4.3 ± 1.1
ABS-250°C	446 ± 16	4.3 ± 0.5
ABS-260°C	190 ± 16	3.1 ± 0.7

patterns was not changed significantly up to 250°C. From 260°C upward, they were gathering together and became more irregular in shape indicated that the phase separation of ABS was accelerated at 260°C.

It can be also found that yellowing of the ABS specimen started at 250°C because it was attacked by heating during thermal aging. Kim and kang have investigated the NMR sequence before and after recycling of ABS specimens. The color of SAN may be varied by the sequence of styrene and acrylonitrile (AN). Yellowing occurs according to the content of sequence AN-AN-AN. However, their study shows that the content of the sequence AN-AN-AN was not much changed by recycling, and the total molecular weight of ABS was not changed [18]. Therefore, the PB component affects the yellowing process which may introduce alterations in the chemical structure of the polymeric chains, resulting in more brittle material with lower ductility [19].

As the thermal aging time increased, phase separation of the PB phase seems to become progressively more significant (Figure 4). After thermal aging at 260°C for 10 min, surface of ABS sample appeared to have a noticeable increase in scratches (Figure 4(b)). With the aging time, the scratches were gathering together and forming wrinkle patterns (Figure 4(c)). Finally, they are arranged in an

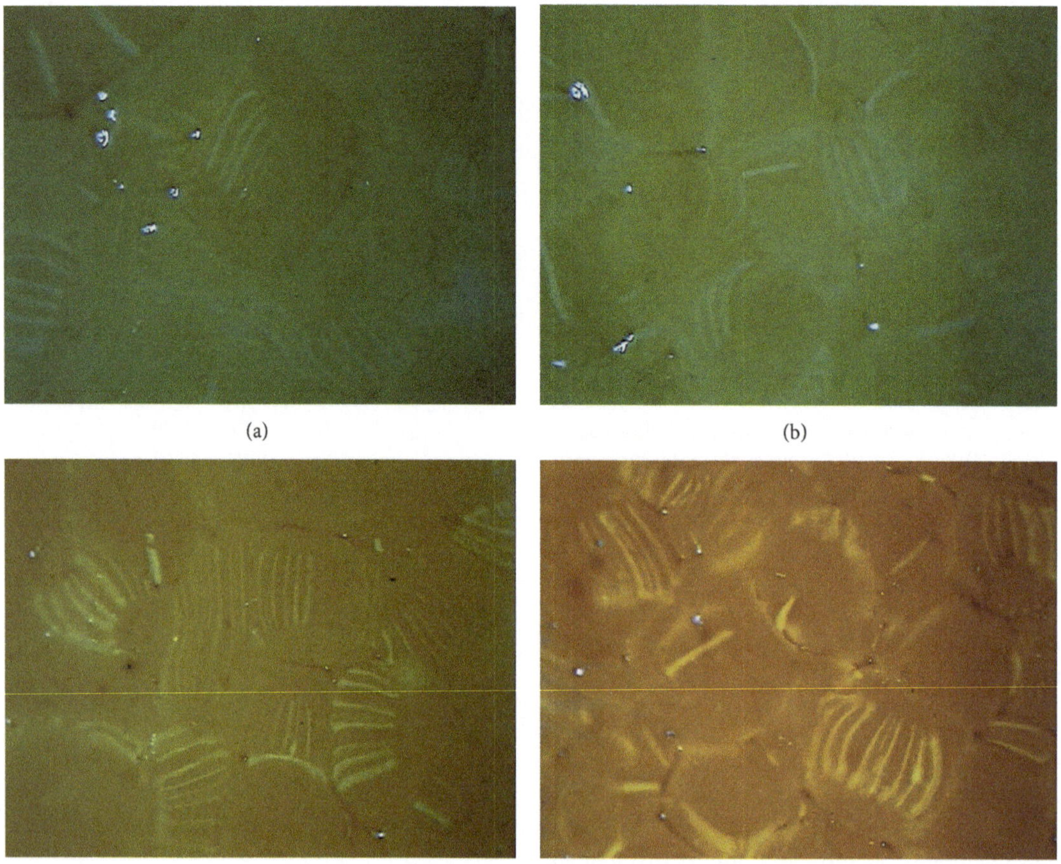

(a) (b) (c) (d)

Figure 3: Phase separation phenomenon of the ABS resin after the thermal aging at (a) 230, (b) 240, (c) 250, and (d) 260°C for 1 h (20x).

orderly repeating pattern extending in all PB rubber regions (Figure 4(d)).

Microphase separation in block copolymers is a possible explanation for the pattern formation. In the case of diblock copolymers, consisting of two subchains (a) and (b) made of different monomers A and B, respectively, even a weak repulsion between unlike monomers induces a strong repulsion between sub-chains. As a result, different sub-chains tend to segregate below some temperature, but as they are chemically bonded, even a complete segregation cannot lead to a macroscopic phase separation as in mixtures of two polymers. Only a local microphase separation occurs, and microphases rich in A or B are formed [15].

The images in Figure 5 demonstrate the fractured surface of the ABS specimen after hot pressing at 220 and 250°C for 20 min. In the SEM micrographs, the light regions represent the PB phase in the ABS, while the dark region represents SAN matrix. The micrographs show that no significant changes in size, shape, and the distribution of the rubber particles. These different test results indicated that ABS resin little affected the thermal decomposition in processing temperature range.

3.3. Viscosity Model. The viscosity is also another important function that needs to provide into the injection moulding flow analysis. The viscosity behavior of ABS resin is obtained by using the Cross-WLF model. Cross-WLF is a well-known rheological model used to predict the viscosities of polymer melts at different temperatures, pressures, and shear rates. This model accounts for the effect of temperature, shear rate, and pressure on the viscosity, over a wide temperature range. The Cross-WLF model is appropriate for postfilling stage, because the temperature and pressure sensitivities of the zero-shear-rate viscosity are well represented [20]. The model is given as

$$\eta\left(T,\dot{\gamma}\right) = \frac{\eta_0(T)}{1+\left(\eta_0(T)\,\dot{\gamma}/\tau\right)^{1-n}},$$
$$\eta_0(T) = D1\exp\left[\frac{-A1\left(T-T_r\right)}{A2+\left(T-T_r\right)}\right], \tag{1}$$
$$T_r = D2 + D3P,$$

where η, γ, τ, T, and P are the viscosity, shear rate, relaxation time, and pressure, respectively. T_r is an arbitrary reference temperature and usually chosen as the material T_g. n, $D1$, $D2$, $D3$, $A1$, and $A2$ are model constants.

The ABS resin used in this study has a melt index of 7 (200°C, 5.0 kg), 50 (220°C, 10.0 kg), and 75 g/10 min

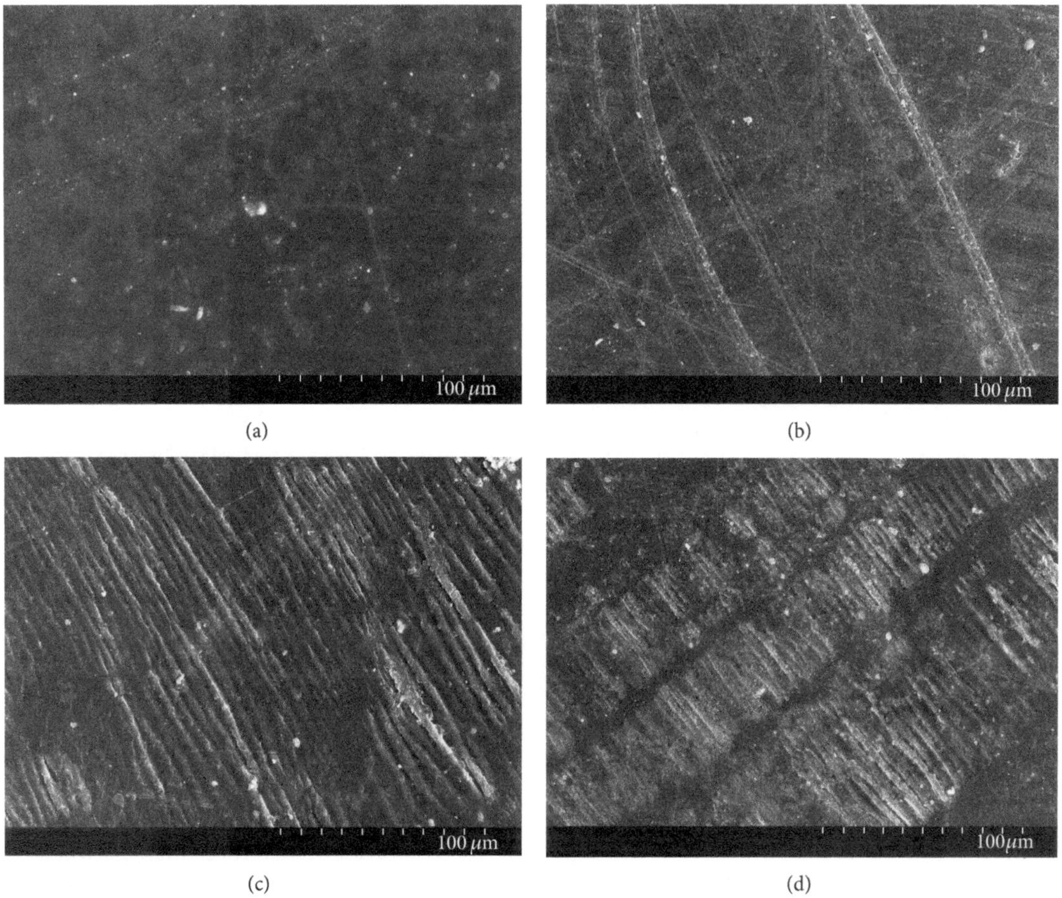

Figure 4: SEM image of the white pattern region of the ABS specimen after thermal aging at 260°C for (a) 0 min, (b) 10 min, (c) 20 min, and (d) 60 min.

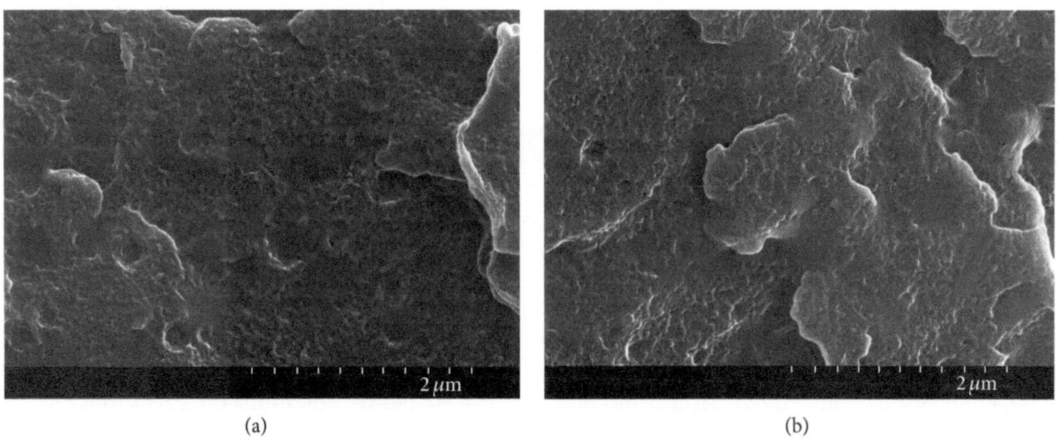

Figure 5: Fractured surface of the ABS specimen after hot pressing at (a) 220 and (b) 250°C for 20 min.

(200°C, 21.6 kg) and a heat distortion temperature of 85°C at 18.6 kg/cm^2 (Table 1). The values of Cross-WLF viscosity model parameters are $n = 0.289$, $\tau^* = 3.48 \times 10^4$ Pa, $D1 = 8.62 \times 10^{11}$ Pa-s, $T^* = 373.15$ K, $A1 = 24.96$, and $A2 = 51.6$ K. The heat capacity, thermal conductivity, and thermal expansion were assumed to be constant at 1847 J/kg·°C, 0.192 W/m·°C, and 0.192×10^{-5}°C^{-1}, respectively.

The viscosity change of pristine ABS at different temperatures is presented in Figure 6. Viscosity data at temperature of 200°C has significantly lower values of viscosity in the whole range of shear rate compared to the flow behavior of ABS at 260°C. It can be also found that shear viscosity of ABS decline with increasing shear rate for each temperature indicating a preservation of shear thinning behavior typical for a

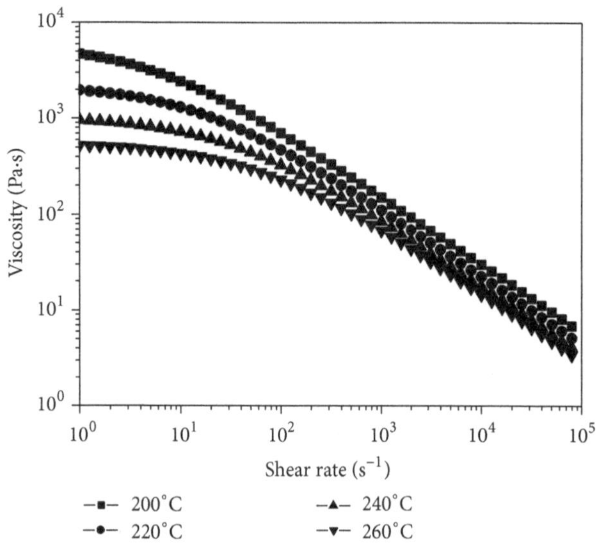

FIGURE 6: Viscosity change of pristine ABS at different temperatures.

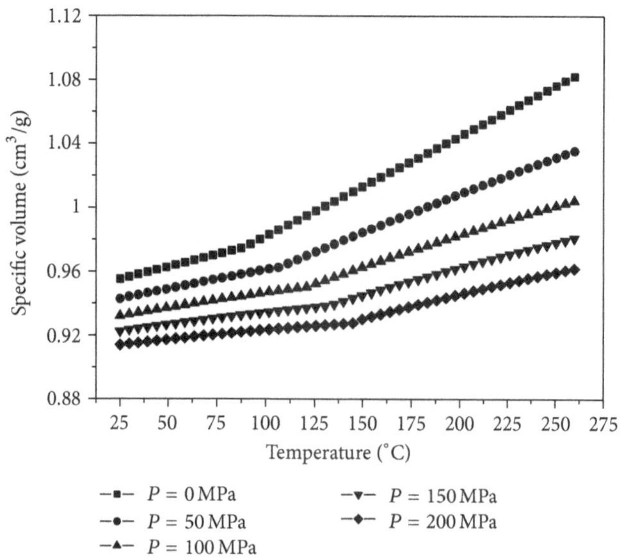

FIGURE 7: PVT behavior of the ABS resin.

common thermoplastic polymer. Thus, it can be concluded that the temperature increase in the tested ABS melt did not influence the shear thinning behavior while it effects only the magnitude of viscosity. Most polymers exhibit two regimes of flow behavior, the Newtonian and shear-thinning. The Newtonian flow occurs at low shear rates, but with increasing shear, the viscosity tends to fall away in what is termed shear-thinning behavior. The requirement viscosity model is very vital because it represents the observed behavior of polymer melts. These requirements that relate to viscosity-temperature space are that the viscosity should decrease while the temperature is increasing. For the requirements that relate to viscosity-shear rate, the viscosity should decrease while the shear rate should increase. Therefore, these requirements must be applied to ABS on the filling phase and observe its behavior while melting.

3.4. PVT Behavior of ABS Resin. The PVT behavior of the ABS resin at 50, 100, 150, and 200 MPa was demonstrated as in Figure 7. This PVT curve is used to obtain information about the compressibility and volumetric expansion of polymeric materials. The kink in the curve characterizes the glass-transition temperature (T_g), which is a function of pressure. The T_g is the temperature below which the physical properties of amorphous materials vary in a manner similar to those of a solid phase (glassy state) and above which amorphous materials behave like liquids (rubbery state), by compression.

The T_g of former-liquid polymer is increased because of the compression. The slopes of the V-T curve denote the bulk thermal-expansion coefficients in the solid and liquid phases, respectively. This type of observations is very important in melt processing, especially where part consistency and minimum thermal shrinkage are important requirements; PVT relationships are used therefore to optimize the packing process cycle in injection molding. When the temperature of the material is increased, its specific volume (the inverse of density) also increases due to thermal expansion. The rate of increase becomes higher at the T_g, because the molecules have more freedom to move and they occupy more space. Above the T_g, the effect of chaining pressure has a significant effect on the specific volume-temperature relationships due to the compressible nature of high polymers; an increase in pressure decreased the specific volume at any given temperature.

3.5. Analysis of Injection-Molding Defects Using Moldflow Simulation. In order to analyse injection defects, Moldflow simulation was conducted using a single cavity mold with hot runner system. The processing parameters are shown in Table 2. The 3D-CAD drawing of cover mold was exported to the FE software using STEP file format and modeled using MF/FLOW module of Moldflow. The surfaces of part were converted to a midsurface mesh via Moldflow Preprocessing (Figure 8). The results of the converting were summarized in Table 4.

When designing plastic parts for the injection-molding process, the important element to understand is how the plastic is filling in the mold. In the mold injection filling stage, molten plastic is injected into the cavity until the cavity is just filled. As plastic flows into the cavity, the plastic in contact with the mold wall quickly freezes and this will create a frozen layer of plastic between the mold and the molten plastic.

Figure 9 shows the filling pattern of plastic flow for inferior cover part. The simulation was performed using measured thickness of inferior cover and design of runner gate for actual mold as shown in Figure 10(a).

Filling pattern results determine that all flow paths fill at the same time, balanced flow, hesitation, and over-packing. This result can be used to implement other parameter changes to improve filling such as, higher or lower injection rate, moving gate location and changes to the parts geometry. The color diagrams show the contour colors that represent

TABLE 4: Converting results of the midsurface mesh via Moldflow preprocessing.

Processing parameters	Unit	Simulation values
Mesh Type	—	Midplane
Total number of nodes	No.	23254
Total number of injection location nodes	No.	1
The injection location node labels	—	1
Total number of elements	No.	46428
Number of part elements	No.	46360
Number of sprue/runner/gate elements	No.	68
Number of channel elements	No.	0
Number of connector elements	No.	0
Average aspect ratio of triangle elements	—	2.0344
Maximum aspect ratio of triangle elements	—	16.7113
Element number with maximum aspect ratio	—	44141
Minimum aspect ratio of triangle elements	—	1.1553
Element number with minimum aspect ratio	—	6855
Total volume	cm^3	5447.5958
Volume filled initially	cm^3	245.5855
Volume to be filled	cm^3	5202.0103
Part volume to be filled	cm^3	5089.9284
Sprue/runner/gate volume to be filled	cm^3	112.0817
Total projected area	cm^3	6144.9301

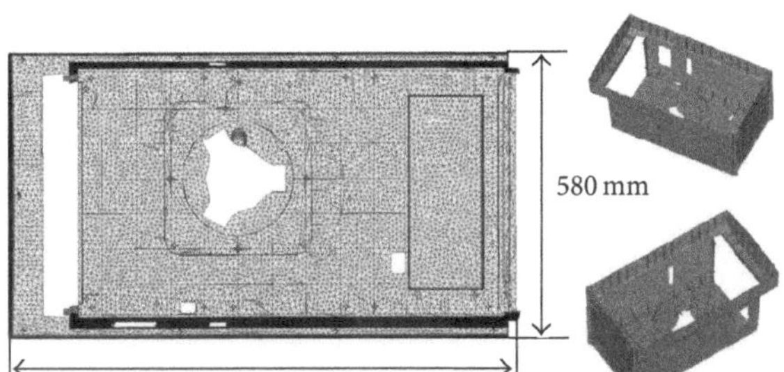

FIGURE 8: Result of the midsurface mesh conversion via Moldflow preprocessing.

the flow of plastic into the part. The result is dark blue at the start of the injection, and the last areas to fill are red. As the percentages of yellow and red increase, the difficulty in molding the part will increase and the part quality will decrease. If the part is a short shot, the section that did not fill has no color. As expected, the cover part filled very unequally and showed the formation of unfilled sections and burrs. The incomplete filling is one type of the defect that normally will appear either a lack of fluidity in the molten polymer or cross-section of mold that is too narrow. Hesitation can lead to asymmetrical and unpredictable flow patterns. It can reduce part quality due to variation in surface appearance, poor packing, high stresses, and nonuniform orientation of the polymer molecules. Alternatively, when the hesitation allows the flow front to freeze completely, part of the cavity may remain unfilled.

3.6. Two Different Case Studies. To avoid the actual mold defects, an other possible 5-gate system was simulated. The hot runners allowed the location of gates in the most favorable positions of cavities. In the procedure of injection molding, it is very important to determine the injection location inside a mold. It is because location at proper gate positions can minimize the flow hesitation at thin areas and this will be useful to avoid the occurrence of defects [21]. By designing the correct gating system, it is useful to have a thin section as the last point to fill, and so all the injection pressure is applied at this point. The configuration of each runner gate and thickness of components were shown in Figures 10(b) and 11(a), respectively.

Figure 12 shows the plastic flow filling pattern of modified 5-gate system. This system had some drawbacks too. The left corner of cover part was very slowly filling and totally shows poor flow behavior. Unfilled section appeared because the component thickness was reduced from 3.2 to 3.0 mm in bottom directions. In fact, thin sections offer more resistance to flow than thick sections [3]. This can result in molten polymer in the thin section stopping or slowing significantly. Once the polymer starts to slow down, it will cool more rapidly, and so the viscosity will increase. This higher viscosity will inhibit flow further causing even faster cooling, and so the problem is self-propagating. Hesitation can occur in thin section of parts that have significant changes in wall thickness.

Based on the previous results, the use of 5-gate configuration was considered inadequate for product fabrication. The simulation study was extended to 6-gate system to eliminate unfilled sections. New identification numbers of each runner gate and thickness of component were shown in Figure 10(c) and Figure 11(b), respectively. The wall thickness of the injection-molded parts has an intense influence in its quality; the flow resistance is big when the wall thickness is too small, which result in unfilled defects in the cavity for large and complicated parts. This is because restricted areas in the mold, such as thin sections, small runners, and long flow lengths, require a larger pressure gradient and therefore a higher pressure to fill. ABS is a two-phase system; the rheology depends on the morphology and mutual interaction of the phases during the flow. Two different morphological models have been proposed to explain the rheological behavior of ABS flow behavior [22]. At lower shear rates, interfacial control is more pronounced where the deforming rubber particles embedded in SAN continuous phase. The rubber particles deform easily, so, various phenomena were observed. At higher shear rates, laminar deformational behavior dominates where both phases deform simultaneously and viscosity decreases rapidly, comparatively. SAN component behaved more Newtonian by increasing the temperature, which is usual for homogenous single phase thermoplastics [23]. Moreover, multiple gates are normally used for the fabrication of complex-shaped parts that all the cavities are filled simultaneously.

The achievement of balanced fill is to make the distance of the material travels from the sprue to each cavity the same. This approach will work as long as the material flows directly from the sprue to the gate of the part. An effective way of

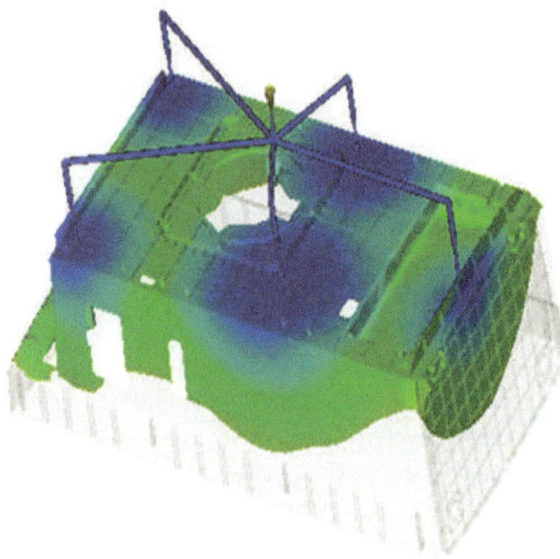

(a)

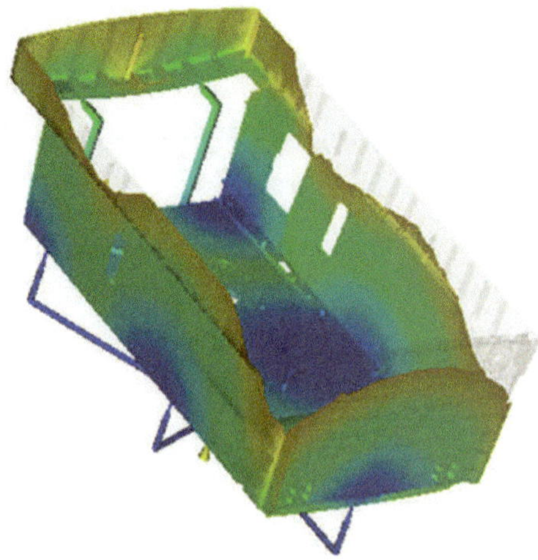

(b)

FIGURE 9: Filling pattern of the molten plastic flow for inferior cover part at (a) 70% filling and (b) 86% filling stage.

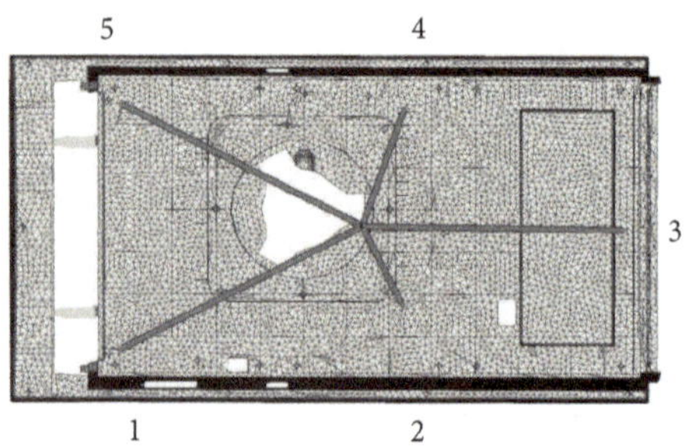

Gate1: (–210, –400)

Gate2: (130, –66)

Gate3: (430, 0)

Gate4: (200, 70)

Gate5: (–210, –400)

(a)

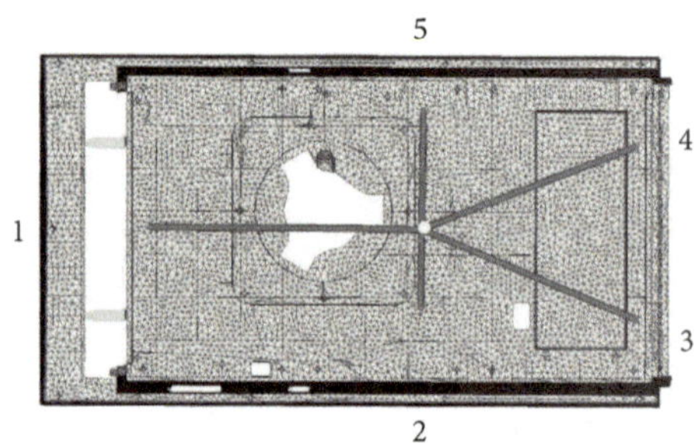

Gate1: (–470, 0)

Gate2: (130, 0)

Gate3: (365, –150)

Gate4: (365, 140)

Gate5: (0, 200)

(b)

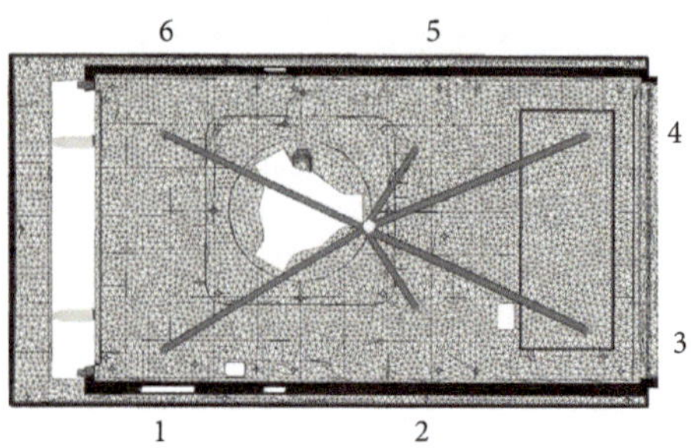

Gate1: (–340, 200)

Gate2: (80, –130)

Gate3: (365, –165)

Gate4: (365, 155)

Gate5: (80, 135)

Gate6: (–340, 155)

(c)

FIGURE 10: Three different hot runner gate systems ((a) actual gate, (b) 5-gate, and (c) 6-gate configurations).

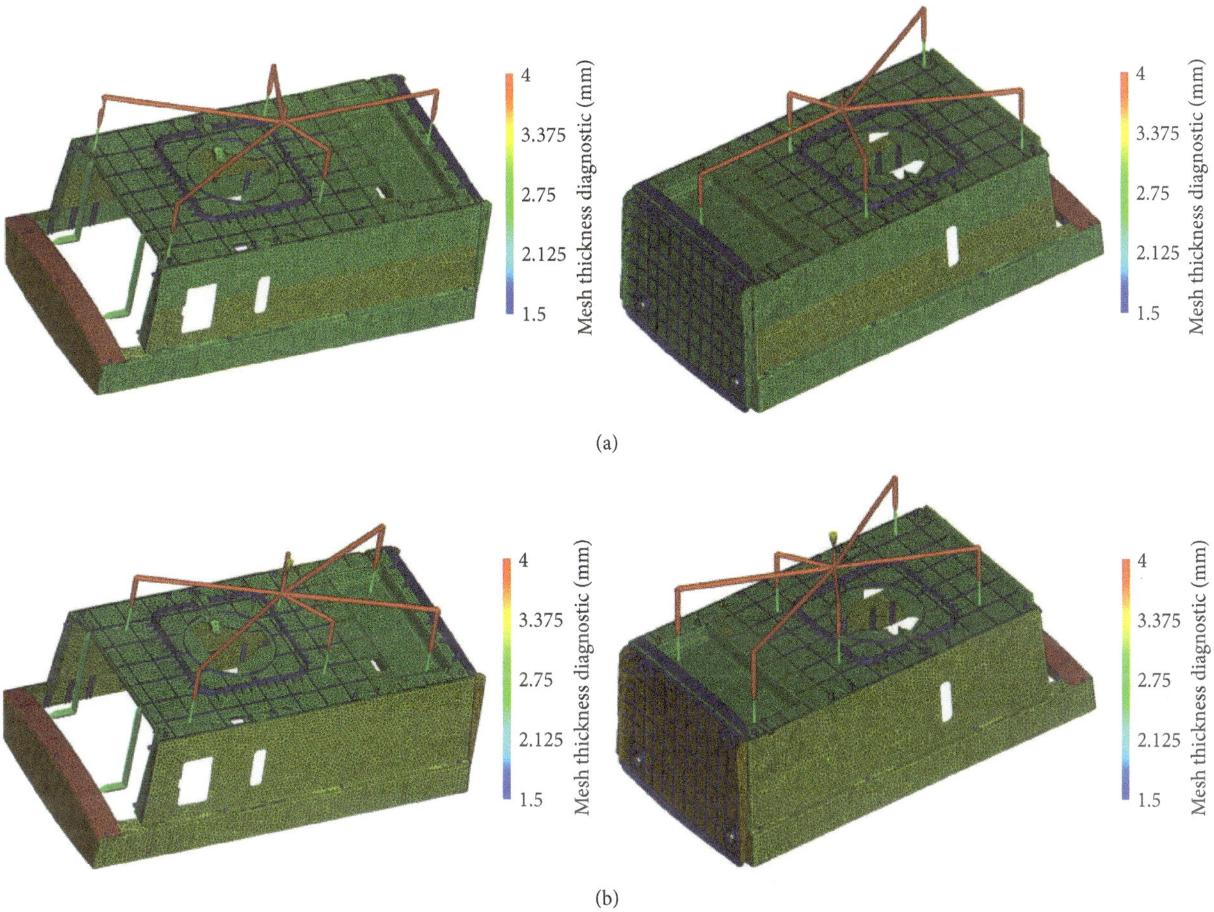

FIGURE 11: Thickness of the component for (a) 5-gate and (b) 6-gate system.

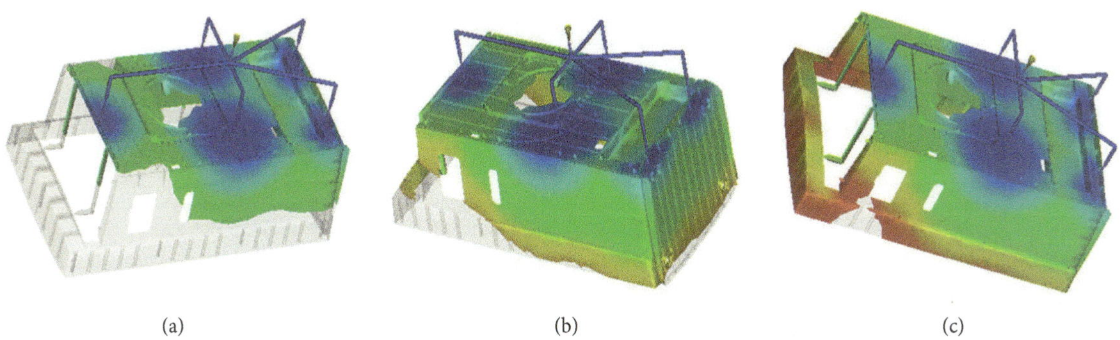

FIGURE 12: Plastic flow filling pattern of the 5-gate system at (a) 60% filling, (b) 88% filling stage, and (c) 98% filling stage.

balancing the fill is to have one main runner that extends from the last cavity of one end of the mold to the last cavity on the opposite end, with subrunners feeding the individual cavities. It can be found that the filling pattern for 6-gate system is fairly uniform without any hesitation effect and all regions with the same color are filled simultaneously (Figure 13).

The injection pressure result in Figure 14, which is produced by a fill analysis, shows the maximum injection pressure value obtained before the velocity/pressure switch-over occurs during the filling stage. At the beginning of filling, the pressure is zero in the absolute pressure scale throughout the mold. The pressure at a specific location starts to increase only after the melt front reaches that location. The pressure continues to increase as the melt front moves past, due to the increasing flow length between this specific location and the melt front. The color at each place on the model represents

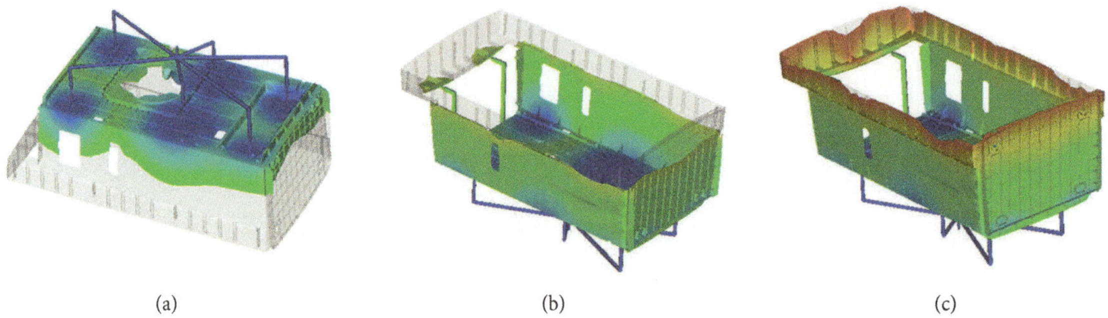

FIGURE 13: Plastic flow filling pattern of the 6-gate system at (a) 50% filling, (b) 72% filling stage and (c) 98% filling stage.

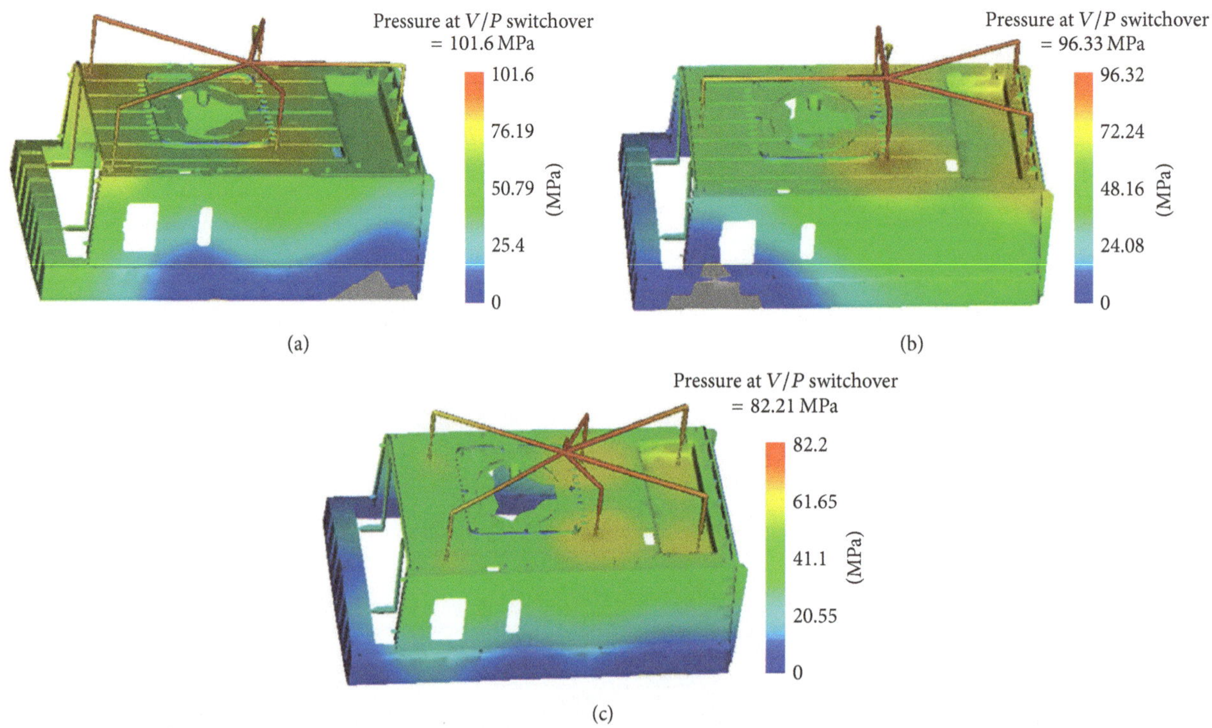

FIGURE 14: Results of the injection pressure analysis for three different gate systems ((a) actual gate, (b) 5-gate, and (c) 6-gate system).

the pressure at the place on the model. Two colors show the highest pressure (red) and lowest pressure (blue). The pressure difference from one location to another is the force that pushes the polymer melt to flow during filling. The pressure gradient is the pressure difference divided by the distance between two locations. The thermoplastic polymer always moves in the direction of the negative pressure gradient, from higher pressure to lower pressure; therefore, the maximum pressure occurs at the polymer injection locations and the minimum pressure occurs at the melt front during the filling stage. When applying modified 5-gate system, the maximum injection pressure reaches 96.3 MPa at completely mold was filled. It appears that the injection pressure was decreased approximately 5.5% more than that of actual gate system (101.6 MPa). In case of 6-gate system, the maximum injection pressure reduced by 23% (82.3 MPa), and the injection pressure required to fill is well within the range of the molding equipment. The injection pressure can be used in conjunction with pressure drop result. Even if a section of a part has an acceptable pressure drop, the actual injection pressure in the same area may be too high. High injection can cause over-packing.

Figure 15 compares the clamping force changes of actual and simulated runner systems. The mold is closed within the platen arrangement and clamped using necessary force to hold the mold shut during the plastic injection cycle, thus preventing plastic leakage over the face of the mold. After the required cooling time, the mold is then opened by the clamping motor. In case of actual gate system, maximum clamping force is too high to properly operate on a 2,500-ton injection-molding machine. The maximum value will be over approximately 42% exceeding its capacity. In contrast, the maximum clamping force of 6-gate system (2,610 ton) was reduced about 37% and 19% than that of actual

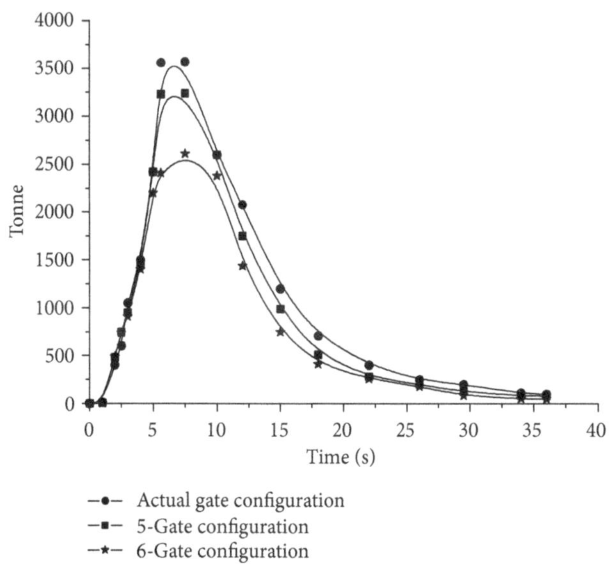

FIGURE 15: Clamping force changes of the actual gate, 5-gate, and 6-gate configurations.

gate (3,570 ton) and modified 5-gate system (3,240 ton), respectively. When the clamping force of the machine is exceeded, the molded parts stretch. This deformation causes wall thickness variation which results in both structural and cosmetic defects.

4. Conclusion

The large-sized cover part for air cleaner was injection molded with ABS resin using a 2,500-ton injection-molding machine. During manufacturing process, many undesired products with unfilled sections and burrs were formed repeatedly. Firstly, to determine the effect of processing temperature on these defects, the phase separation behavior, weight loss, and tensile properties of ABS resin were investigated. The tensile properties, weight loss, and phase separation of ABS resin were not significantly changed until 250°C. These results indicated that the unfilled sections and burr defects were not originated in processing temperature. In order to analyse injection defects, Moldflow simulation was conducted using a single cavity mold with hot runner system. As revealed, inadequate mold design results to a poor plastic filling behavior and a decrease of the filling rate of polymer melts into the mold cavity. This increases injection pressure and leads to higher clamping force resulting in an unfilled sections and burrs. To improve these defects, possible two hot runner gate systems were designed and simulated using Moldflow program. The first configuration consisted of a single manifold with five runners, while the second configuration employed single manifold with six runners. The configuration with 5-gate system caused a higher injection pressure and clamping force than those of the configuration with 6-gate system. The second design avoided the problem of higher injection pressure and clamping force; hence, the defects of the actual mold as described have been solved. This study provides a better understanding of the incomplete filling defects for large-size injection-molding parts and gives a better part quality and reduces rework and scrapping of materials.

Conflict of Interests

This work does not have a direct financial relation with the commercial identities mentioned in the paper that might lead to a conflict of interests for any of the authors.

References

[1] Z. B. Chen and L.-S. Turng, "A review of current developments in process and quality control for injection molding," *Advances in Polymer Technology*, vol. 24, no. 3, pp. 165–182, 2005.

[2] Y. K. Shen, P. H. Yeh, and J. S. Wu, "Numerical simulation for thin wall injection molding of fiber-reinforced thermoplastics," *International Communications in Heat and Mass Transfer*, vol. 28, no. 8, pp. 1034–1042, 2001.

[3] N. M. Neves, A. J. Pontes, and A. S. Pouzada, "Experimental validation of morphology simulation in glass fibre reinforced polycarbonate discs," *Journal of Reinforced Plastics and Composites*, vol. 20, no. 6, pp. 452–465, 2001.

[4] I. S. Dairanieh, A. Haufe, H. J. Wolf, and G. Mennig, "Computer simulation of weld lines in injection molded poly(methyl methacrylate)," *Polymer Engineering and Science*, vol. 36, no. 15, pp. 2050–2057, 1996.

[5] M.-C. Huang and C.-C. Tai, "Effective factors in the warpage problem of an injection-molded part with a thin shell feature," *Journal of Materials Processing Technology*, vol. 110, no. 1, pp. 1–9, 2001.

[6] H. Shin and E.-S. Park, "Analysis of crack phenomenon for injection-molded screw using moldflow simulation," *Journal of Applied Polymer Science*, vol. 113, no. 4, pp. 2702–2708, 2009.

[7] F. Pisciotti, A. Boldizar, M. Rigdahl, and I. Ariño, "Effects of injection-molding conditions on the gloss and color of pigmented polypropylene," *Polymer Engineering and Science*, vol. 45, no. 12, pp. 1557–1567, 2005.

[8] S. Weir, "Predicting surface defects in injection molded PVC components," *Journal of Vinyl and Additive Technology*, vol. 16, no. 4, pp. 231–234, 1994.

[9] T. Boronat, V. J. Segui, M. A. Peydro, and M. J. Reig, "Influence of temperature and shear rate on the rheology and processability of reprocessed ABS in injection molding process," *Journal of Materials Processing Technology*, vol. 209, no. 5, pp. 2735–2745, 2009.

[10] X. Bai, D. H. Isaac, and K. Smith, "Reprocessing acrylonitrile-butadiene-styrene plastics: structure-property relationships," *Polymer Engineering and Science*, vol. 47, no. 2, pp. 120–130, 2007.

[11] B. H. Min, "A study on quality monitoring of injection-molded parts," *Journal of Materials Processing Technology*, vol. 136, no. 1, pp. 1–6, 2003.

[12] P. K. Kennedy, *Practical and Scientific Aspects of Injection Molding Simulation*, chapter 2, Technische Universiteit Eindhoven, Eindhoven, The Netherlands, 2008.

[13] J. Koszkul and J. Nabialek, "Viscosity models in simulation of the filling stage of the injection molding process," *Journal of Materials Processing Technology*, vol. 157-158, pp. 183–187, 2004.

[14] B. E. Tiganis, L. S. Burn, P. Davis, and A. J. Hill, "Thermal degradation of acrylonitrile-butadiene-styrene (ABS) blends,"

Polymer Degradation and Stability, vol. 76, no. 3, pp. 425–434, 2002.

[15] J. B. Adeniyi, "Clarification and discussion of chemical transformations involved in thermal and photo-oxidative degradation of ABS," *European Polymer Journal*, vol. 20, no. 3, pp. 291–299, 1984.

[16] M. G. Wyzgoski, "Effects of oven ageing on ABS," *Polymer Engineering and Science*, vol. 17, pp. 265–269, 1976.

[17] G.İ. Öztürk, *Selected methods of surface engineering applied to materials [M.D. thesis]*, Graduate School of Engineering and Natural Sciences, Sabanc University, 2004.

[18] J. K. Kim and C. K. Kang, "Basic studies on recycling of ABS resin," *Polymer Plastics Technology and Engineering*, vol. 34, no. 6, pp. 875–890, 1995.

[19] D. Salari and H. Ranjbar, "Study on the recycling of ABS resins: simulation of reprocessing and thermo-oxidation," *Iranian Polymer Journal*, vol. 17, no. 8, pp. 599–610, 2008.

[20] R. K. Jena, X. Chen, C. Y. Yue, and Y. C. Lam, "Viscosity of COC polymer (TOPAS) near the glass transition temperature: experimental and modeling," *Polymer Testing*, vol. 29, no. 8, pp. 933–938, 2010.

[21] H. S. Kim, J. S. Son, and Y. T. Im, "Gate location design in injection molding of an automobile junction box with integral hinges," *Journal of Materials Processing Technology*, vol. 140, no. 1–3, pp. 110–115, 2003.

[22] A. A. Qaiser, A. Qayyum, and R. Rafiq, "Rheological properties of ABS at low shear rates: effects of phase heterogeneity," *Malaysian Polymer Journ*, vol. 4, pp. 29–36, 2009.

[23] J.-Z. Liang, "Effects of extrusion conditions on rheological behavior of acrylonitrile-butadiene-styrene terpolymer melt," *Journal of Applied Polymer Science*, vol. 85, no. 3, pp. 606–611, 2002.

9

Growth Kinetics of Monodisperse Polystyrene Microspheres Prepared by Dispersion Polymerization

Fan Li,[1,2] Chong Geng,[1] and Qingfeng Yan[1]

[1] *Department of Chemistry, State Key Laboratory of New Ceramics and Fine Processing, Tsinghua University, Beijing 100084, China*
[2] *Department of Material Science and Engineering, Tsinghua University, Beijing 100084, China*

Correspondence should be addressed to Qingfeng Yan; yanqf@mail.tsinghua.edu.cn

Academic Editor: Alain Durand

Dispersion polymerization has been widely applied to the synthesis of monodisperse micron-sized polymer colloidal spheres. Many efforts have been devoted to studying the influence of initial conditions on the size and uniformity of the resultant microspheres, aiming to synthesize micron-size monodisperse colloidal spheres. However, the inner contradiction between the size and the size distribution of colloidal spheres hinders the realization of this goal. In this work, we drew our attention from the initial conditions to the growth stage of dispersion polymerization. We tracked the size evolution of colloidal sphere during the dispersion polymerization, through which we established a kinetic model that described the relationship between the monomer concentration and the reaction time. The model may provide a guideline to prepare large polymer colloidal spheres with good monodispersity by continuous monomer feeding during the growth stage to maintain the concentration of monomer at a constant value in a dispersion polymerization process.

1. Introduction

As a new functional macromolecule material, monodisperse polymer microspheres have many applications in environmental conservation, biomedicine, colloid science, electronic information material, and many other areas [1–3] due to their superiority of good sphericity, size tunability, large specific surface area, and excellent absorbability [4–6]. Particularly, monodisperse micron-sized spheres are ideal materials as advanced coatings [7], fillers of chromatographic column [8, 9], standard particles of electron microscope and Coulter particle size testers, and spacers in LCD [10], which has drawn increasing interest to synthesize monodisperse micron-sized spheres [11, 12].

The first report on monodisperse polymer colloidal spheres was the polystyrene spheres prepared by Vanderhoff and Brandford [13]. So far, many synthesis strategies have been developed such as emulsion polymerization, suspension polymerization, dispersion polymerization, and seed swelling polymerization. In view of the demand of micron size and monodispersity, the former two methods are excluded because they cannot meet the two requirements at the same time. Emulsion polymerization can only gain monodisperse colloidal spheres with the size below 1 μm. Suspension polymerization may achieve large microspheres with size ranging from 10 μm to 100 μm while the microspheres possess a bad monodispersity [14]. The latter two methods are suitable for the synthesis of uniform colloidal spheres with micron size. Seed swelling polymerization, including the dynamics swelling developed by Okubo et al. [15] and two-step swelling initiated by Ugelstad et al. [16], can attain colloidal spheres with larger sphere diameter. However, in terms of the whole process, seed swelling polymerization is complex. In addition, the monodispersity is determined not only by the swelling process itself, but also by the initial monodispersity of the seeds prepared by dispersion polymerization in advance. Thus, the key point is how to gain larger microspheres with uniform size by dispersion polymerization.

To attain larger microspheres with excellent monodispersity, many efforts have been made to explore the influence of the initial condition of the reaction system on the resultant microspheres, such as the dosage of monomer, initiator, and dispersing agent. Generally, with the increase in the monomer's initial concentration, the sphere diameter increases and the distribution of sphere diameter broadens. Thus, there is a suitable range for monomer's initial concentration to achieve uniform and large microspheres [17, 18]. Similarly, increasing the concentration of initiator leads to higher reaction rate and larger sphere diameter while it may also broaden the distribution of the sphere diameter simultaneously [17, 19, 20]. In addition, the dispersing agent which prevents microspheres from gelatinization and stabilizes the reaction system requires a sufficient supply to guarantee its functions. However, the increase in the amount of the dispersing agent results in a smaller size despite a better monodispersity [21–24]. In this view, the inner contradiction between the size and the size distribution of microspheres hinders the further increase of the size of monodisperse microspheres. It is thus believed that the inner contradiction cannot be solved solely by adjusting the initial conditions.

Recently, Song et al. [25] employed two-stage dispersion polymerization to synthesize large and uniform dye-contained polymer spheres. In this method, the dye-comonomer, which significantly destroyed the final monodispersity if fed into the reaction at the beginning, was added to the reaction system after the nucleation stage. Through this strategy, the influence of the dye-comonomer on the nucleation was avoided and a narrow size distribution of the final colloidal spheres was guaranteed. It was expected that two-stage or multistage dispersion polymerization strategy paved a way to the synthesis of micron-sized large and monodisperse polymer colloidal spheres. However, when and how to feed the extra monomer into the reaction system to achieve this goal remains debatable since most previous works were mainly based on trial and error.

In this work, we focused on the whole process in one-stage dispersion polymerization and tracked the size evolution of colloidal sphere at different reaction time. Based on the experimental results, we established a theoretical model to describe the relationship between the monomer concentration and the reaction time during dispersion polymerization. The model was expected to provide guidance to the two-stage or the multi-stage dispersion polymerization, aiming to synthesize micron-sized large and monodisperse polymer colloidal spheres.

2. Materials and Methods

2.1. Materials. All chemicals, including styrene (St, 99%, Aldrich), anhydrous ethanol (99.95%, Aldrich), polyvinylpyrrolidone (PVP, MW 40000, TCI), and azobis(isobutyronitrile) (AIBN, Sinopharm Chemical Reagent Co., Ltd), were used as received without further purification. Deionized (DI) water (resistivity greater than 18.2 MΩ·cm, Ultra Pure UV, China) was used in all experiments.

2.2. Synthesis of PS Colloidal Microspheres by Using Dispersion Polymerization. Briefly, the synthesis was conducted in a three-neck round-bottom flask equipped with a condenser and immersed in a water bath with a preset temperature. Firstly, the monomer of 17 mL styrene and the stabilizer of 1.5 g PVP were dissolved in 98 mL anhydrous ethanol and then added to the flask. The solution was stirred with a rate of 200 rpm and the temperature was kept at 70°C while nitrogen gas was continuously led into the flask to avoid the oxidation of the initiator AIBN in the whole process. After 30 minutes, an initiator solution of 0.15 g AIBN dissolved in 28 mL ethanol was introduced into the system. The reaction continued for 24 h and then was stopped by cooling at ambient conditions. To track the evolution of sphere size during dispersion polymerization, the mixture in the reaction flask was sampled at specific intervals and dried on a glass substrate. The sphere size was measured using an optical microscope.

2.3. Characterization. The optical microphotographs of the PS colloidal microspheres were taken using an optical microscope (Olympus, BX51TRF), which was connected to a CCD camera (Pixelink-B742) and a computer for real-time image recording.

3. Results and Discussion

3.1. The Morphology of PS Colloidal Spheres Prepared by Dispersion Polymerization. As can be seen from the optical microscope photograph shown in Figure 1, micron-size PS colloidal spheres were obtained by using the one-stage dispersion polymerization method. The PS colloidal spheres were spontaneously organized into a closely packed hexagonal lattice on the glass substrate upon evaporation of solvent, indicating the high monodispersity of the PS spheres. To reduce the error in the measurement of the size of a single PS sphere, we chose the closely packed area and measured the direct distance between the several tangent microspheres in line and calculated the average distance between neighboring two spheres. According to the measured data in Figure 1, the average diameter of the resultant PS colloidal spheres was calculated to be 2.72 μm.

3.2. Growth Kinetics of PS Spheres during Dispersion Polymerization. The process of dispersion polymerization can be basically divided into several stages. At the beginning of the reaction, monomer, dispersing agent, and initiator are all dissolved in the solvent. When the temperature reaches the decomposition temperature of the initiator, the initiator decomposes and produces radicals. The radicals trigger the polymerization process, which results in formation of the oligomers. When the chain length of the oligomers exceeds a critical value, nucleation occurs and the oligomers precipitate from the solvent. At the same time, dispersing agents are absorbed by the nuclei and form graft copolymers with the polymer chains, which makes microspheres stable in the solvent. Then the synthesis enters the stage of growth. Monomer and copolymer in the solvent are persistently captured by polymer nuclei to provide the motive power of the growth, which keeps on until termination of the reaction.

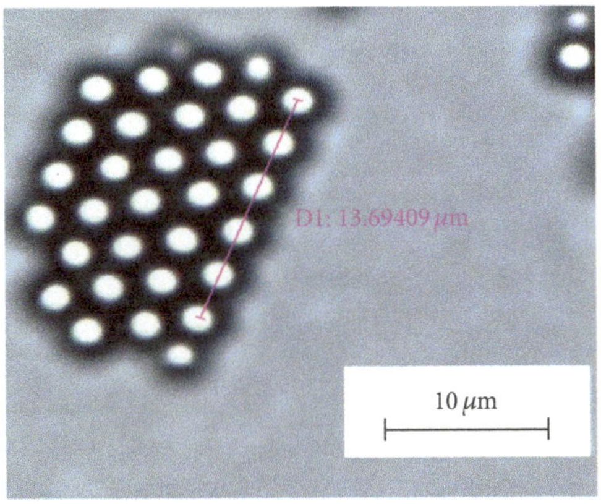

FIGURE 1: Optical microphotograph of PS colloidal spheres prepared by dispersion polymerization.

In the process of dispersion polymerization, the nucleation stage is a transition period for the system from homogenous phase to heterogeneous phase. This period determines the number of microspheres in the system which should be constant during the polymerization to guarantee the uniformity of final microspheres [21, 22, 26–28].

Knowing the evolution of monomer concentration during dispersion polymerization is of importance especially for synthesis of large and uniform colloidal spheres. Considering the reaction mechanism involved in dispersion polymerization, however, a large proportion of the monomers exist in the solvent in the form of oligomers whose chain length does not exceed the critical value for nucleation. Thus, direct measurement of the monomer concentration is not feasible. Herein, we establish a theoretical model to describe monomer concentration at different reaction time based on the diameter of microspheres measured at different time intervals.

The following are the major assumptions used in formulating the model. Firstly, in view of the good monodispersity of the final microspheres, the reaction is considered as an ideal dispersion polymerization, in which nucleation number is just determined by the nucleation stage, and secondary nucleation does not occur in the later growth stage. It is also assumed that the number of microspheres remains during growth stage equal to the nucleation number at the beginning of reaction. Secondly, although the microspheres in the growth stage may mostly seize the oligomer chains rather than the monomers, it is considered that the increasing rate of the microspheres' weight or volume is proportional to the equivalent concentration of the monomers, some of which are in the form of free oligomers. Thirdly, considering the actual slight volume change during the mixture process of styrene and ethanol, which can hardly be observed in the experiment, the deviation of volume is neglected in our model to reduce the complexity of the calculation. Thus, a simple approximation was adopted by the replacement of the total volume of the mixture of several liquids with the sum of their own volumes.

According to the conservation of mass, the relationship between the product and the reactant at any time t is given by

$$N\rho_{\text{sphere}} V_{\text{sphere}} = m_0 - CV, \tag{1}$$

where N is the number of microspheres in the system at time t, ρ_{sphere} is the density of polystyrene microspheres, m_0 is the initial mass of styrene monomer added into the system, C is the present concentration of surplus styrene in the system, and V is the total volume of the system at time t. V_{sphere}, the average volume of one sphere, can be described by the formula of spheroidal volume

$$V_{\text{sphere}} = \frac{\pi}{6} d^3, \tag{2}$$

where d is the average diameter of microspheres at time t.

The concentration of styrene in the system at time t can be expressed using

$$C = \frac{\rho_{\text{styrene}} V_{\text{styrene}}}{V} = \frac{\rho_{\text{styrene}} \left(V - V_{\text{ethanol}}\right)}{V}, \tag{3}$$

where V_{styrene} stands for the surplus volume of styrene in the system, while V_{ethanol} stands for the surplus volume of ethanol, which is equal to the initial volume with the loss in the reaction ignored.

By putting together the three equations above, out comes the relationship between the concentration of monomer and the average diameter of microspheres at time t, which can be given by

$$C = \frac{\rho_{\text{styrene}} \left(m_0 - N\rho_{\text{sphere}} \cdot (\pi/6)\, d^3\right)}{m_0 - N\rho_{\text{sphere}} \cdot (\pi/6)\, d^3 + \rho_{\text{styrene}} V_{\text{ethanol}}}. \tag{4}$$

Based on our experimental data, we input relevant parameters. The initial parameters were the same as the data described in the experimental section, while the density of microspheres was taken as that of polystyrene. According to the former assumption, the number of microspheres was a constant value which is equal to the nucleation number. Thus, we set the time t at the end of the reaction with the microspheres' diameter d taken as the final size which was 2.72 μm in our work, while the final concentration C_{final} was assumed to be one percent of the initial value. In this way, the number of microspheres was approximately calculated as follows:

$$N = \frac{\rho_{\text{styrene}} V_0 - C_{\text{final}} V_{\text{ethanol}}}{\rho_{\text{sphere}} \cdot (\pi/6)\, d_{\text{final}}^3} \approx 1.347 \times 10^{12}. \tag{5}$$

Thus, with the input of the initial parameters and the estimated number of microspheres N, the relationship between the monomer concentration and the microspheres diameter at any time was attained.

To introduce the time dimension, we measured the diameter of the PS spheres sampled from the reaction system at different time intervals. According to the relationship expressed in (4), we gained the corresponding monomer concentration, which is shown in Table 1.

Table 1: The sphere diameters and monomer concentration at different time.

Reaction time (h)	Sphere diameter (μm)	Monomer concentration (g/mL)
0	0	0.105
1	1.170	0.098
1.5	1.364	0.094
2	1.360	0.094
3	1.760	0.080
4	1.960	0.069
5	2.114	0.060
7	2.376	0.039
9	2.410	0.036
11	2.652	0.010
23	2.720	0.001

In order to further deduce the relationship between the monomer concentration and the reaction time, we calculated different function forms of monomer concentration versus reaction time and processed the data with linear fitting. The results are shown in Figure 2.

As can be seen in Figure 2, the linear fitting of $\ln C$ and t has the largest correlation coefficient of 0.965. As a consequence, we consider that the natural log of the monomer concentration is linearly associated with the reaction time. In other words, there exists an exponential relationship between the monomer concentration and the reaction time, which can be expressed using

$$C(t) = C_0 e^{-kt}, \quad (6)$$

where the initial monomer concentration C_0 can be attained by the initial parameters in our experiment. Then, a program was written to find optimal value of k which minimized the sum of squared relative errors between the theoretical curve and the experimental points (t_i, C_i). That was one value that made the value of $\sum_i [(C_i - C(t_i))/C(t_i)]^2$ minimum. The final result was obtained as follows:

$$C = 0.105 \times e^{-0.10t}. \quad (7)$$

Combining (4) with (7), the relationship between the colloidal sphere diameter and the reaction time could be derived. The resultant curve is shown in Figure 3 in comparison to the experimental data. As can be seen from Figure 3, the theoretical curve matches the experimental data well.

3.3. Application of the Kinetic Model on Synthesis of Large and Monodisperse Colloidal Spheres via Dispersion Polymerization. Through the deformation of (3), the total volume of the reaction system can be obtained as follows:

$$V = \frac{\rho_{\text{styrene}} V_{\text{ethanol}}}{\rho_{\text{styrene}} - C}. \quad (8)$$

Table 2: Different monomer feeding rates required at various beginning time of continuous monomer feeding.

t (h)	C (g/mL)	dC/dt (g/mL/h)	V (mL)	ΔV (mL/h)
1	0.0950	−0.00950	140.9	1.48
2	0.0860	−0.00860	139.4	1.32
3	0.0778	−0.00778	138.0	1.18
4	0.0704	−0.00704	136.8	1.06
5	0.0637	−0.00637	135.7	0.95
6	0.0576	−0.00576	134.7	0.86
7	0.0521	−0.00521	133.9	0.77
8	0.0472	−0.00472	133.1	0.69
9	0.0427	−0.00427	132.4	0.62
10	0.0386	−0.00386	131.8	0.56

ΔV means the monomer feeding rate needed if the beginning of the continuous monomer feeding is set at time t.

With the relationship between the monomer concentration and the reaction time described in (7), the monomer consumption rate at any time of the reaction can be easily deduced by taking the derivative of the monomer concentration with respect to time:

$$\Delta V = \frac{1}{\rho_{\text{styrene}}} \cdot \frac{dC}{dt} \cdot V. \quad (9)$$

Thus, assuming that the initiator does not lose efficacy during the reaction, the monomer concentration will be held at a constant value if one continuously feeds extra monomer into the system according to the consumption rate described in (9). This will be helpful to keep the system at a high reaction rate, leading to the enlargement of the diameter of the final PS colloidal microspheres. In fact, the initiator efficiency will inevitably descend during the process of dispersion polymerization; so in actual process, we can meet the requirement of constant efficiency by replenishing initiator during the reaction.

According to the analysis above, we calculated the relationship between the beginning time of the monomer feeding and the monomer feeding rate within the present system of our experiment, which is shown in Table 2. In this way, even if an extra monomer is added into the reaction system, the monomer concentration will still be low enough to avoid new nucleation as long as one begins the continuous feeding of extra monomer in the growth stage rather in the nucleation state. As a consequence, the monodispersity of the colloidal spheres is ensured. At the same time, continuous feeding of extra monomer inevitably leads to colloidal spheres with large size.

4. Conclusion

In summary, a kinetic model has been established to describe the exponential relationship between the monomer concentration and the reaction time during synthesis of colloidal spheres by using dispersion polymerization. Accordingly, one may gain the quantitative relationship between the

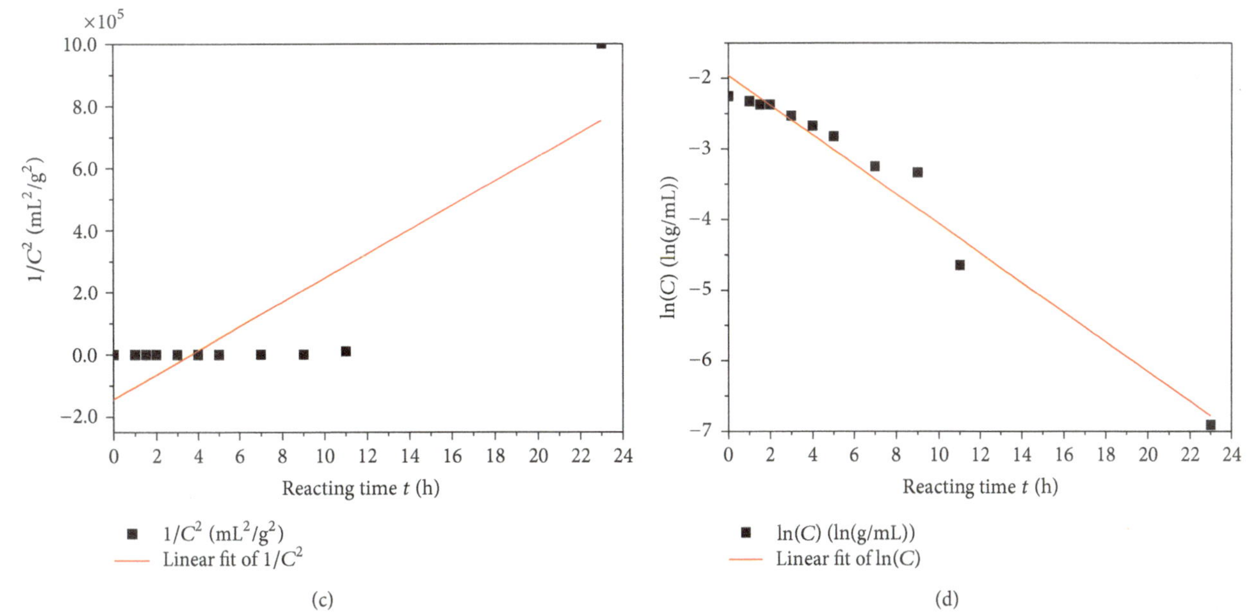

FIGURE 2: Linear fitting results of different function forms of concentration C with reaction time t. (a) linear fitting of C and t, $r = 0.800$ (b) linear fitting of $1/C$ and t, $r = 0.763$ (c) linear fitting of $1/C^2$ and t, $r = 0.703$ (d) linear fitting of $\ln C$ and t, $r = 0.965$.

monomer consumption rate and the reaction time. Taking advantage of this model, it is possible to attain larger colloidal spheres by continuously feeding monomer into the reaction system during the growth stage according to the monomer consumption rate. In this way, the reaction rate is kept at a high constant all the time until the reaction is artificially ended. Furthermore, secondary nucleation can be effectively avoided because both the average concentration and the local concentration of monomer in the reaction system are less than the critical value. Ultimately, large and uniform colloidal spheres can be prepared using the dispersion polymerization technique.

Conflict of Interests

The authors declare no competing financial interests.

Acknowledgments

This work was supported by the Natural Science Foundation of China (Project nos. 50903046 and 51173097) and the National Key Basic Research Program of China (no. 2013CB632902). The Tsinghua University Initiative Scientific Research Program and the Tsinghua National Laboratory

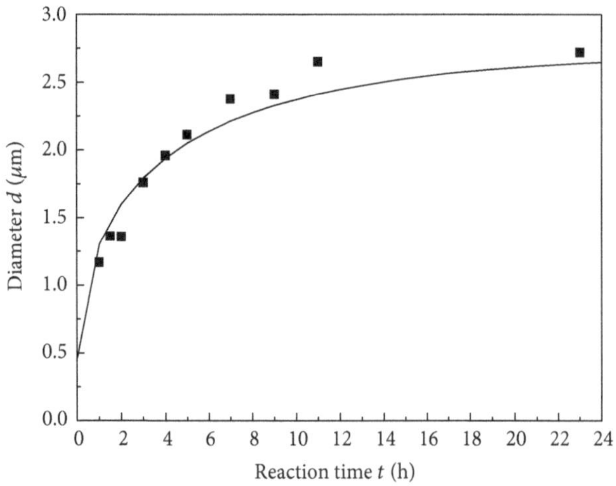

FIGURE 3: The evolution of spheres diameter with extended reaction time. The squares stand for the experimental points.

for Information Science and Technology (TNList) Cross-discipline Foundation are also acknowledged for partial financial support.

References

[1] S. K. Mi, H. L. Gyu, H. Jae-Min, and L. Hyunjung, "Synthesis of monodisperse PS-co-PDMS microspheres by dispersion polymerization," *Materials Science and Engineering*, vol. 27, no. 5–8, pp. 1247–1251, 2007.

[2] S. O. Kyoung, B. won, and K. Hwayoung, "Dispersion polymerization of N-vinylcarbazole using siloxane-based and fluorine-based surfactants in compressed liquid dimethyl ether," *Polymer*, vol. 48, no. 6, pp. 1450–1454, 2007.

[3] I. Miraballes-Martínez and J. Forcada, "Synthesis of latex particles with surface amino groups," *Journal of Polymer Science A*, vol. 38, pp. 4230–4237, 2000.

[4] R. P. A. Dullens, E. M. Claesson, and W. K. Kegel, "Preparation and properties of cross-linked fluorescent poly(methyl methacrylate) latex colloids," *Langmuir*, vol. 20, no. 3, pp. 658–664, 2004.

[5] H. Kawaguchi, "Functional polymer microspheres," *Progress in Polymer Science*, vol. 25, no. 8, pp. 1171–1210, 2000.

[6] R. M. Santos and J. Forcada, "Acetal-functionalized polymer particles useful for immunoassays," *Journal of Polymer Science A*, vol. 35, no. 9, pp. 1605–1610, 1997.

[7] K. Ishii, "Synthesis of microgels and their application to coatings," *Colloids and Surfaces A*, vol. 153, no. 1-3, pp. 591–595, 1999.

[8] E. Caro, R. M. Marce, P. A. G. Cormack, D. C. Sherrington, and F. Borrull, "Molecularly imprinted solid-phase extraction of naphthalene sulfonates from water," *Journal of Chromatography A*, vol. 1047, no. 2, pp. 175–180, 2004.

[9] Y. Watebe, K. Hosoya, N. Tanaka, and T. Kubo, "Shielded molecularly imprinted polymers prepared with a selective surface modification," *Journal of Polymer Science A*, vol. 43, no. 10, pp. 2048–2060, 2005.

[10] S. Omi, M. Saito, T. Hashimoto, M. Nagai, and G.-H. Ma, "Preparation of monodisperse polystyrene spheres incorporating polyimide prepolymer by dispersion polymerization in the presence of L-ascorbic acid," *Journal of Applied Polymer Science*, vol. 68, no. 6, pp. 897–907, 1998.

[11] K. Ishizu and N. Tahara, "Microsphere synthesis by emulsion copolymerization of methyl methacrylate with poly(methacrylic acid) macromonomers," *Polymer*, vol. 37, no. 13, pp. 2853–2856, 1996.

[12] N. B. Viswanthan, P. A. Thomas, J. K. Pandit, M. G. Kulkarni, and R. A. Mashelkar, "Preparation of non-porous microspheres with high entrapment efficiency of proteins by a (water-in-oil)-in-oil emulsion technique," *Journal of Controlled Release*, vol. 58, no. 1, pp. 9–20, 1999.

[13] J. W. Vanderhoff and E. B. Brandford, *Polymer Colloid I*, Plenum Press, New York, NY, USA, 1971.

[14] D.-H. Lee, J.-W. Kim, and K.-D. Suh, "Monodisperse micron-sized polymethylmethacrylate particles having a crosslinked network structure. V," *Journal of Materials Science*, vol. 35, no. 24, pp. 6181–6188, 2000.

[15] M. Okubo, M. Shiozaki, M. Tsujihiro, and Y. Tsukuda, "Preparation of micron-size monodisperse polymer particles by seeded polymerization utilizing the dynamic monomer swelling method," *Colloid & Polymer Science*, vol. 269, no. 3, pp. 222–226, 1991.

[16] J. Ugelstad, P. C. Mork, R. Schmid, T. Ellingsen, and A. Berge, "Preparation and biochemical and biomedical applications of new monosized polymer particles," *Polymer International*, vol. 30, no. 2, pp. 157–168, 1993.

[17] C.-M. Tseng, M. S. El-Aasser, and J. W. Vanderhoff, "Optimization of polymer-solvent interaction in a multisolvent system by the unifac group-contribution method," *Journal of Applied Polymer Science*, vol. 32, no. 5, pp. 5007–5019, 1986.

[18] H. Bamnolker, B. Nitzan, T. Tennenbaum, B. BarToov, M. Hinz, and H. Seliger, "Scientific and Clinical Applications of Magnetic Carriers ," in *Proceedings of the 1st International Conference on Scientific and Clinical Applications of Magnetic Carriers*, pp. 37–51, Rostock, Germany, September 1996.

[19] A. Tuncel, R. Kahraman, and E. Piskin, "Monosize polystyrene lattices carrying functional-groups on their surfaces," *Journal of Applied Polymer Science*, vol. 51, pp. 1485–1498, 1994.

[20] K. Hu, V. L. Dimonie, E. D. Sudol, and M. S. El-Aasser, "Monodisperse poly(butadiene/styrene) particles by dispersion polymerization," *Journal of Applied Polymer Science*, vol. 55, no. 10, pp. 1411–1415, 1995.

[21] B. Thomson, A. Rudin, and G. Lajoie, "Dispersion copolymerization of styrene and divinylbenzene: synthesis of monodisperse, uniformly crosslinked particles," *Journal of Polymer Science A*, vol. 33, no. 3, pp. 345–347, 1995.

[22] B. Thomson, A. Rudin, and G. Lajoie, "Dispersion copolymerization of styrene and divinylbenzene. II: effect of crosslinker on particle morphology," *Journal of Applied Polymer Science*, vol. 59, no. 13, pp. 2009–2028, 1996.

[23] M. Holderle, G. Bar, and R. Mulhaupt, "Synthesis and characterization of oxazoline-functional polymer particles by free radical nonaqueous dispersion polymerization," *Journal of Polymer Science A*, vol. 35, no. 13, pp. 2539–2548.

[24] D. Horak, F. Svec, and J. M. Frechet, "Preparation and control of surface properties of monodisperse micrometer size beads by dispersion copolymerization of styrene and butyl methacrylate in polar media," *Journal of Polymer Science A*, vol. 33, no. 14, pp. 2329–2338, 1995.

[25] J.-S. Song, F. Tronc, and M. A. Winnik, "Two-stage dispersion polymerization toward monodisperse, controlled micrometer-sized copolymer particles," *Journal of the American Chemical Society*, vol. 126, no. 21, pp. 6562–6563, 2004.

[26] H. Kurioka, H. Uyama, and S. Kobayashi, "Peroxidase-catalyzed dispersion polymerization of phenol derivatives," *Polymer Journal*, vol. 30, no. 6, pp. 526–529, 1998.

[27] M. Okubo, T. Yamashita, and E. Ise, "Thermodynamics for preparation of micron-sized, monodispersed, monomer-adsorbed polymer particles having snowman shape by utilizing the dynamic swelling method and the seeded polymerization," *Proceedings of the Japan Academy B*, vol. 75, no. 7, pp. 195–200, 1999.

[28] M. Okubo, Y. Konishi, M. Takebe, and H. Minami, "Preparation of micron-sized, monodispersed,anomalous polymer particles by utilizing the solvent-absorbing/releasing method," *Colloid and Polymer Science*, vol. 278, no. 10, pp. 919–926, 2000.

10

Pharmaceutical Properties and Applications of a Natural Polymer from *Grewia mollis*

Elijah I. Nep, Patricia O. Odumosu, Ndidi C. Ngwuluka, Patrick O. Olorunfemi, and Nelson A. Ochekpe

Biomaterials and Drug Delivery Research Group, Faculty of Pharmaceutical Sciences, University of Jos, PMB 2084, Jos 930001, Nigeria

Correspondence should be addressed to Elijah I. Nep; elijah.nep@biodrudel.com

Academic Editor: Alain Durand

The use of naturally occurring biocompatible materials has been the focus of recent research activity in the design of dosage forms for immediate and controlled release formulations. Grewia gum is an intracellular gum obtained by extraction from the inner stem bark of the shrub *Grewia mollis* (Malvaceae). It grows abundantly (wild or cultivated) in the middle belt region of Nigeria, and the mucilage has been used by indigenes of this belt as thickener in soups. Grewia gum has been investigated for potential applications in pharmaceutical dosage forms. The industrial extrapolation of the applications of the gum has, however, been slowed by the limited structural, toxicological, and stability data available on the gum. This paper highlights ethnobotanical uses of *G. mollis* shrub and discusses the structural features, functional properties, and applications of grewia gum with emphases on its pharmaceutical potentials.

1. Introduction

Plant materials are playing increasing role as alternatives to chemical food additives and synthetic pharmaceutical excipients. Natural polysaccharide gums swell to form highly viscous solutions or dispersions in aqueous media. They have the advantage of biocompatibility, low cost, and relative abundance [1] compared to their synthetic counterparts. They are widely used in the pharmaceutical industry as polymers in various drug delivery systems [2–7].

Grewia polysaccharide gum is a natural resource that could be used as an excipient in the pharmaceutical industry in Nigeria to reduce the costs of pharmaceutical products. It may provide a suitable alternative to the synthetic counterparts which are expensive and mostly imported. It is obtained by extraction (maceration in cold or hot water) of the inner stem bark of the edible plant *G. mollis*, Juss (Malvaceae). In Nigeria, *G. mollis* from which the gum is extracted grows abundantly (wild or cultivated) in the middle belt region of the country where it is used as thickener in local delicacies.

This review delineates the ethnobotanical uses, physicochemical properties, structural properties, and the potential applications of the gum in drug delivery systems.

2. *Grewia mollis* Plant

G. mollis belongs to the flowering plant genus, *Grewia*. It was formerly placed in the family Tiliaceae or Sparrmanniaceae. Today most authors place the genus in the mallow family Malvaceae.

2.1. Botany. The plant has been described [8, 9] as a shrub or small tree growing to attain a height of 10.5 m with young branches densely stellate-pubescent. The young branches turn dark grey to black or reddish brown when older.

The leaves of the plant are elliptic to elliptic-oblong, usually between 2.0 and 14.5 cm long and 0.7 and 5.5 cm wide. They are acute to slightly acuminate at the apex, broadly rounded or obliquely truncate at the base. The leaf margins are coarsely and sharply serrate, more or less glabrous to sparsely minutely stellate-pubescent above, but densely and finely greyish to brownish white-pubescent beneath [9].

The flowers are yellow and bisexual; sepals 6–10 mm long; petals obviate to oblong, 4–6 mm long, ±2 mm wide, and sometimes notched at the apex. The ovary is 1.5–2 mm long and densely hairy [9].

Table 1: Ethnomedicinal uses of *G. mollis* plant.

Use	Plant part	Preparation	References
Wound healing	Bark, leaves	Mucilages are applied to ulcers, cuts, sores, and snake bites	[9, 13]
Cough	Bark, roots	Decoction	[9, 13]
Fever	Bark Leaves, and fruit	Extract of bark and/or leaves. Fruits are also eaten for this purpose	[9, 13]
Diarrhoea	Stem bark	Decoction	[9, 24]
Child birth Pain	Stem bark, leaves	Maceration, decoction	[9, 13]
Colic	Bark, leaves	Infusion	[9]
Constipation	Bark	Mucilage	[9]
Rickets	Leaves	Decoction	[9]
Palpitation	Roots, leaves	Decoction	[9]
Sore eyes	Root shavings	Sap squeezed out and applied under the eyelid	[9]
Rheumatism	Roots	Paste	[9, 13]
Anal prolapse	Bark	Pounded and dissolved in water	[9, 24]

Figure 1: *G. mollis* plant. The figure is adapted from http://www.prota4u.org/ [10].

The fruit has been described as unlobed and globose drupe of about 4–7 mm long and 5–7 mm wide, covered with a fine whitish tomentellum. The fruits are green when younger and turning yellow when older [8, 9]. A digital image of the aerial part of the plant is shown in Figure 1.

2.2. Propagation. *G. mollis* grows in the wild and can be propagated by seed or seedlings. The seeds are collected from the dried fruits that may have fallen on the ground [9].

3. Uses of the Aerial Parts of *G. mollis*

The different parts of *G. mollis* have been used for different purposes across the regions and places where the plant grows or is cultivated.

3.1. Mucilage from Leaves and Stem Bark. In the Democratic Republic of Congo, the bark is kneaded with water into a viscous substance that is added to sauces while, in Gabon, the inner bark is sometimes eaten as food. In Nigeria, the inner stem bark is used as thickener in soups and in local cakes made from beans or corn flour commonly called "Kosai" and "Punkasau" in Hausa (Nigeria), respectively [11, 12].

The infusion of the bark obtained by cold or hot maceration in water is applied to give a smooth surface to mud walls and floors [13] while the wood ash is used as salt substitute including the ash of the leaves, stems, and roots [9].

3.2. Vegetables and Fruits. The flowers, buds, and young shoots are added to soups and sauces as garnishing while, in Sudan, the young leaves are cooked and eaten as vegetables. The fruit is eaten raw or boiled [9].

3.3. Ethnomedicinal Uses. Several ethnomedicinal uses of the infusion, decoction, maceration, or mucilage from the leaves, roots, or stem bark of *G. mollis* have been documented. These ethnomedicinal uses are described in Table 1.

4. Biological Evaluation and Chemical Constituents of *G. mollis*

Several scientists have investigated the histopathological and toxicological effects of extracts of *G. mollis* on laboratory animals.

Onwuliri et al. [14] conducted toxicological and histopathological studies in rats using the ethanolic extract of the stem bark of *G. mollis*. Tannins, saponins, flavonoids, glycosides, balsam, phenols, terpenes, and steroids were isolated from the stem bark, but alkaloids were absent. The extract exhibited toxic properties at the lethal dose (LD_{50}) of 1500 mg/kg body weight. The convoluted tubules of the kidney but no structural effects on the liver and heart were observed suggesting that the extract may be safe in humans but should be used with caution on patients with renal failure.

Studies by Obidah et al. [12] on the toxic effects of the stem bark of *G. mollis* showed that the addition of the pulverized stem bark to the normal diet of male Wister rats at concentrations of 0, 1, 5, and 10% given to them as feed for 4 weeks, resulted in no deaths, and remarkable changes in appearance were not observed in the treated animals. However, rats fed with 10% dietary level showed significant ($P < 0.05$) increases in serum transaminases activities which was accompanied by decreased food intake. There was no observed effect on serum alkaline phosphatase activity, urea, creatinine, triglycerides, cholesterol, glucose concentrations, and body and organ weights in their study. These workers concluded that dietary exposure of rats to *G. mollis* stem bark powder at high concentrations (10%) may cause some adverse effects, especially liver injury.

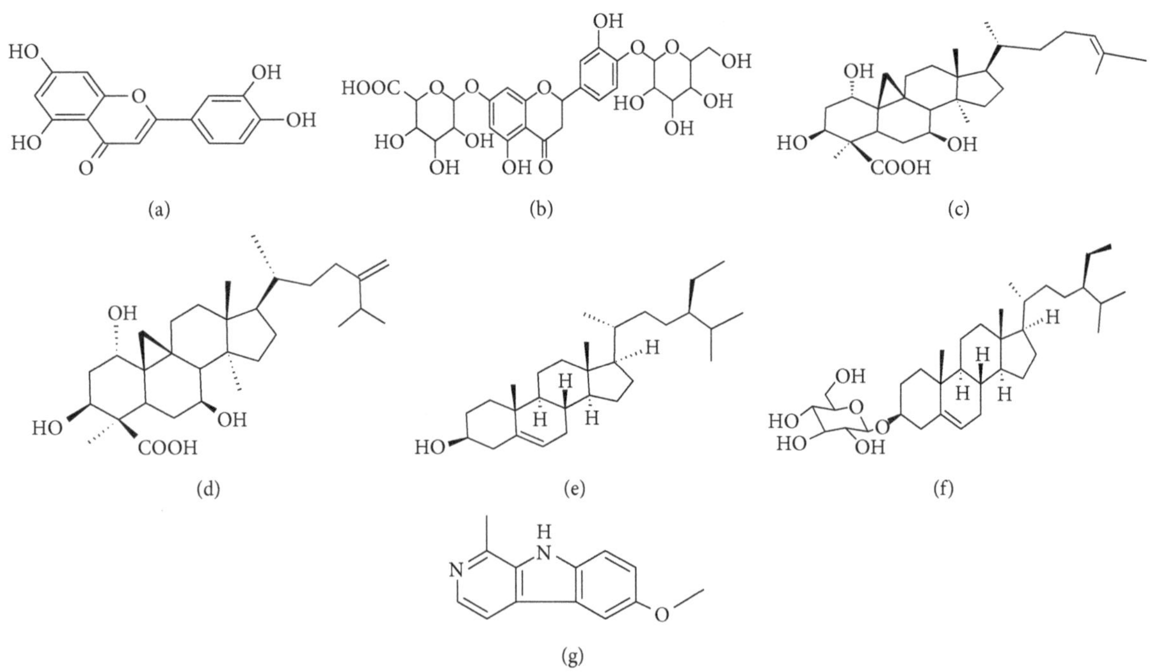

Figure 2: Molecular structure of isolates from aerial parts of *G. mollis*. (a) Luteolin, (b) 7-(1-O-β-D-galacturonide) 4′-(1-O-β-glucopyranosyl)-3′4′,5,7-tetrahydroxyflavone, (c) 7β-hydroxy-23-enedeoxojessic acid, (d) 7β-hydroxy-23-deoxojessic acid, (e) β-sitosterol, (f) β-sitosterol-3-O-glucoside, and (g) 6-Methoxyharmane.

It may be argued however that gum extracted from the stem bark of *G. mollis* when used as excipient is incorporated into formulations in smaller proportions usually lower than the 10% proportion added to the feed of the animals in the foregoing study. Consequently, it may be expected that the low use level of *G. mollis* gum as excipient in formulations should not cause any adverse effects in humans.

Asuku et al. [15] evaluated the methanolic extract of *G. mollis* leaves for its antioxidant and hepatoprotective properties by an *in vivo* procedure. They reported significant ($P < 0.05$) hepatoprotective potential evidenced by the lowering of serum levels of bilirubin, aspartate aminotransferase, and alanine aminotransferase and decreasing of malondialdehyde levels in rats pretreated or posttreated with carbon tetrachloride (CCl_4). They concluded that *G. mollis* leaves contain potent antioxidant compounds that could offer protection against hepatotoxicity and ameliorate preexisting liver damage and oxidative stress conditions.

Saleem et al. [16] assessed the nutritive value of the leaves and fruits of three *grewia* species under semiarid environment, and the results of the study indicated that the three species could be introduced as a source of fodder in animal production farms and silvopastoral systems.

4.1. Compounds Isolated from G. mollis. Al-Youssef et al., [17] evaluated the methanolic extract of the aerial part of *G. mollis* and reported isolation of a number of phytochemical constituents such as flavonoids (luteolin and 7-(1-O-β-D-galacturonide)-4′-(1-O-β-glucopyranosyl)-3′4′,5,7-tetrahydroxyflavone), triterpenoids (7β-hydroxy-23-enedeoxojessic acid and 7β-hydroxy-23-deoxojessic acid), and steroids (β-sitosterol and β-sitosterol-3-O-glucoside). The molecular structures of these isolates are shown in Figure 2.

Rosler et al. [18] earlier reported the isolation of 6-methoxyharmane (Figure 2(g)), a Harman alkaloid from *G. mollis*. Harman alkaloids belong to the class of β-carbolines. They bind strongly to benzodiazepine receptors in the brain to induce inverse agonist effects to benzodiazepines such as convulsive, anxiogenic, and memory enhancing effects [19–21].

Efiom and Oku [22], working on the n-hexane fraction of *G. mollis* root extract collected in Niger State, Nigeria, isolated two triterpene compounds, lup-20-en-3-ol and 1,3-hexyloxacyclotridec-10-en-2-one.

4.2. Phytochemical Screening, Antibacterial and Anti-Inflammatory Properties. Shagal et al. [23] reported the presence of saponins and phenolic compounds in the ethanol fraction of *G. mollis* leaves while glycosides were additionally present in the ethanolic stem bark extracts. Tannins, volatile oils, and flavonoids were also observed in the ethanolic root extracts.

Antibacterial studies on the ethanolic leaf and root extracts were reported to exhibit inhibitory activity against *Staphylococcus aureus* and *Escherichia coli* while the stem bark extract exhibited activity against *S. aureus* and *Salmonella typhi*. The aqueous root extracts showed no inhibition against the same organisms except for the aqueous leaf extract which showed some inhibitory activity against *Streptococcus* spp. and *E. coli*. The aqueous extract of the stem bark also exhibited inhibition against *S. typhi* which possibly justifies the use of the plant in the treatment against diarrhoea and dysentery.

Al-Youssef et al. [17] investigated the *in vitro* antibacterial activity of the aerial part of *G. mollis* collected in Akba-Al-Abar, Saudi Arabia, against various strains of bacteria such as *Staphylococcus epidermidis* (ATCC 9615), *Bacillus subtilis* (ATCC 1174), *S. aureus* (ATCC 25175), *Klebsiella pneumonia* (ATCC 33495), *Pseudomonas aeruginosa* (ATCC 12453), and *E. coli* (ATCC 10536). Moderate inhibitory activity was exhibited by the methanolic extract against *S. epidermidis, B. subtilis, and S. aureus* when compared with the standard antibiotics cloxacillin (5 mcg/disc), amoxicillin (10 mcg/disc), and ciprofloxacin (5 mcg/disc). The extract also showed inhibitory activity against oral cavity pathogen, *S. aureus,* at MIC value of 64 mcg/ml and 128 mcg/ml against methicillin resistant *S. aureus* (MRSA, ATCC 15187), respectively. The authors reported a time and concentration dependent killing by the methanolic extract of *G. mollis* against *S. aureus* (ATCC 25175).

The development of a new class of polyurethane nanofibers containing aqueous extract of aerial parts of *G. mollis* by one step electrospinning was reported by Musarat et al. [25] for its potential biomedical application as an antimicrobial fiber. They incorporated the aqueous extract into the polymer media in order to influence the morphology and size of the polyurethane nanofibers. Further testing of the extract loaded nanofibers was also observed to inhibit the growth of *E. coli* (ATCC 52922) and *S. aureus* (ATCC 29231).

Al-Youssef et al. [17] also investigated the anti-inflammatory activity of the same extract at 500 mg/kg which showed pronounced anti-inflammatory effect after two hours. Their findings provide preliminary evidence suggesting that fractions of *G. mollis* exhibit inhibitory effects in inflammatory process and support the use of this plant for the treatment of arthritis, dermatitis, and wounds in African and Saudi Arabian traditional medicine.

5. Grewia Polysaccharide Gum

Interest in grewia polysaccharide gum as a pharmaceutical excipient has been on for over a decade [26]. This growing interest is informed by the viscous property of the gum extracted from the inner stem bark of the *G. mollis* plant. Consequently, the gum has been extracted and some of its physicochemical properties evaluated.

5.1. Extraction and Purification. Several approaches to the extraction of the gum from the inner stem bark of the shrub have been reported [27–30]. Briefly, the dried and pulverized inner stem bark of the *G. mollis* shrub is dispersed in demineralized water using an impeller. The fibrous material from the dispersed mucilage is removed by straining through a muslin cloth. Thereafter the mucilage is centrifuged before extraction of the gum with 96% ethanol. The extracted gum is redispersed in water and reextracted to get a beige-coloured gum which is then dried in an oven at 50°C for 8 h.

The gum can be further purified by treatment with 0.1 M sodium hydroxide or hydrochloric acid or with sodium chloride followed by extraction with 96% ethanol. It must be noted that treatment of the natural material with dilute alkali, dilute acid, or electrolytes could result in modification of the parent material with consequent variations in the physicochemical properties of the resultant material.

5.2. Some Physicochemical Properties of Grewia Gum. The gum contains traces of metals (Ca, K, Na, Mg, Zn, and Fe) and a viscosity-average-molecular weight of 316,000 [28]. Metals constitute about 6.1% of the gum, while proteins and lipids account for 1.24 and 0.021%, respectively. It has a high intrinsic viscosity with a swelling capacity greater than tragacanth and methylcellulose.

Grewia polysaccharide gum slowly hydrates and swells in water. The gum has aqueous solubility of about 0.2 mg/mL [29]. The low solubility of the gum was attributed to insoluble cell-wall materials making up a larger proportion of the gum. However, the presence of acetyl groups has been reported to account for the insolubility of certain gums in water [31]. Grewia polysaccharide gum has been shown from Fourier-transformed infrared (FTIR) spectroscopic studies to contain acetyl groups which may account for the low solubility of the gum in water [29]. Generally, water poorly soluble gums due to presence of acetyl groups are improved by deacetylation with dilute ammonia [31–33].

The rheological properties of the gum dispersion and the water vapour permeability of aqueous-based grewia gum films have been reported [27, 34]. The aqueous dispersions of the gum at different concentrations exhibited pseudoplastic flow behaviour. The viscosities of the gum dispersions decrease with increase in temperature. The effect of incorporation of electrolytes to dispersions of the gum was also investigated. The result showed a decrease in the viscosity of the gum brought about by electrolyte which is proportional to the concentration as well as the valence of the cation, and these findings have been confirmed by other workers [29].

5.3. Chemistry of Grewia Gum. Okafor [27] using paper chromatography reported that the primary structure of grewia polysaccharide gum may comprise of d-glucose, d-rhamnose and galacturonic acid. Nep and Conway [29, 35] determined the monosaccharide composition of the gum using gas chromatography. These workers reported that the primary structure of grewia gum may comprise five neutral sugars: d-rhamnose, d-arabinose, d-xylose, d-galactose, and d-glucose (Figure 3). The mean relative concentrations of these sugars in the polysaccharide were determined as glucose (67.14%), rhamnose (6.2%), xylose (2.72%), galactose (9.61%), and arabinose (12.71%).

The FTIR spectral analysis of the gum exhibited the typical bands and peak characteristics of polysaccharides [29] as shown in Figure 4. These authors attributed the broad band occurring at about 3425 cm^{-1} to the presence of hydroxyl (–OH) groups while the peak obtained at about 2927 cm^{-1} was attributed to stretching modes of the C–H bonds of methyl groups ($-CH_3$) of the rhamnose. Absorption bands around 1618 and 1420 cm^{-1} were attributed to carboxylate groups of uronic residues of galacturonic acid while the region between 1500 and 1800 cm^{-1} and absorption peaks at 1735 cm^{-1} and 1256 cm^{-1} were linked to the presence of carboxylic and acetyl groups, respectively. The absorption

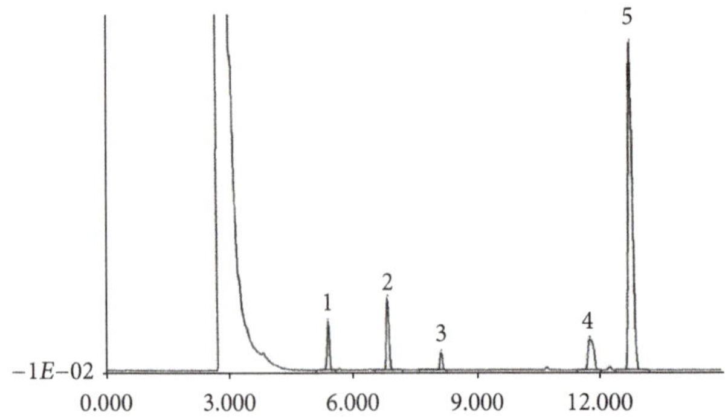

FIGURE 3: GC chromatograms of alditol acetates from grewia polysaccharide gum after hydrolysis with 2 N H_2SO_4 (Adapted from Nep and Conway [29]).

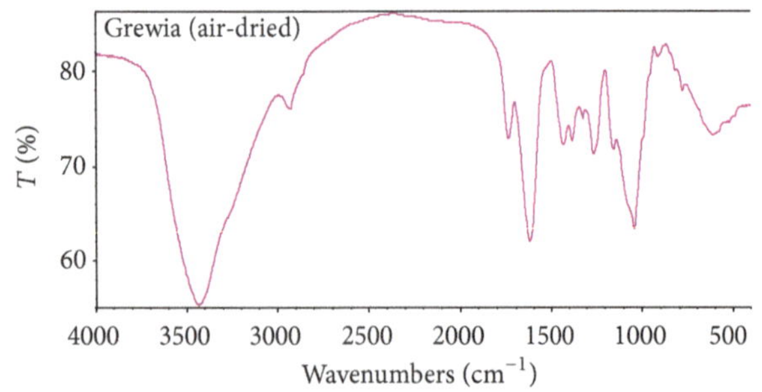

FIGURE 4: FT-IR spectra of grewia gum (adapted from Nep and Conway [29]).

band between 500 and 900 cm^{-1} may represent the finger print region for grewia polysaccharide gum.

The gum has also been characterized using other techniques such as scanning electron microscopy (SEM), gel permeation chromatography (GPC), differential scanning calorimetry (DSC), and thermogravimetric analysis of the extracted sample. Also, spectroscopic techniques such as X-ray photoelectron spectroscopy (XPS), solid-state nuclear magnetic resonance (NMR), and 1H and 13C NMR techniques [29] have been employed in the characterization of grewia gum. Based on findings from the various techniques, the gum has been described as a typically amorphous polysaccharide gum of high thermal stability with an average molecular weight of 5925 kDa expressed as the pullulan equivalent.

The effect of drying methods (air-drying, freeze-drying, and spray-drying) on the physicochemical properties of the gum has been investigated using the techniques outlined above [35]. Solid-state NMR results indicated that drying technique had little effect on the structure of the polysaccharide gum, but X-ray photoelectron spectroscopy (XPS) showed that surface chemistry of the gum varied with drying methods. Thermogravimetric analyses showed that oxidation onset varied according to the drying method. The workers concluded that, for industrial extrapolation, air-drying of the gum may be preferable to spray-drying and freeze-drying when relative cost, product stability, and powder flow are the priority, in tablet formulation, for instance.

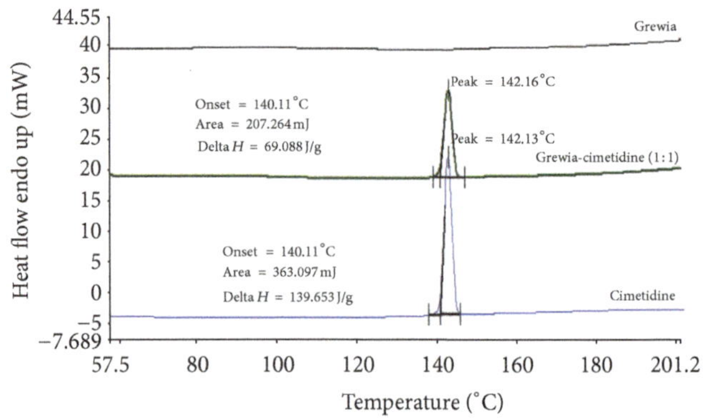

FIGURE 5: DSC traces for cimetidine, grewia gum, and cimetidine/grewia gum physical mixture (1 : 1) under nitrogen atmosphere and a scan rate of 10°C/min up to 200°C (adapted from Nep and Conway [39]).

6. Pharmaceutical Values of Grewia Polysaccharide Gum

6.1. Preformulation Studies. The assessment of possible interactions between a drug and different excipients used in formulation usually precedes large scale development trials of solid dosage forms [36, 37]. Physical and chemical interactions between excipients and active pharmaceutical ingredients (API) are common place [38] and necessitate the screening of novel excipients for possible incompatibilities. In view of the suitability of grewia polysaccharide gum as formulation excipient in solid or liquid dosage forms, studies on the potential interaction or incompatibility of the gum with pharmaceutical actives and/or excipients used in solid dosage formulation are imperative.

Nep and Conway [39] investigated potential interactions between grewia gum and cimetidine, ibuprofen, and some standard excipients used in the formulation of tablets (Figure 5). The thermal and molecular behaviours of mixtures of grewia gum with cimetidine, ibuprofen, and the standard excipients (lactose monohydrate, magnesium stearate, colloidal silicone dioxide, and microcrystalline cellulose) were analyzed using differential scanning calorimetry and FT-IR spectroscopy to assess potential interactions. The results obtained indicated that grewia gum is an inert natural polymer which can be used alone or in combination with other excipients in the formulation of pharmaceutical dosage forms.

6.2. Pharmaceutical Applications. Investigations into the pharmaceutical application of gum from *G. mollis* were first reported in the early 2000s [27, 28, 34, 40, 41]. The gum was reported to possess excellent binding property in sodium salicylate tablets (Okafor and Chukwu [40]). It was found to be as effective as gelatin when employed as a binder in concentrations of 2–6% w/w. At the same concentration, the gum was more effective as a binder than maize starch or acacia gum.

Audu-Peter and Gokum [42] found that the method of incorporating the gum into tablet formulation had effect

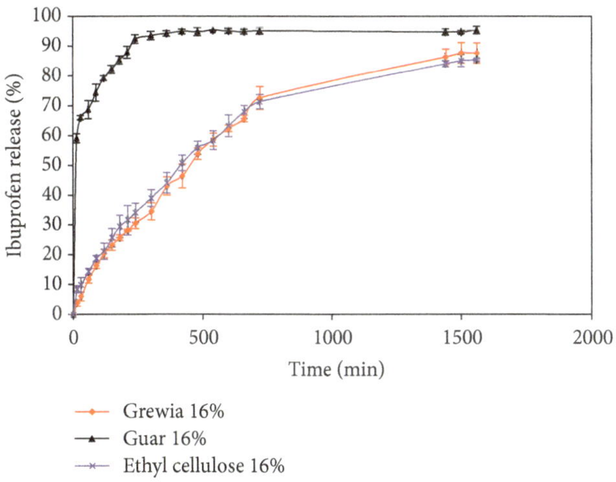

FIGURE 6: Release profiles of ibuprofen from the matrices containing 16% w/w of grewia, guar, or ethyl cellulose in phosphate buffer solution at 37 ± 1°C (n = 3, mean ± sd) (adapted from Nep and Conway [44]).

on tablet properties. They discovered that tablets with better tablet properties such as hardness and friability were produced when the gum was incorporated by activation with water than by wet granulation or direct compression. Acid and thermal treatment of the gum resulted in improved drug release from tablets attributable to reduced viscosity of the gum (Audu-Peter and Isah [43]).

Emeje et al. [28] in 2008 investigated the binding property of the gum. These workers compared the binding property of the gum with polyvinyl pyrolidone in paracetamol tablet formulations and analysed the compression properties of the formulations using density measurements and application of Heckel and Kawakita equation. They found that grewia gum compares favourably with the standard binder PVP and may be a useful substitute binder in paracetamol formulations.

Grewia gum has been investigated as a sustained release polymer matrix in ibuprofen [44]. It was used at a concentration of 16%, 32%, or 48% to formulate tablets of ibuprofen. This was compared with similar formulations prepared using hydroxypropyl methylcellulose (HPMC), guar gum, or ethyl cellulose as the polymer matrix. The results indicated that grewia gum at these concentrations was capable of sustaining the release of ibuprofen tablets up to 24 hours (Figure 6). Similarly, these scientists demonstrated the potential of grewia gum to sustain the release of a water-soluble drug in tablet formulations [45]. Tablets of cimetidine containing 40% of grewia gum were prepared by direct compression.

Similar formulations containing HPMC, gum Arabic, carboxy methylcellulose, or ethyl cellulose were prepared for comparison. The results indicated that grewia gum was superior to hydrophilic matrices of HPMC, carboxy methylcellulose, and gum Arabic in sustaining the release of cimetidine from tablets. In both instances, the workers concluded that grewia gum may be a useful excipient when used alone or in combination with other polymers to modify the release of soluble drugs or poorly soluble drugs from polymeric matrices.

Grewia gum fractions obtained by centrifugation at 4500 rpm for 30 minutes with average molecular weights between 230 and 235 kDa have been shown to demonstrate improved aqueous solubility that was useful in delivering more solids to the substrate when used as a film coating agent [26, 30].

Grewia gum has also been evaluated as a mucoadhesive in tablets [46–48]. Nep and Okafor [47] showed that the gum compared favourably with tragacanth when used as bioadhesive in tablet formulation. Nep and Conway [48] compared the mucoadhesive performance of compacts or gels of grewia polysaccharide gum with those of guar gum, carboxymethylcellulose, hydroxypropyl methylcellulose, and carbopol 971P. A software-controlled penetrometre, TA.XTPlus texture analyzer (Stable Microsystems, UK) was used to measure the detachment force of the compacts while the polymer gels were evaluated for hardness, stickiness, work of cohesion, and work of adhesion. The results showed that grewia gels had a significantly greater work of adhesion than carboxymethylcellulose gels ($P < 0.05$) and HPMC gels ($P < 0.001$). The mucoadhesive performance of grewia compacts was comparable to those of HPMC and carbopol 971P compacts ($P > 0.05$). It was concluded that grewia polysaccharide gum should be suitable for the formulation of retentive drug delivery devices.

The application of the gum as a suspending agent in paediatric formulations has been reported [49–51]. The suspending ability of air-dried or freeze-dried grewia gum in ibuprofen suspension was compared with similar suspension formulations containing xanthan gum, sodium carboxymethylcellulose, or acacia gum as suspending agents [49]. The stability of the ibuprofen suspension formulations was assessed using parameters such as appearance, pourability, viscosity, rheology, sedimentation volume and redispersibility. Other evaluation parameters used include degree of flocculation, zeta potential, and microbial load. The result from the study indicated that air-dried or freeze-dried grewia gum may provide a suitable alternative as suspending agent in paediatric suspension formulations.

7. Conclusion

Grewia gum is a natural resource whose potential pharmaceutical values have been established. Our literature survey revealed that, pharmaceutically, grewia gum has been well investigated for its application as binder, disintegrant or mucoadhesive in solid dosage forms and as stabilizer or suspending agent in liquid formulations. However, the scientific research on grewia gum is yet to be translated into industrial applications due to the absence of detailed physical and chemical characterization, stability data, and scientifically validated bioactive properties. Although the use of grewia gum as thickener in local delicacies spans generations, it may be necessary to reconcile the ethnomedicinal use of extract of the aerial parts of the plant with the absolute requirement that excipients are pharmacologically inert. Detailed scientific investigations that will fill in these details will not only help in furthering the scientific understanding of grewia gum, but importantly it will assist in realizing its commercial potential as pharmaceutical excipient. Furthermore, grewia gum should be explored in design and development of other drug carriers such as micro- and nanomaterials, membranes, capsules, and hot-melt extruded carriers and in wound healing applications. Grewia gum holds promise as an excipient to be explored for the unmet needs in drug delivery.

Conflict of Interests

The authors certify that there is no conflict of interests with any financial organization regarding the material discussed in the paper.

References

[1] C. W. Vendruscolo, C. Ferrero, E. A. G. Pineda et al., "Physicochemical and mechanical characterization of galactomannan from Mimosa scabrella: effect of drying method," *Carbohydrate Polymers*, vol. 76, no. 1, pp. 86–93, 2009.

[2] H. E. Huber and G. I. Christenson, "Utilization of hydrophilic gums for control of drug release from tablet formulation. 1: disintegration and dissolution behaviour," *Journal of Pharmaceutical Sciences*, vol. 1, pp. 59–66, 1996.

[3] T. R. Bhardwaj, M. Kanwar, R. Lal, and A. Gupta, "Natural gums and modified natural gums as sustained-release carriers," *Drug Development and Industrial Pharmacy*, vol. 26, no. 10, pp. 1025–1038, 2000.

[4] J. Varshosaz, N. Tavakoli, and S. A. Eram, "Use of natural gums and cellulose derivatives in production of sustained release metoprolol tablets," *Drug Delivery:* , vol. 13, no. 2, pp. 113–119, 2006.

[5] O. K. Udeala and A. Chukwu, "The binding property of mucuna gum in sulphadimidine and chloroquine phosphate tablets," *Nigerian Journal of Pharmaceutical Sciences*, vol. 1, pp. 59–66, 1985.

[6] M. R. Bilany, "Suspensions and emulsions," in *Pharmaceutics: The Design and Manufacture of Medicines*, M. E. Aulton, Ed., pp. 383–405, Churchill Livingstone, Philadelphia, Pa, USA, 3rd edition, 2007.

[7] D. L. Middleton, S. H. S. Leung, and J. R. Robinson, "Ocular bioadhesive delivery systems," in *Bioadhesive Drug Delivery Systems*, V. Leenaerts and R. Gurny, Eds., pp. 179–202, CRC Press, Boca Raton, Fla, USA, 1990.

[8] C. Whitehouse, M. Cheek, S. Andrews, and B. Verdcourt, "Tiliaceae and Muntingiaceae," in *Flora of Tropical East Africa*, H. J. Beentje, Ed., p. 120, A.A. Balkema, Rotterdam, Netherlands, 2001.

[9] D. Louppe, A. A. Oteng-Amoako, and M. Brink, *Plant Resources of Tropical Africa 7(1). Timber 1*, PROVA Foundation, Wageningen, Netherlands; Backhuys Publishers, Lenden, Netherlands; CTA, Wageningen, Netherlands, pp. 298–300, 2008.

[10] "Digital image of *Grewia mollis* plant," http://www.prota4u.org/protav8.asp?h=M11,M15,M17,M18,M25,M26,M27,M34,M36,M4,M6,M7,M8,M9&t=Grewia,mollis&p=Grewia+mollis#Description.

[11] L. S. Gill, *Ethno-Medical Uses of Plants in Nigeria*, University of Benin Press, Benin, Nigeria, 1992.

[12] W. Obidah, J. L. Godwin, J. Z. Fate, and M. A. Madusolumuo, "Toxic effects of *Grewia mollis* stem bark in experimental rats," *Journal of American Science*, vol. 6, no. 12, pp. 1544–1548, 2010.

[13] J. M. Dalziel, *The Useful Plants of West Africa*, Crown Agents, London, UK, 1937.

[14] F. C. Onwuliri, J. D. Mawak, D. L. Wonang, and E. A. Onwuliri, "Phytochemical toxicological and histo-pathological studies of some medicinal plants in Nigeria," *International Journal of Natural and Applied Sciences*, vol. 2, pp. 225–229, 2006.

[15] O. Asuku, S. E. Atawodi, and E. Onyike, "Antioxidant, hepatoprotective, and ameliorative effects of methanolic extract of leaves of *Grewia mollis* Juss. on carbon tetrachloride-treated albino rats," *Journal of Medicinal Food*, vol. 15, no. 1, pp. 83–88, 2012.

[16] N. A. Saleem, M. El-Nour, and A. A. M. Khalid, "The nutritive value of leaves and fruits of three grewia species under semi-arid environment," *Journal of Agricultural Science and Technology*, vol. B2, pp. 956–964, 2012.

[17] H. M. A. Al-Youssef, M. Amina, and A. M. El-Shafae, "Biological evaluation of constituents from *Grewia mollis*," *Journal of Chemical and Pharmaceutical Research*, vol. 4, no. 1, p. 508, 2012.

[18] H. Rosler, H. Framm, and R. N. Blomster, "The isolation of 6-hydroxyharmane from *Grewia mollis*," *Journal of Natural Products*, vol. 41, no. 4, pp. 383–384, 1978.

[19] W. Pfau and K. Skog, "Exposure to β-carbolines norharman and harman," *Journal of Chromatography B*, vol. 802, no. 1, pp. 115–126, 2004.

[20] P. K. Goyal, "Phytochemical and pharmacological properties of the genus *Grewia*: a review," *International Journal of Pharmacy and Pharmaceutical Sciences*, vol. 4, no. 4, pp. 72–78, 2012.

[21] P. Venault and G. Chapouthier, "From the behavioral pharmacology of beta-carbolines to seizures, anxiety, and memory," *The Scientific World Journal*, vol. 7, pp. 204–223, 2007.

[22] O. O. Efiom and E. Oku, "Isolation and characterisation of triterpene lup-20-en-3-ol and 1, 3-hexyloxacyclotridec-10-en-2-one from the root of *Grewia mollis*," *New York Science Journal*, vol. 5, no. 11, pp. 138–141, 2012.

[23] M. H. Shagal, D. kubmarawa, and Z. Idi, "Phytochemical screening and antimicrobial activity of roots, stembark and leaf extracts of *Grewia mollis*," *African Journal of Biotechnology*, vol. 11, no. 51, pp. 11350–11353, 2012.

[24] N. Idika and M. Niemogha, "The diversity of uses of medicinal plants in Nigeria," in *A Textbook of Medicinal Plants of Nigeria*, T. Odugbemi, Ed., p. 63, University of Lagos Press, Lagos, Nigeria, 2008.

[25] A. Musarat, H. M. Al-Youssef, T. Amna et al., "Polyurethane/G.mollis composite nanofibers for biomedical applications," *Journal of Nanoengineering and Nanomanufacturing*, vol. 2, pp. 85–90, 2012.

[26] I. J. Ogaji, E. I. Nep, and J. D. Audu-Peter, "Advances in natural polymers as pharmaceutical excipients," *Pharmaceutica Analytica Acta*, vol. 3, article 146, 2012.

[27] I. S. Okafor, "The rheological properties of grewia gum," *Nigeria Journal of Polymer Science and Technology*, vol. 2, pp. 169–175, 2001.

[28] M. Emeje, C. Isimi, and O. Kunle, "Effect of Grewia gum on the mechanical properties of Paracetamol tablet formulations," *African Journal of Pharmacy and Pharmacology*, vol. 2, pp. 1–6, 2008.

[29] E. I. Nep and B. R. Conway, "Characterization of Grewia Gum, a potential pharmaceutical excipient," *Journal of Excipients and Food Chemicals*, vol. 1, no. 1, pp. 30–40, 2010.

[30] I. Ogaji, "Characterization and application of grewia gum as a film coating agent in theophylline hydrochloride tablets," *Pharmaceutics and Pharmaceutical Technology*, vol. 308, 2011.

[31] A. Setia, S. Goyal, and N. Goyal, "Applications of gum karaya in drug delivery systems: a review on recent research," *Der Pharmacia Lettre*, vol. 2, no. 5, pp. 39–48, 2010.

[32] D. Le Cerf, F. Irinei, and G. Muller, "Solution properties of gum exudates from Sterculia urens (Karaya gum)," *Carbohydrate Polymers*, vol. 13, no. 4, pp. 375–386, 1990.

[33] A. Imeson, "Exudate gums," in *Thickening and Gelling Agents For Food*, A. Imeson, Ed., pp. 66–97, Chapman and Hall, London, UK, 1992.

[34] I. S. Okafor and A. Chukwu, "Water vapour permeability of aqueous-based grewia gum film," *Nigeria Journal of Polymer Science and Technology*, vol. 2, pp. 176–182, 2003.

[35] E. I. Nep and B. R. Conway, "Physicochemical characterization of grewia polysaccharide gum: effect of drying method," *Carbohydrate Polymers*, vol. 84, no. 1, pp. 446–453, 2011.

[36] G. Bruni, V. Berbenni, C. Milanese, A. Girella, and A. Marini, "Drug-excipient compatibility studies in binary and ternary mixtures by physico-chemical techniques," *Journal of Thermal Analysis and Calorimetry*, vol. 102, no. 1, pp. 193–201, 2010.

[37] M. J. Peres-Filho, M. P. N. Gaeti, S. R. De Oliveira, R. N. Marreto, and E. M. Lima, "Thermoanalytical investigation of olanzapine compatibility with excipients used in solid oral dosage forms," *Journal of Thermal Analysis and Calorimetry*, vol. 104, no. 1, pp. 255–260, 2011.

[38] M. E. Brown, E. M. Antunes, B. D. Glass, M. Lebete, and R. B. Walker, "DSC screening of potential prochlorperazine-excipient interactions in preformulation studies," *Journal of Thermal Analysis and Calorimetry*, vol. 56, no. 3, pp. 1317–1322, 1999.

[39] E. I. Nep and B. R. Conway, "Preformulation studies on grewia gum as a formulation excipient," *Journal of Thermal Analysis and Calorimetry*, vol. 108, no. 1, pp. 197–205, 2012.

[40] I. S. Okafor and A. Chukwu, "The binding property of Grewia gum in sodium salicylate tablets," *West African Journal of Biological Sciences*, vol. 14, pp. 9–21, 2003.

[41] I. S. Okafor and I. M. Danat, "The influence of granulating solvents on drug release from tablets containing grewia gum," *Journal of Pharmacy and Bioresources*, vol. 1, pp. 76–83, 2004.

[42] J. D. Audu-Peter and B. G. Gokum, "Effect of methods of incorporating grewia gum as binder on tablet properties," *Nigerian Journal of Pharmaceutical Research*, vol. 4, pp. 68–73, 2005.

[43] J. D. Audu-Peter and S. Isah, "Evaluation of grewia gum as binder in paracetamol tablet," *Journal of Pharmacy and Bioresources*, vol. 4, pp. 68–73, 2007.

[44] E. I. Nep and B. R. Conway, "Grewia polysaccharide as a pharmaceutical excipient in matrix tablets," *Journal of Excipients and Food Chemicals*, vol. 2, no. 1, pp. 3–15, 2011.

[45] E. I. Nep and B. R. Conway, "Polysaccharide gum matrix tablets for oral controlled delivery of cimetidine," *Journal of Pharmaceutical Sciences and Research*, vol. 2, no. 11, pp. 708–716, 2010.

[46] E. I. Nep and I. S. Okafor, "Evaluation of the bioadhesive property of Grewia Gum in mebendazole tablet formulation 1: in pig gastric mucus," *Nigerian Journal of Pharmaceutical Research*, vol. 4, no. 2, pp. 52–58, 2005.

[47] E. I. Nep and I. S. Okafor, "Evaluation of the bioadhesive property of Grewia gum in indomethacin tablet formulation in pig gastric mucus," *Journal of Pharmacy and Bioresources*, vol. 3, no. 2, pp. 62–69, 2006.

[48] E. I. Nep and B. R. Conway, "Grewia gum 2: mucoadhesive properties of compacts and gels," *Tropical Journal of Pharmaceutical Research*, vol. 10, no. 4, pp. 393–401, 2011.

[49] E. I. Nep and B. R. Conway, "Evaluation of Grewia polysaccharide gum as a suspending agent," *International Journal of Pharmacy and Pharmaceutical Sciences*, vol. 3, no. 2, pp. 168–173, 2011.

[50] J. O. Kadiri and I. S. Okafor, "Evaluation of the suspending property of Grewia gum in metronidazole suspension," *Nigerian Journal of Pharmaceutical Research*, vol. 8, no. 1, pp. 247–254, 2010.

[51] I. J. Ogaji and S. W. Hoag, "Effect of Grewia gum as a suspending agent on ibuprofen pediatric formulation," *AAPS PharmsciTech*, vol. 12, no. 2, pp. 507–513, 2011.

11

Effect of Plasticizers on Structural and Dielectric Behaviour of [PEO + $(NH_4)_2C_4H_8(COO)_2$] Polymer Electrolyte

Kamlesh Pandey,[1] Nidhi Asthana,[1] Mrigank Mauli Dwivedi,[1] and S. K. Chaturvedi[2]

[1] *National Center of Experimental Mineralogy and Petrology, University of Allahabad, Allahabad, UP 211002, India*
[2] *Department of Env. Science, Chitrakoot Gramodaya Vishwavidyalay, Chitrakoot, MP 485334, India*

Correspondence should be addressed to Kamlesh Pandey; kp542831@gmail.com

Academic Editor: Han-Do Kim

Improvements in ion transport property of polyethylene-oxide- (PEO-) based polymer electrolytes have been investigated, using different types of plasticizers. The effects of single and coupled plasticizers [i.e., EC, (EC + PC), and (EC + PEG)] on structural and electrical behavior of pristine electrolyte were studied by XRD, SEM technique, and impedance spectroscopy. The electrical conductivity of the best plasticized system was found to be 4×10^{-6} S/cm. Argand plots show dispersive nature of relaxation time or inhomogeneous space charge polarization of plasticized polymer electrolyte.

1. Introduction

Recently, plasticized polymer electrolytes have been extensively used in the electrochemical devices, like secondary batteries, fuel cells, sensors, and smart windows. In order to improve their electrical and electrochemical properties for technical and industrial application, structural modification through plasticized polymer electrolytes is extensively used. Various techniques (like plasticization, copolymerization, etc.) have been adopted to achieve the desired objective of high ionic conduction and better stability of polymer composite electrolytes [1–3]. Large number of host polymer [e.g., polyethylene oxide (PEO), polymethylene methacrylate (PMMA), polypropylene oxide (PPO), polyvinyl alcohol (PVA), polyvinyl formal (PVF), etc.] [4–6] with different alkali and alkaline salts have been investigated in the past four decades. Among these polymers, PEO is the most studied polymeric system. The electrical conductivity of PEO is found relatively low at ambient temperature in comparison to the existing conventional liquid/hybrid electrolytes. The ionic conductivity of polymer electrolyte mainly depends upon (i) crystallinity of the material (ii) simultaneous cation and anion motion, and (iii) the ion-pair formation (anion complex-cation interaction). To overcome these problems, various approaches have been made to modify the structure of polymer electrolyte in the last three decades. Ammonium adipate $(NH_4)_2C_4H_8(COO)_2$ contains smaller size cation (H^+) with larger size anion which can easily dissociate and provide higher protonic conduction [7].

The addition of plasticizers in polymer electrolyte is one of the most effective approaches to improve the mobility of ionic or/and the interfacial interaction among ionic and polar groups in polymer chains. The choice of plasticizer and its concentration can affect film permeability and mechanical properties [8, 9]. The other advantage of plasticization is to modify the polymeric matrix and provide the enhanced conductivity. In general, low molecular weight and high dielectric constant additive, such as polyethylene glycol (PEG), ethylene carbonate (EC), or propylene carbonate (PC) are used to modify the permeability of the polymer film [10]. The addition of plasticizer in polymer matrix tends to decrease the glass transition temperature and increase deformability, viscosity, elasticity, abrasion resistance, and elastic recovery in the electrolyte system. Use of plasticizers also increases the salt-solvating power which leads to an increase in amorphous content of the polymer matrix and tends to dissociate ion-pairs into free cations and anions thereby leading to an overall enhancement in conductivity. However, many theories have been proposed to account for the mechanism of plasticizer action, which leads to the lowering of processing temperature [11]. The "Lubricity Theory" suggests that the plasticizers act as a lubricant to facilitate movements of the charge carrier over each other, where as the "Gel Theory" considers the disruption of polymer-polymer interaction (Hydrogen bond and Vander Waals or Ionic forces). The "Free Volume Theory" states that a study of plasticization is the way to increase free volume and was useful to clarify the lowering of the glass transition temperature (T_g) by a plasticizer [12–14].

In the present study, we outline the effect of plasticizers that is (single/coupled) (EC, EC + PC and EC + PEG) on electrical properties of [$(PEO)_{0.9}$ + $(C_6H_{16}N_2O_4)_{0.1}$] electrolyte that is ionic conduction and ionic behavior (electrical and dielectric response).

2. Experimental

Plasticized polymer nanocomposite electrolyte films were prepared using phase inversion technique. The polyethylene oxide [(PEO) (M.W. 6×10^5, ACROS organics)] and Ammonium adipate ($C_6H_{16}N_2O_4$) (Rankem, India) of AR grade were used for the synthesis of composite polymer electrolyte (CPE). The stoichiometric ratio of PEO and vacuum dried ($C_6H_{16}N_2O_4$) was dissolved in deionized (DI) water at 40°C and mixed together. This solution was stirred for 12 h continuously. This gelatinous polymer solution was finally casted in polypropylene petri dish. The solution casted film was finally dried at room temperature in a constant humidity condition (RT = 30°C and RH = 60%) for obtaining free standing films of CPE. To study the effect of plasticizer, we have used different concentrations of the plasticizer by wt (%) of EC, PC, and PEG, respectively. Following systems were synthesized for further studies:

(a) ES-1 $(PEO)_{0.9} + (C_6H_{16}N_2O_4)_{0.1}$,

(b) ES-2 $[(PEO)_{0.9} + (C_6H_{16}N_2O_4)_{0.1}]_{0.99} : [EC]_{0.01}$,

(c) ES-3 $[(PEO)_{0.9} + (C_6H_{16}N_2O_4)_{0.1}]_{0.99} : [EC + PC]_{0.01}$,

(d) ES-4 $[(PEO)_{0.9} + (C_6H_{16}N_2O_4)_{0.1}]_{0.99} : [EC + PEG]_{0.01}$.

Structural behavior of [$(PEO)_x$ + $(C_6H_{16}N_2O_4)_{1-x}$]: plasticizer$_y$ where y = 0 and 0.10] system was studied with XRD, SEM, and IR spectroscopic investigations. The XRD pattern was recorded between 2θ = 15–60° at room temperature using Phillips X-pert diffractometer. The SEM image was obtained by JOEL, EPMA model JXA-8100. The stub was coated with graphite to a thickness of 50 Å under vacuum using JEOL high vacuum evaporator. The coated samples were randomly scanned, and photographs were taken by EPMA. The infrared spectrum was recorded on Perkins Elmer IR spectrophotometer in a range of 4000–400 cm^{-1}. The electrical conductivity was evaluated using complex impedance plot obtained by computer controlled Hioki (JAPAN)-LCZ HI Tester (model 3520-01) in the frequency range of 1 Hz to 100 KHz. The other dielectric parameters, modulus spectra (real and imaginary part), and a.c. conductivity were derived from impedance spectroscopic data. Mobility of the mobile ionic species was evaluated using the dielectric parameters.

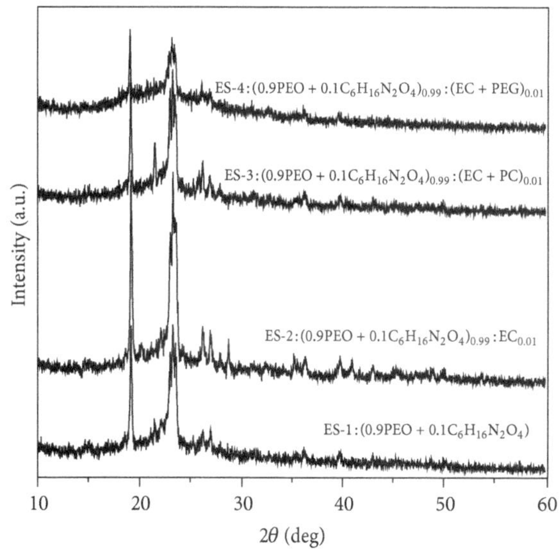

FIGURE 1: XRD pattern of different polymer electrolyte systems.

3. Result and Discussion

XRD pattern of polymer electrolytes [$(PEO)_x$ + $(C_6H_{16}N_2O_4)_{1-x}$]: plasticizer$_y$ where y = 0 and 0.10] are shown in Figure 1. Two strong diffraction peaks at 2θ = 19° (120) and 23° (112) (i.e., characteristic peak of PEO) are recorded in all composite electrolyte samples. A comparative look of XRD pattern shows that the plasticized composite electrolytes have relatively sharper peak at 19° and more broadness of 23° peak with respect to the pristine sample. This change indicates the polymer-salt-plasticizer interaction as well as the decrease in crystalline behaviour of the composite electrolyte. The increase in amorphosity supports the high conductivity in PCPE. W-H plot, shows that the size of the average crystallites is decreasing ES-2 (29 nm) > [ES-3] (27 nm) > [ES-4] (24 nm). This effect clearly reveals that the addition of plasticizer in the pristine electrolyte affects the viscosity ratio and it controls the particle size.

Infra red spectrum of pristine and PCPE with different plasticizers (EC, EC + PC, or EC + PEG) electrolyte films are shown in Figure 2. The absorption peak of $(NH_4)_2C_4H_8(COO)_2$ shows various peaks between 4000 cm^{-1} and 400 cm^{-1}. Some peaks related to skeleton vibration of salt with deformation mode of C–CN and bending vibrations of N=N=N are appeared in the lower wave number region (upto 600 cm^{-1}). In ES-1 [PEO-$(NH_4)_2C_4H_8(COO)_2$] electrolyte film, the absorption peak

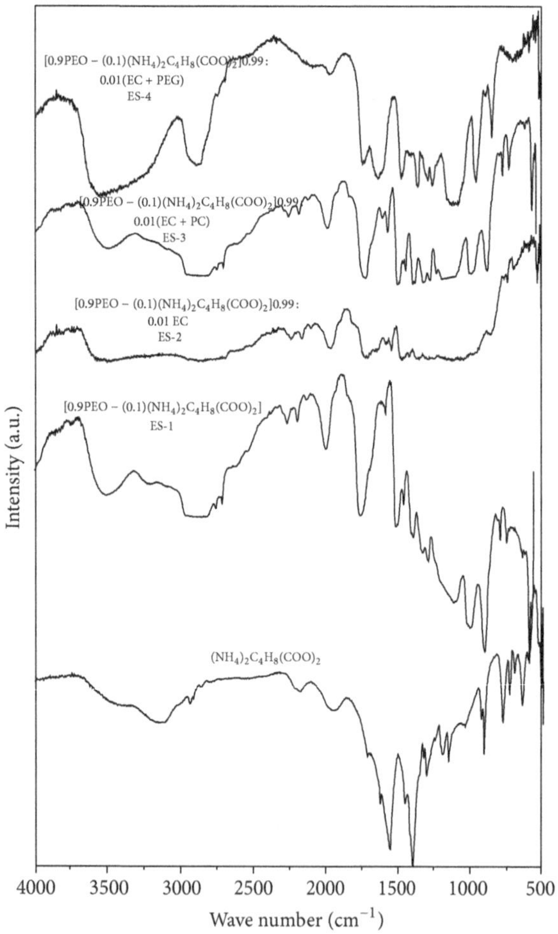

FIGURE 2: The infrared spectrum of different electrolyte system.

at 3502 cm^{-1} is due O–H stretching vibration. The peaks at 2894, 2236, and 1945 cm^{-1} related to the C–H stretching vibration, C–C stretching in PEO, and asymmetric C=C=C stretching vibration are recorded, respectively. After the doping of salt in PEO matrix, few new peaks at 1720, 1274 cm^{-1} were observed due to C=O stretching vibration of $(NH_4)_2C_4H_8(COO)_2$ in polymer chain. Furthermore, addition of salt in polymer the peaks at 1303 and 1400 cm^{-1} were shifted toward higher wave numbers 1342 cm^{-1} and 1465 cm^{-1} due to polymer salt complexation. Similarly, the peak at 1035 cm^{-1} is shifted to 1054 cm^{-1}. The peaks at 943, 840, 734, 690, 578, and 534 cm^{-1} were assigned to C–H deformation vibration in PEO, CH_2 rocking vibration, and C–C skeleton vibration of polymer chain.

In the plasticized polymer electrolyte system, ethylene carbonate (EC) gives new absorption peaks at 1382 cm^{-1} and 3490 cm^{-1} related to characteristic peak of EC. It shifts the peak at 2172 cm^{-1} towards lower wave number side that is 2162 cm^{-1}. The shift in $(NH_4)_2C_4H_8(COO)_2$ peak (1303 cm^{-1} and 1400 cm^{-1}) at higher wave number indicates C=C=C bending vibration with higher frequency. The peak at 3502 cm^{-1} of O–H vibration is disappearing in the presence of EC, which indicates the removal of O–H vibration in electrolyte system. The presence of plasticizer increases the frequency of vibration of polymer chain and facilitates more flexibility.

Propylene carbonate (PC) as plasticizer in pristine electrolyte shows few new absorption peaks at 2694, 2858, 1697, and 1243 cm^{-1} related to characteristic peak of PC. The combination of EC and PC in the pristine polymer electrolyte shows a minor shift with respect to their individual peaks. In this case, the reappearance of –OH-related peak took place which is an indication of interaction of PC in the electrolyte matrix. The other combinations of poly ethylene glycol (PEG) with ethylene carbonate (EC) give some new peaks at 2086, 1984, 1741, 1631, 1249, and 1297 cm^{-1}. These peaks are related with the presence of PEG in the plasticized electrolyte. The reappearance of C–C stretching vibration peak at 1097 cm^{-1} is an indication of main chain vibration in the presence of PEG. The new peaks at 1631, 1741, and 1984 cm^{-1} are CH deformation vibration, CH_2 symmetric deformation vibration, and C=O stretching vibration of PEG. The peak at 2858 cm^{-1} of ES-1 of N–H stretching vibration with small increment in wave number is indicating highly frequent vibration. The addition of plasticizer in the electrolyte provides more flexible path for the ionic conduction in electrolyte system. Such type of complexation can also be seen in XRD pattern of plasticized polymer electrolyte.

The SEM images of ES-1, that is, $[(PEO)_{0.9} + (C_6H_{16}N_2O_4)_{0.1}]$ electrolyte system with different plasticizers were recorded to assess the morphology of electrolyte system (shown in Figure 3). In ES-1, several heterogeneous distributions of polycrystalline domain are observed. The addition of plasticizers drastically changes the microstructure of the pristine electrolyte. It clearly shows heterogenous morphology with modified crystalline domain. The addition of plasticizers with electrolyte results in an the induced delay in phase relation or lower surface energy. A comparative look of SEM images of plasticized system with pristine electrolyte provides more smoothness with more grains. The film plasticized with combination of (PEG + EC) has a higher degree of surface smoothness and smaller pin hole due to enhanced flexibility. The plasticizer is believed to act as a spacer between molecules of polymer by forming links with them. It was mentioned that dipole interaction occurs between polar groups in the plasticizer. In films with EC + PC, smooth and better structure was obtained. But EC + PEG containing film gives more smooth morphology of films and enhancement in visible grains. The developments of grains are very useful for interfacial contact between the polymer and salt. The plasticized film with EC + PEG has a higher degree of surface smoothness and smaller pinhole in the cross section.

The variations of a.c. conductivity (σ) with frequency for different plasticizers are shown in Figure 4(a). In pristine polymer electrolyte, the conductivity of the system due to dissociation of salt facilitates the free charge carriers/ions which enhances ionic conduction in electrolytes [15]. Incorporation of plasticizers in the pristine electrolyte system provides better mechanical strength as well as higher conductivity.

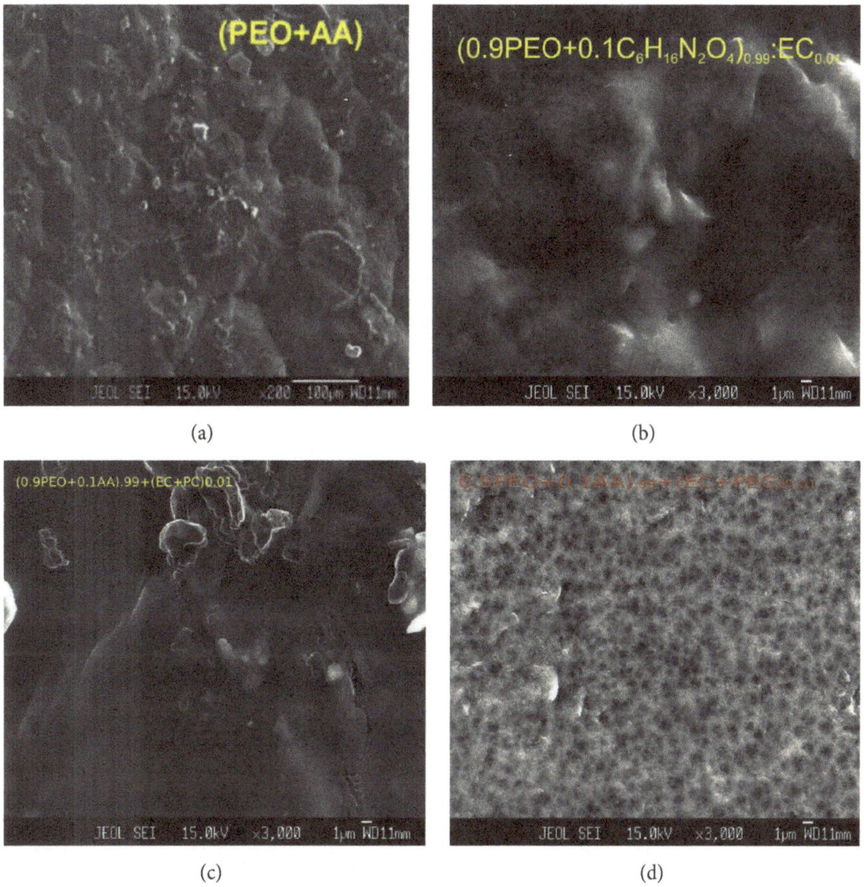

FIGURE 3: The SEM image of (a): ES-1, (b): ES-2, (c): ES-3, and (d): ES-4 polymer electrolyte systems.

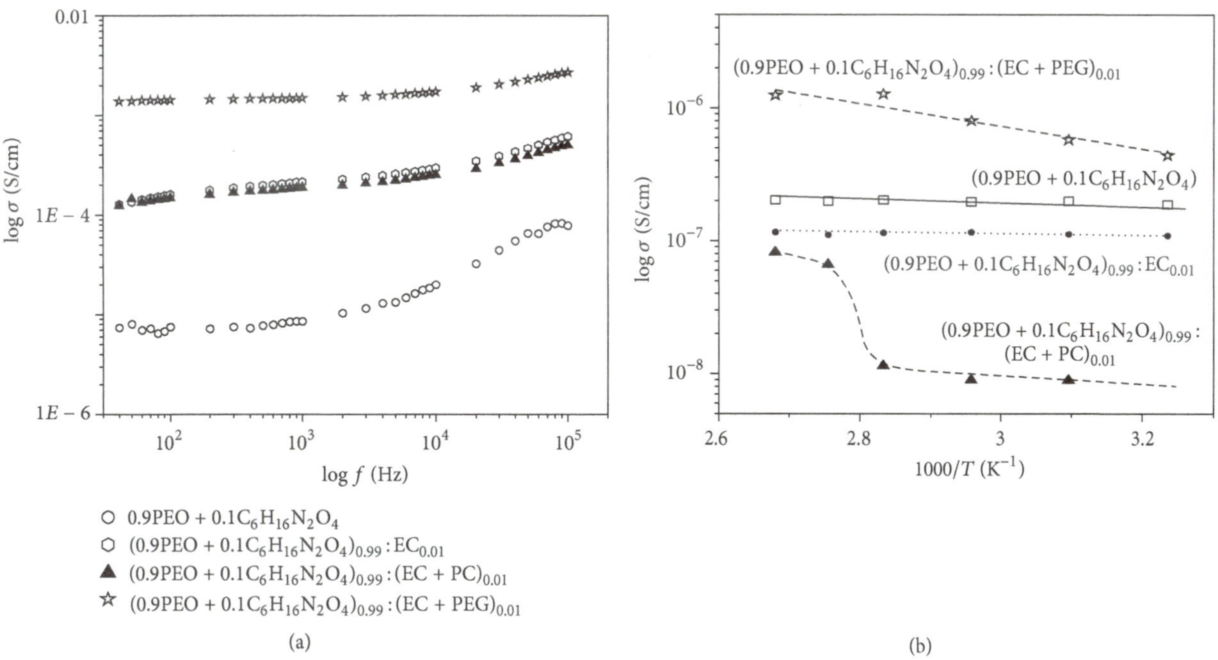

FIGURE 4: (a) The variations of a.c. conductivity with frequency at room temperature. (b) The variations of conductivity with temperature for different polymer electrolyte systems.

Table 1: Transport parameters for different electrolyte systems.

Electrolyte system	σ ($S{\cdot}cm^{-1}$)	n (cm^{-3})	μ ($cm^2\,v^{-1}\,s^{-1}$)	D ($cm^2\,s^{-1}$)
$(PEO)_{0.9} + (C_6H_{16}N_2O_4)_{0.1}$	1.87×10^{-7}	1.32×10^{13}	0.88×10^{-1}	2.29×10^{-3}
$[(PEO)_{0.9} + (C_6H_{16}N_2O_4)_{0.1}]_{0.99} : [EC]_{0.01}$	1.09×10^{-7}	2.19×10^{13}	0.31×10^{-1}	0.81×10^{-3}
$[(PEO)_{0.9} + (C_6H_{16}N_2O_4)_{0.1}]_{0.99} : [EC + PC]_{0.01}$	0.15×10^{-7}	3.26×10^{13}	0.27×10^{-1}	7.18×10^{-3}
$[(PEO)_{0.9} + (C_6H_{16}N_2O_4)_{0.1}]_{0.99} : [EC + PEG]_{0.01}$	4.37×10^{-7}	0.93×10^{13}	2.93×10^{-1}	7.59×10^{-3}

This enhancement in conductivity is possibly due to better percolation path. In (EC + PC) and (EC + PEG), combination of plasticizer shows better effect than pristine/and with single plasticizer. The reason behind this is that PEG polymer adopts a random coil configuration in solution, acting as if it was a sphere with an effective radius. The highest conductivity in this case was also explained as the more solubility of the system which facilitates more linear H-bond (extra) in the system which provides the extra percolation path for the mobile charge carriers. It has been also reported by other workers that the frequency-dependent conductivity and relaxation dynamics are both sensitive to the motion of charge species and dipoles of the polymer electrolyte [16, 17]. The temperature-dependent conductivity of pristine and plasticized polymer electrolyte is shown in Figure 4(b). In pristine electrolyte system, conductivity increases with temperature linearly up to 70° then suddenly jumps and again increases regularly up to 100°C. But in other three plasticized systems, a.c. conductivity linearly increases with temperature. The trend of conductivity is ES-1 < ES-2 < ES-3 < ES4 valid from room temperature to 373 K. The plots of log σ versus $10^3/T$ are found to be linear a limited temperature range and well fit in the Arrhenius equation

$$\sigma_T = \sigma_{\circ} \exp\left(\frac{-E_a}{KT}\right), \tag{1}$$

where $\sigma_{\circ}$ is the preexponential factor, E_a is the activation energy, K is the Boltzmann constant, and T is the absolute temperature. The activation energy (E_a) of each system was calculated from the slop of the curve shown in Figure 4(b). The increase in conductivity with temperature is also explained by the Rice and Roth model [18] that expresses the ionic conductivity σ as

$$\sigma = \frac{2}{3}\left[\frac{(Ze)_2}{KBT_m}\right] nE_a\tau \exp\left(\frac{-E_a}{KT}\right), \tag{2}$$

where Z is valency, m the mass of the conducting ion, e electronic charge, and τ is time of flight between the sites. Using (2), number of charge carriers (n), ionic mobility (μ), and diffusion coefficient (D) can be identified.

The ionic conductivity is represented by

$$\sigma = \sum n_i e_i \mu_i, \tag{3}$$

where n_i, e_i, and μ_i are the charge carriers density of ionic charge of the ith ion, respectively. In general, an increase of ionic conductivity of the complex system is due to an increase of number of mobile charge carriers introduced/produced in the material with a change of constituent concentration. The calculated values of these constants are given in Table 1. On the basis of the transport parameters, calculated by Rice and Roth model, it is interesting to observe that the combination of plasticizers provide a better medium for mobility in the matrix. The plasticized polymer electrolyte (namely, ES-4 with conductivity 4E-6) is found to be the highest mobility. The number of charge carriers is even low for this electrolyte. Here, it is clear that the highest conductivity is due to higher mobility of ionic species.

The frequency-dependent a.c. conductivity at different temperatures of all four electrolytes (namely, ES-1, ES-2, ES-3, and ES-4) is shown in Figure 5(a) and their scaling behaviour in Figure 5(b). In Figure 5(a) conductivity increases with frequency at different temperatures (T = 35°C, 45°C, 55°C, 70°C, and 85°C). In all cases, the conductivity pattern shows a frequency independent plateau, with an indication of electrode polarization. Scaling of a.c. conductivity is an important feature for the polymer electrolyte. The a.c. conductivity follows scaling law [19]

$$\frac{\sigma}{\sigma_{\circ}} = F\left(\frac{f}{f_{\circ}}\right); \qquad f(x) \cong 1 + x^n, \tag{4}$$

where $f_{\circ}$ is the characteristics frequency, but arbitrarily determined frequency, and x is % of salt concentration. All the measured conductivity data with respect to frequency at different temperatures were scaled according to $f_{\circ} = \sigma_{\circ} T$ and shown in Figure 5(b). In the first two electrolyte systems (ES-1: PEO with ammonium adipate salt) (ES-2: Pristine with EC plasticizer), the scaled plot is scattered, but in the other two cases (with combination of plasticizer EC + PC and EC + PEG) the data collapse almost in single master curve. This suggests that in later cases, the relaxation mechanism is independent of temperature, and it also represents their ion-hopping motion behavior. Slight dispersion in the higher frequency regime is possibly due to enormous diffusion phenomenon. The relative dielectric constant ε' determines the maximum energy that can be stored in

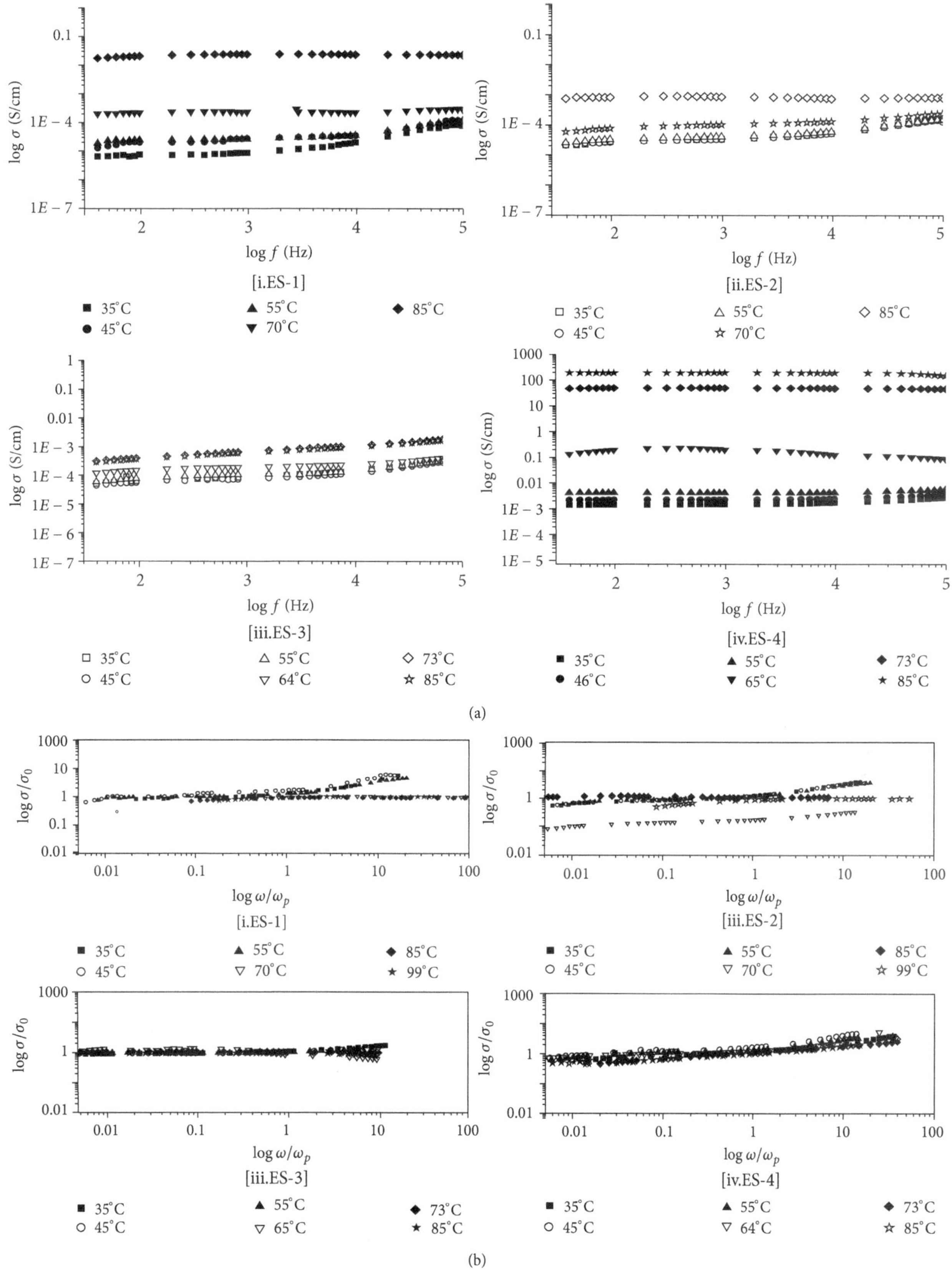

FIGURE 5: (a) The variations of a.c. conductivity with frequency at different temperatures. (b) Variation of scaled a.c. conductivity with respect to scaled frequency for different electrolyte systems.

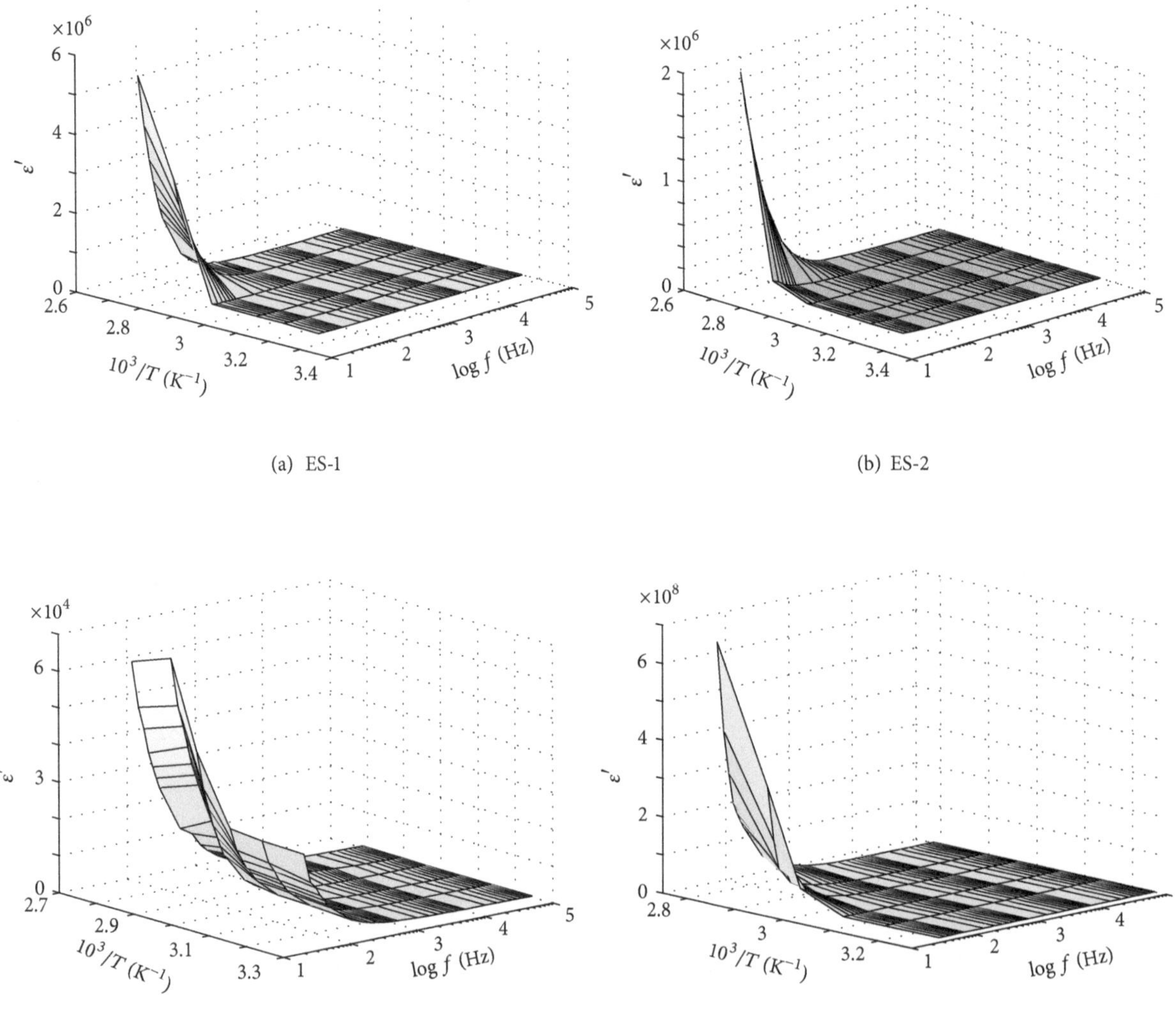

(a) ES-1 (b) ES-2

(c) ES-3 (d) ES-4

FIGURE 6: Variation of the dielectric constant (ε') with frequency and temperature of pristine and plasticized electrolytes.

the material. However, the relative loss factor ε'' evaluates the absorption of electrical energy by a dielectric material that is subjected to an alternating electromagnetic field. Furthermore, the dissipation factor $\tan\delta = \varepsilon''/\varepsilon'$ determines the absorbance of electromagnetic field in material. The relaxation can be seen in dielectric materials particularly in polymer electrolytes when they are subjected to an a.c. electric field. The electronic and ionic relaxations are correlated to very high rapid oscillations of weak dipoles, but after plasticization of polymer electrolytes, these dipoles relaxations subsequently appear at near audio frequencies. This relaxation arises from the facts that free charges/ions were present due to doping of salt in polymer. These free charges/ions are mobiles in the applied magnetic field. Further, these free carriers are then blocked at interface between the two media of different conductivity and permittivity. An interfacial or Maxwell-Wagner-Sillars (MWS) polarization can occur depending on the frequency of applied field. This type of polarization can be seen in heterogeneous or multiphase system/composite system at high temperature and low frequency region. However, it is difficult to discriminate between MWS relaxations and d.c. conduction. The MWS effect has been found to be possibly responsible for the low frequency and high temperature dielectric losses in composites system of dispersion of fillers in polymer matrix or by copolymerization or by the addition of low molecular weight polymer in host matrix. This type of relaxation has been studied by various groups of researchers. The dielectric constant (ε') and loss (ε'') of pristine and plasticized electrolytes are shown in Figures 6 and 7. These

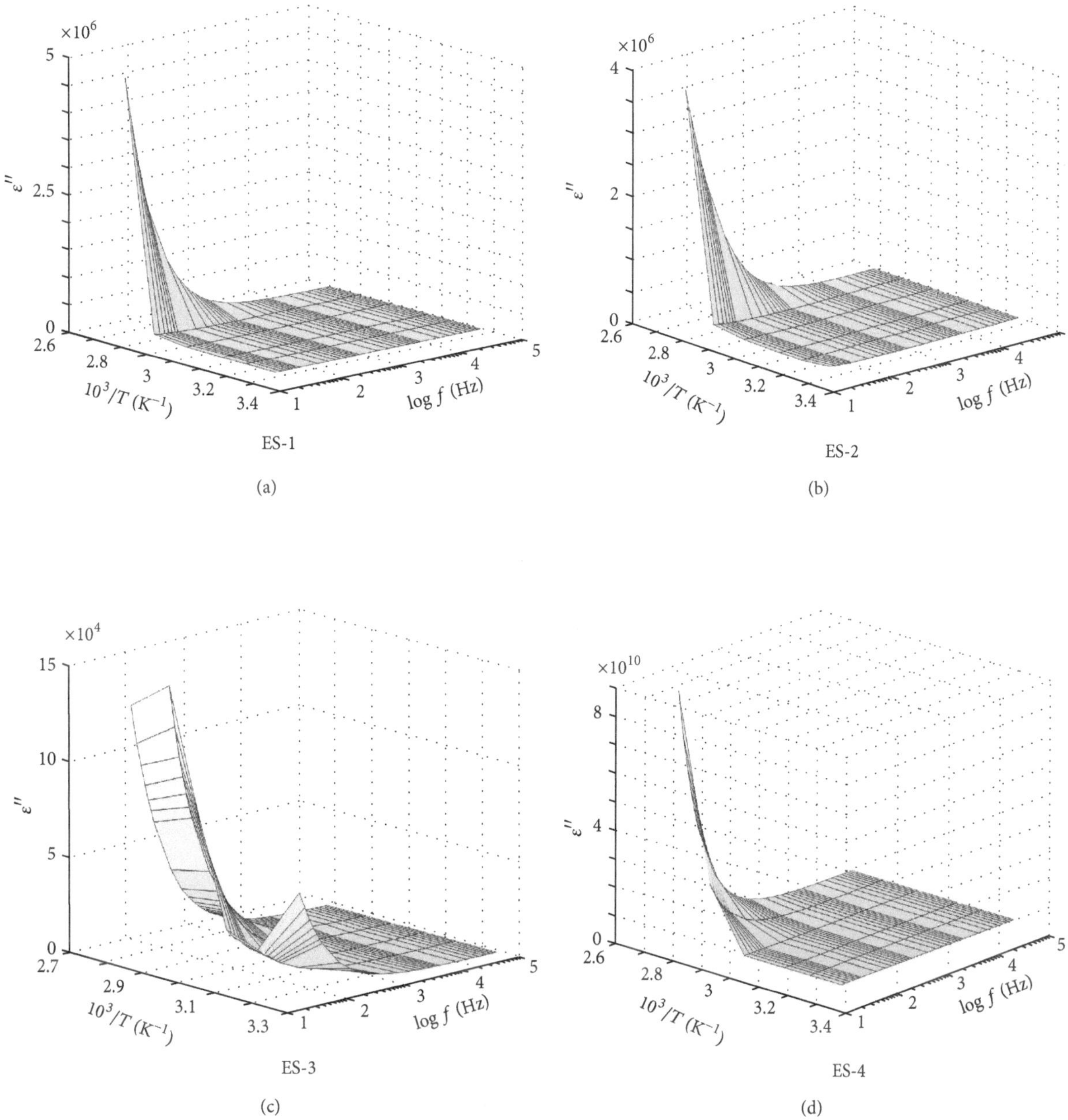

FIGURE 7: Variation of the dielectric loss (ε'') with frequency and temperature of pristine and plasticized electrolytes.

curves give a peculiar nature of the polymeric substance due to low frequency dispersion. The high value at low frequency region is an indication of electrode polarization. It can be deduced that plasticizers with high dielectric constant allows greater dissolution of the electrolyte salt resulting in an increased number of charge carriers. The apparent role of plasticizer in a host polymer is to change the viscosity of electrolyte and assist in the dissociation of the salt thereby increasing the number of charge carriers. The value of real and imaginary part of dielectric relaxation shows a regular increase in the plasticized sample relative to the pristine electrolyte. This increase in ε' and loss ε'' value is directly based on the production of more charge carriers/dipoles in the matrix. The variation of (ε') and (ε'') with different frequency does not provide any new insight. But the temperature-dependent tangent loss variation clearly gives a single peak at 363 K in the pristine electrolyte system (Figure 8(a)), whereas the plasticized systems show two peaks (Figures 8(b), 8(c), and 8(d)). The new peak appears at the lower temperature side near

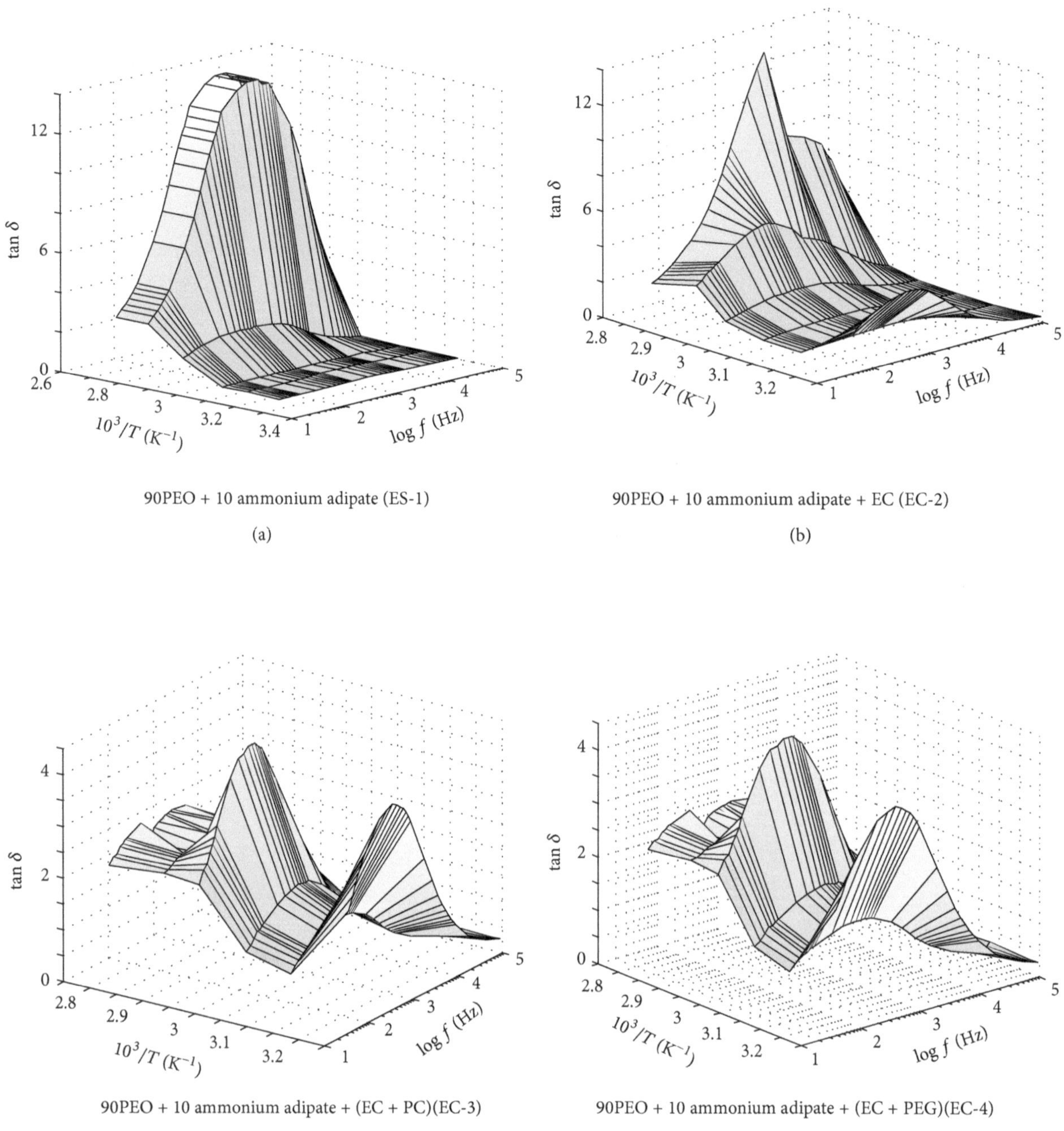

90PEO + 10 ammonium adipate (ES-1)

(a)

90PEO + 10 ammonium adipate + EC (EC-2)

(b)

90PEO + 10 ammonium adipate + (EC + PC)(EC-3)

(c)

90PEO + 10 ammonium adipate + (EC + PEG)(EC-4)

(d)

FIGURE 8: Variation of the temperature-dependent tangent loss of pristine and the plasticized electrolyte system.

10 K frequency and becomes relatively intense in EC, (EC + PC) then (EC + PEG) systems. The peak at the higher frequency region is due to β relaxation peak of the host polymer, and second peak may infer that plasticizer EC or EC/PC or PEG has dissociated more salt and generated more ions. The increase in intensity of peak is related to the more salt dissociation/dissolution in the two plasticizers. The Argand plots between the real and imaginary part of the dielectric relaxation for all the four systems (pristine and plasticized) were presented in Figure 9. All the curves are incomplete half semicircles. The incomplete half semicircle in all the curves cannot be explained by Debye-Model (i.e., single relaxation time) and suggest multiple relaxation times in the electrolyte system. The formation of shapes as ellipsoid in these system is due to various factors like hopping nature, space charge polarization, or inhomogeneously of

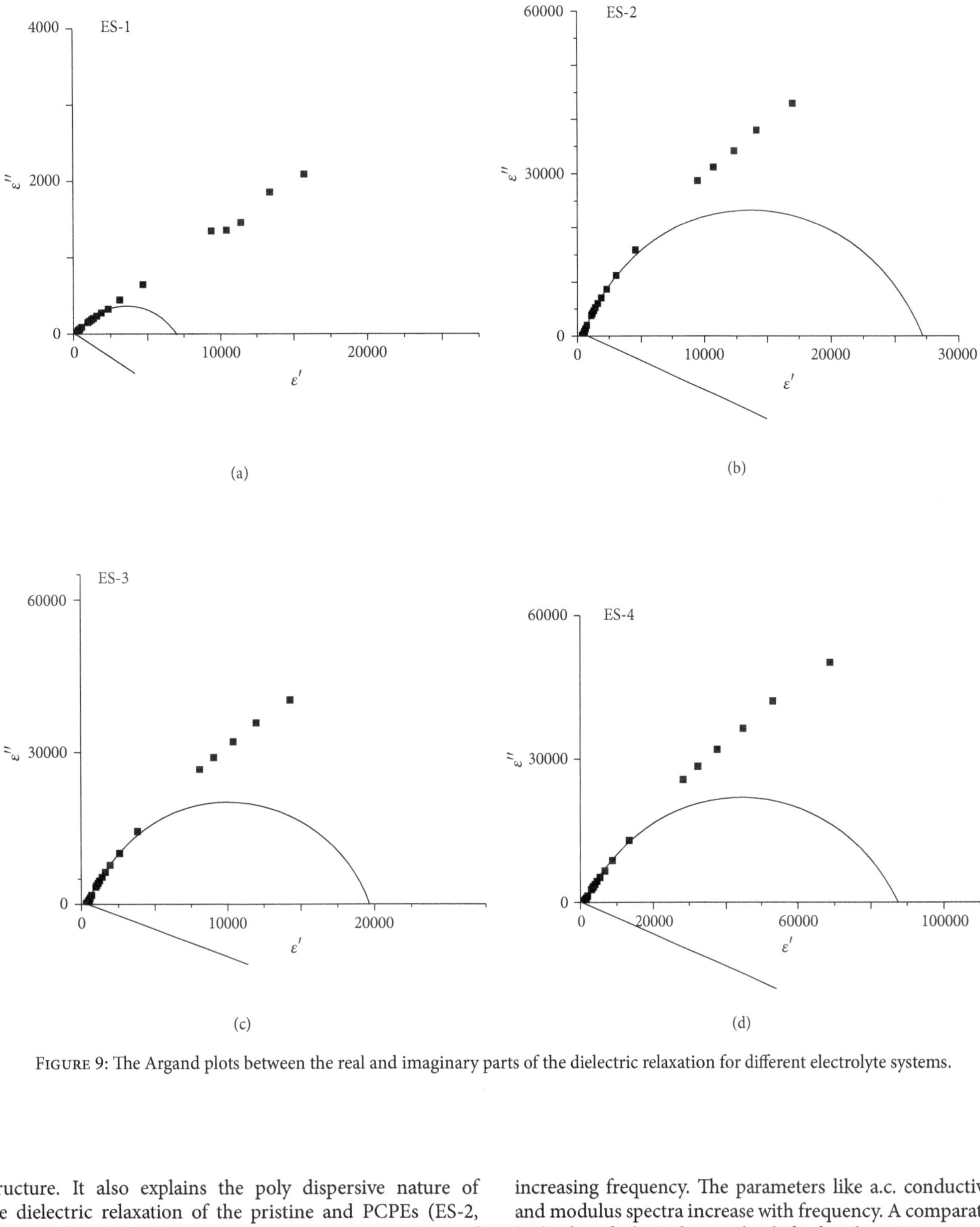

FIGURE 9: The Argand plots between the real and imaginary parts of the dielectric relaxation for different electrolyte systems.

structure. It also explains the poly dispersive nature of the dielectric relaxation of the pristine and PCPEs (ES-2, ES-3, and ES-4). The master curve representation of various dielectric parameters (ε^*, σ^*, M^* and $\tan\delta$) with frequency is shown in Figure 10 for pristine (marked as "A" in Figure 10) and best conductive plasticized electrolytes ES-4 (marked as "B" in Figure 10). From Figure 10, it is obvious that relative permittivity and tangent loss decrease with increasing frequency. The parameters like a.c. conductivity and modulus spectra increase with frequency. A comparative look of $\tan\delta$ plot indicates the shift of peak position towards the lower frequency side with respect to pristine electrolyte. In the same plot other parameters like σ^*, M^*, and ε^* recorded some specific changes in plasticized sample with respect to the pristine system. In plasticized sample (ES-4), a clear intercept is observed in the real and imaginary part

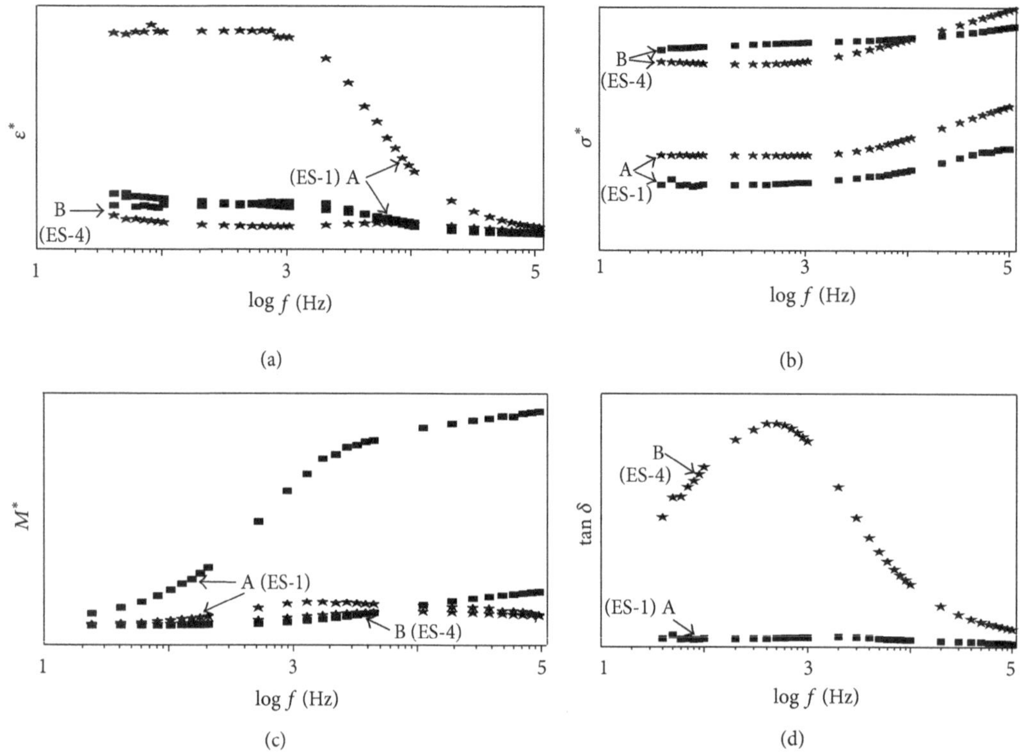

FIGURE 10: The master curve representation of various dielectric parameters (ε^*, σ^*, M^*, and $\tan\delta$) with frequency for ES-1 and ES-4 plasticized electrolytes.

of scaled value of σ^*, M^*, and ε^* on frequency level, and it is noted as relaxation frequency. This intersection is also an indication of better ionic conduction relaxation mechanism.

4. Conclusion

A group of plasticized polymer nanocomposite electrolytes were prepared and characterized. Effect of plasticization on dielectric behaviour, electrical conductivity, and other physical property was clearly visible on the pristine electrolyte. XRD study emphasizes that the increase of amorphous content is relatively high after plasticization. The better electrical conductivity depends on the combination of plasticizers. The maximum conductivity was obtained with (EC + PEG). The addition of PEG modifies the matrix and provides an excellent path for conduction. This enhancement in electrical conductivity of the films agrees well with the changes in the local microstructure/structure on plasticizer addition. Scaled curve shows single transport mechanism of ion conduction in the plasticized polymer electrolytes.

Acknowledgments

The authors thank the BRNS, Department of Atomic energy, Government of India and the CST Lucknow, India for the financial support.

References

[1] F. M. Gray, *Solid Polymer Electrolyte Fundamental and Application*, New York, NY, USA, 1991.

[2] A. Chandra, P. C. Srivastava, and S. Chandra, "Ion transport studies in PEO:NH_4I polymer electrolytes with dispersed Al_2O_3," *Journal of Materials Science*, vol. 30, no. 14, pp. 3633–3638, 1995.

[3] K. Pandey, M. M. Dwivedi, Asthana, Nidhi, M. Singh, and S. L. Agrawal, "Structural and ion transport studies in (100-x) PVdF + xNH_4SCN Gel electrolyte," *Materials Sciences & Applications*, vol. 2, no. 7, pp. 721–728, 2011.

[4] P. K. Khare, P. L. Jain, and R. K. Pandey, "Charging transient in polyvinyl formal," *Bulletin of Materials Science*, vol. 24, no. 4, pp. 401–406, 2001.

[5] M. Kono, E. Hayashi, and M. Klantanabey, "Network polymer electrolytes with free chain ends as internal plasticizer," *Journal of The Electrochemical Society*, vol. 145, no. 5, pp. 1521–1527, 1998.

[6] Y. Kang, H. J. Kim, E. Kim, B. Oh, and J. H. Cho, "Photocured PEO-based solid polymer electrolyte and its application to lithium-polymer batteries," *Journal of Power Sources*, vol. 92, no. 1-2, pp. 255–259, 2001.

[7] M. A. S. Azizi Samir, A. M. Mateos, F. Alloin, J.-Y. Sanchez, and A. Dufresne, "Plasticized nanocomposite polymer electrolytes based on poly(oxyethylene) and cellulose whiskers," *Electrochimica Acta*, vol. 49, no. 26, pp. 4667–4677, 2004.

[8] M. H. Sheldon, M. D. Glasse, R. J. Latham, and R. G. Linford, "The effect of plasticiser on zinc polymer electrolytes," *Solid State Ionics*, vol. 34, no. 1-2, pp. 135–138, 1989.

[9] A. Roy, A. Ghosh, S. Datta et al., "Effects of plasticizers and surfactants on the film forming properties of hydroxypropyl methylcellulose for the coating of diclofenac sodium tablets," *Saudi Pharmaceutical Journal*, vol. 17, no. 3, pp. 233–241, 2009.

[10] A. Bhide and K. Hariharan, "Ionic transport studies on $(PEO)_6$:$NaPO_3$ polymer electrolyte plasticized with PEG_{400}," *European Polymer Journal*, vol. 43, no. 10, pp. 4253–4270, 2007.

[11] M. Egashira, H. Todo, N. Yoshimoto, and M. Morita, "Lithium ion conduction in ionic liquid-based gel polymer electrolyte," *Journal of Power Sources*, vol. 178, no. 2, pp. 729–735, 2008.

[12] C. A. Entwistle and R. C. Rowe, "Plasticization of cellulose ethers used in the film coating of tablets," *Journal of Pharmacy and Pharmacology*, vol. 31, pp. 269–272, 1979.

[13] J. K. Sears and J. R. Darby, "Mechanism of plasticizer action," in *The Technology of Plasticizers*, J. K. Sears and J. R. Darby, Eds., pp. 35–77, Wiley-Interscience, New York, NY, USA, 1982.

[14] W. Barenswaard, V. M. Litvinov, F. Socrren, R. L. Scherreenberg, C. Gondard, and C. Colemonts, "Crystallinity and microstructure of plasticized poly(vinyl chloride). A ^{13}C and ^{1}H solid state NMR study," *Macromolecules*, vol. 32, no. 1, pp. 167–180, 1999.

[15] K. Pandey, M. M. Dwivedi, M. Tripathi, M. Singh, and S. L. Agrawal, "Structural, thermal and ion transport studies on nanocomposite polymer electrolyte-(PEO + SiO_2):NH_4SCN system," *Ionics*, vol. 14, no. 6, pp. 515–523, 2008.

[16] H.-M. Lai and G. W. Padua, "Properties and microstructure of plasticized zein films," *Cereal Chemistry*, vol. 74, no. 6, pp. 771–775, 1997.

[17] N. Baskaran, "Conductivity relaxation and ion transport processes in glassy electrolytes," *Journal of Applied Physics*, vol. 92, no. 2, p. 825, 2002.

[18] M. J. Rice and W. L. Roth, "Ionic transport in Superionic conductors: a theoretical model," *Journal of Solid State Chemistry*, vol. 4, no. 2, pp. 294–310, 1972.

[19] D. L. Sidebottom, "Universal approach for scaling the a.c. conductivity in ionic glasses," *Physical Review Letters*, vol. 82, no. 18, pp. 3653–3656, 1999.

12

Effect of MWCNT on Thermal, Mechanical, and Morphological Properties of Polybutylene Terephthalate/Polycarbonate Blends

C. P. Rejisha, S. Soundararajan, N. Sivapatham, and K. Palanivelu

Department of Plastics Technology, Central Institute of Plastics Engineering and Technology (CIPET), Chennai, Tamil Nadu 600 032, India

Correspondence should be addressed to S. Soundararajan; bssrajan@yahoo.co.in

Academic Editor: Cornelia Vasile

This paper evaluated the effect of multiwall carbon nanotube (MWCNT) on the properties of PBT/PC blends. The nanocomposites were obtained by melt blending MWCNT in the weight percentages 0.15, 0.3, and 0.45 wt% with PBT/PC blends in a high performance corotating twin screw extruder. Samples were characterized by tensile testing, dynamic mechanical analysis, thermal analysis, scanning electron microscopy, and X-ray diffraction. Concentrations of PBT and PC are optimized as 80 : 20 based on mechanical properties. A small amount of MWCNT shows better increase in the thermal and mechanical properties of the blends of PBT/PC nanocomposite when compared to nanoclays or inorganic fillers. The ultimate tensile strength of the nanocomposites increased from 54 MPa to 85 MPa with addition of MWCNT up to 0.3% and then decreased.The tensile modulus values were increased to about 60% and the flexural modulus was more than about 80%. The impact strength was also improved with 20% PC to about 60% and with 0.15% MWCNT to about 50%. The HDT also improved from 127°C to 205°C. It can be seen from XRD result that the crystallinity of PBT is less affected by incorporating MWCNT. The crystallizing temperature was increased and the MWCNT may act as a strong nucleating agent.

1. Introduction

Polyesters like PBT and PET are semicrystalline and possess high stiffness, hardness, dimensional stability, good bearing strength, good electrical properties, excellent flow characteristics, and good resistance to chemicals but have low impact strength. PC has excellent impact strength.

Polymer blends are developed in an effort to meet cost/performance required in many engineering applications. In these blends, it is important to establish some level of interfacial adhesion between the components in order to achieve the necessary toughness. The required bonding is achieved in some commercial blends, for example, polybutylene terephthalate/polyethylene terephthalate (PBT/PET), polybutylene terephthalate/polycarbonate (PBT/PC), as a result of the partial miscibility between the blend components [1–5]. PC-PBT is useful for car bumpers, front ends by Bayer (trade name: Macro Blend). DSM also recently announced PC-PET blends. The PBT/PC finds use in many engineering applications like automotive body panels, in outdoor power, or recreational equipment, appliance housings, telecommunication, and so forth. The PC will improve the impact strength and elongation of PBT since PBT like PET has lower impact strength [6–8]. The various blends of PBT and the properties improvement are also given in [8].

The properties of the high polymer blends can be further improved by compounding with glass fibre or carbon fibre. The recently found carbon nanotube (single-/mutliwall) has the potential to further improve the properties due to their high strength (~130 GPa) and has high aspect ratio (length: few micrometres; diameter: 1 nm) [9–12]. The positive effect of nanotubes to polymer properties can be derived directly not only from the MWCNT properties and their volume fraction but also from their influence on morphology, crystallinity [13], and glass transition temperature [14–16]. Though it imparts good properties, it is very difficult to achieve these fundamental properties in experimental composites due to the need to disperse the individual nanotubes into the matrix and to ensure sufficient interfacial stress transfer between nanotubes and matrix [17]. The main issue in the production of carbon nanotube reinforced polymers is to avoid the agglomeration of nanotubes in the polymer matrix [18].

Recent studies on MWCNT with ethylene vinyl acetate (EVA) show improved flame retardancy and thermal properties [19]. Studies on glass fibers-carbon nanotube epoxy composites have been reported [20]. Novel approach for the conductive nanocompounds by using polypyrrole MWCNT also has been published in the literature [21]. Applications of MWCNT composites in aviation industry with various characteristics of CNT and MWCNT are given in this reference [22]. Comparison and analysis of physical properties of carbon nanomaterial doped polymer composites were also reported [23]. Highly electrical conducting PS/MWCNT nanocomposites properties by *in situ* polymerization were reported [24]. Studies on PC/ABS blends with MWCNT were recently reported [25].

The MWCNT in blends may be useful as an ultraviolet (UV) stabilizer in the outdoor applications like automotive applications with higher mechanical and thermal properties, though the electrical insulation may be little lower. In the present paper, PBT is melt-blended with PC in three different formulations (10, 20, and 30 wt%). In 80% : 20% ratio PBT and PC, the MWCNTs were melt-dispersed. The influence of MWCNT on the mechanical and thermal properties of the blend was evaluated.

2. Experimental

2.1. Materials. PBT (Crastin, CNC010), melting temperature 225°C, was purchased from DuPont (Chennai, India) and PC (CORNING), melting temperature 220°C, was procured from Sigma Aldrich (Chennai, India). The multiwall nanotube was supplied by Sun Nanotech Co. Ltd., China.

2.2. Twin Screw Compounding. Melt processing is carried out in Berstorff high performance corotating twin screw extruder (L/D ratio −30 : 1, capacity 5–30 Kg/hr). In the first stage, PBT/PC blends containing different volume percentages (90 : 10, 80 : 20, and 70 : 30%) were prepared at a temperature range of 180–250°C with a screw speed of 110 rpm. In the second stage, the nanocomposite was prepared by compounding different concentrations of MWCNT (0.15, 0.3, and 0.45%) with the optimized composition of PBT/PC (80 : 20%) blend. PBT and PC were dried at 100°C for 8 hrs in an air-circulated oven before compounding. Finally, these blended materials were molded at an injection molding machine as per American Society for Testing and Materials (ASTM) standard.

2.3. Testing of PBT/PC Blends and Their MWCNT Composites. Tensile properties were investigated using Universal testing machine (Shimadzu Autograph, model: AG 50 KN, Japan) according to ASTM D 638 with a cross-head speed of 50 mm/minute. The impact strength test is carried out according to ASTM D 256. The test is carried out using an Izod Impact tester (atsfaar, Italy). The specimens have dimensions 63.5 × 12.7 × 3 mm, with "V" notch depth of 2.54 mm and notch angle 45°. The flexural strength and modulus were done as per ASTM D 790 using the same universal testing machine.

Nonisothermal crystallization analysis was performed by means of a differential scanning calorimeter (PerkinElmer, Pyris Diamond DSC). The standard procedure performed was as follows: samples of about 5–10 mg were heated from 5°C to 250°C at a scan rate of 10°C/min and held for 2 min in order to eliminate any thermal history of the material. Subsequently, the samples were cooled to 50°C using scan rate of 10°C/min. The heat distortion temperature (HDT) (atsfaar, Italy) of the nanocomposites was measured under load (1.8 MPa) according to ASTM D 648.

Dynamic mechanical analysis measurement was carried out on a thermal analyser (TA Instrument). The materials were cut into strips with length of 60 mm and thickness of 3 mm. Tests were performed at a frequency of 10 Hz between 30°C and 220°C with a ramp of 3°C/min. The characteristics of the fracture surface of the samples were investigated after slight gold sputtering, using a field emission scanning electron microscope (SEM-FEG). The wide angle X-ray diffraction (WAXD) analysis was performed with a BRUKER D8 ADVANCE X-ray diffractometer using Ni-filtered Cu Kα X-ray, and the diffracting intensities were recorded at a step of every 2θ over the range of $10–40\theta$. The microstructure of the nanocomposite and the CNT distribution in the polymer blend were observed.

3. Result and Discussion

3.1. Dispersion of MWCNTs. Dispersion of MWCNTs is critical in order to improve the mechanical properties. Poor dispersion of MWCNT leads to localized micron-sized agglomerates which can negatively influence the strengthening mechanism of MWCNTs such as crack deflection or crack bridging. If the MWCNTs are well dispersed, they form a three-dimensional physical network thereby effectively enhancing the overall performance of the material.

3.2. Mechanical Properties

3.2.1. Effect of Incorporation of PC on the Mechanical Properties of PBT. Mechanical properties of virgin PBT and PBT/PC at different concentrations (90 : 10, 80 : 20, and 70 : 30%) are summarized in Table 1. It is observed that on increasing the PC concentration in the blend up to 20%, the mechanical properties increase. This is due to the toughening effect of PC. Above 20% of PC, the blend shows a decrease in tensile strength and flexural strength. This is due to the higher polarity of PC and semicompatibility with PBT. But, in the case of impact strength, as the concentration of PC increases, the impact strength also increases because the PC has an excellent impact strength compared to that of PBT.

3.2.2. Effect of Loading MWNT on the Mechanical Properties of PBT/PC Blend. Table 2 shows mechanical properties of various MWCNT reinforced composites. It shows that the tensile and flexural strength and modulus (Figure 1) increase on the

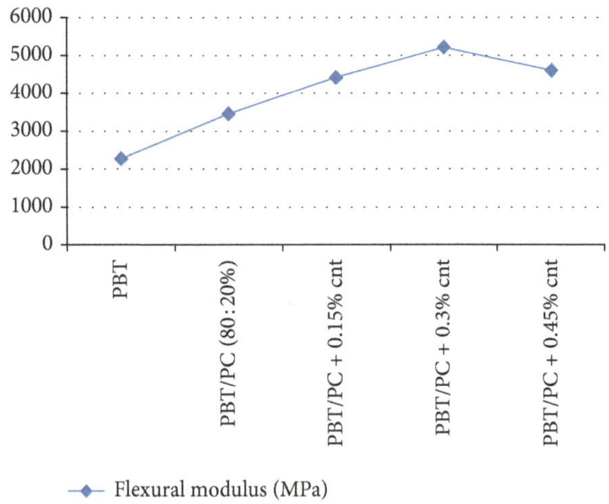

FIGURE 1: Flexural modulus of PBT/PC with MWCNT.

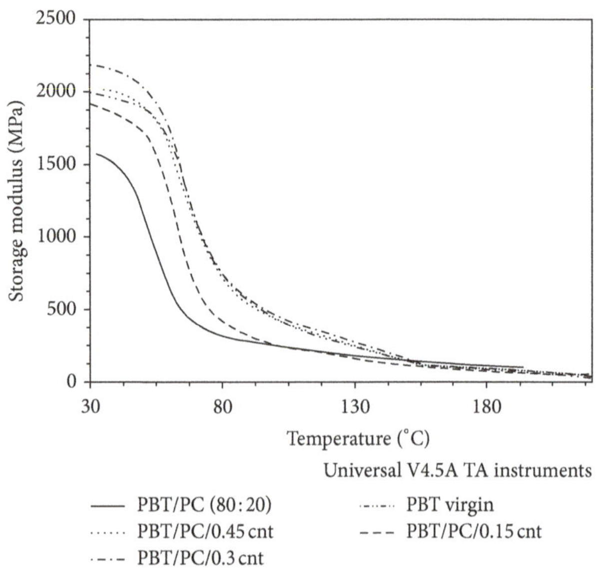

FIGURE 2: Storage modulus of PBT, PBT/PC blends and MWCNT composites.

addition of MWCNT. It can be observed that with an increase from 0.15 wt% to 0.3 wt % of MWCNT concentration, the Young's modulus and tensile strength increase enormously. This may be attributed to the better reinforcement of the MWCNT when incorporated in the PBT/PC matrix and hence better performance was observed. On the other hand, the tensile strength reduces by incorporation of 0.45 wt% of MWCNT. Similar trend was observed with flexural strength and modulus. It may be caused by the disturbance of the PBT crystallization and saturation in this sample. The incorporation of MWCNT on the matrix first increases for 0.15% and then reduces the impact strength value. This is due to the reinforcing effect of MWCNT on the matrix. The PBT/PC MWCNT composites show higher impact strength values than those of virgin PBT.

TABLE 1: Mechanical properties of PBT and PBT/PC blend at various concentrations.

Sample ID	Tensile strength (MPa)	Tensile modulus (GPa)	Flexural strength (MPa)	Flexural modulus (GPa)	Impact strength (J/m)
PBT (100%)	54	2.21	72	2.27	53
PBT/PC (90 : 10)	61	2.39	79	2.91	72
PBT/PC (80 : 20)	65	2.51	87	3.45	85
PBT/PC (70 : 30)	63	2.44	83	3.02	91

3.3. Thermal Properties

3.3.1. Heat Distortion Temperature (HDT). The HDT plays an important role in determining the performance of engineering plastics at the elevated temperature. The HDT can be influenced by various factors such as the polymer melt, mold temperature, the nucleating agent, and various processing conditions, which is often related to the mechanical and thermal behavior of the polymer composite. The variations in the HDT values for the PBT, PBT/PC blend, and the nanocomposites are shown in Table 3. The HDT values of the nanocomposites increased enormously with the MWCNT content which is due to the enormous increase in the flexural modulus of the nanocomposites. In the HDT measurement, the ability of the polymeric material to retain stiffness with the increasing temperature is important for a high HDT value. The nanocomposites containing 0.3% MWCNT show the highest HDT value.

3.3.2. Differential Scanning Calorimetry (DSC). The thermal parameters, melting temperature (T_m), crystallization temperature (T_c), and melting enthalpy (H_m) were obtained from the DSC thermogram. Table 4 shows the DSC thermogram readings.

The melting temperature of PBT is lower in the blends (PBT/PC, 80 : 20) compared to the virgin PBT. The lowering of this temperature for PBT/PC is due to the decreasing concentration of PBT. Incorporation of MWNT causes an increase in delta H_m value. The increase is up to 0.3% of MWCNT. This may be due to the stiffening effect of MWCNT on the matrix. Above 0.3%, the crystallinity of PBT may be affected and hence the reduced H_m value.

The heat of fusion values shows very interesting behavior. The ΔH_m value for the PBT/PC blend is lower than the value expected. But this value is less than that of virgin PBT. The PBT/PC blend does not show much variation in T_m compared to virgin PBT. This may be due to less disturbance of the crystallinity of PBT in the blend. The nanocomposite shows a higher value than the blend. The crystallization temperature for all nanocomposites is little higher than that of the virgin PBT. Faster cooling can be achieved. The crystallite or spherulites sizes may be reduced and transparency may be improved.

TABLE 2: Mechanical properties of PBT, PBT/PC blend, and MWCNT composites.

Sample ID	Tensile strength (MPa)	Tensile modulus (GPa)	Flexural strength (MPa)	Flexural modulus (GPa)	Impact strength, Izod (J/m)
PBT (100%)	54	2.21	72	2.27	53
PBT/PC (80%/20%)	65	2.51	87	3.45	85
PBT/PC + 0.15% MWCNT	74	3.2	112	4.42	79
PBT/PC + 0.3% MWCNT	85	4.0	137	5.21	72
PBT/PC + 0.45% MWCNT	79	3.5	129	4.60	64

TABLE 3: HDT values of PBT, PBT/PC blend, and MWCNT composites.

Concentration of MWCNT (%)	Concentration of PBT (%)	HDT [at 0.46 MPa stress] (°C)
0%	100	127
0%	80 : 20	130
0.15%	80 : 20	187
0.3%	80 : 20	205
0.45%	80 : 20	196

3.4. Dynamic Mechanical Analysis (DMA). The viscoelastic nature of the various nanocomposites was characterized by dynamic mechanical analysis. Glass transition temperature (T_g), storage modulus (E^{I}), loss modulus (E^{II}), and tan δ were obtained from DMA. Figure 2 shows the temperature verses storage modulus E^{I} of PBT, PBT/PC blend, and various nanocomposites. It can be observed that the storage modulus of the nanocomposites is higher than that of the PBT/PC blend. In all nanocomposites, the incorporation of MWCNT causes a measurable increase in stiffness. The storage modulus of PBT/PC is increased by the stiffening effect of the nanotubes. This effect is significant at temperature below T_g of PBT. This shows that the reinforcing effect of MWCNT is mainly active in the amorphous phase.

The tan δ curves of the various blends and MW nanocomposites are shown in Figure 3. It can be seen that tan δ peak of the nanocomposites shifted to slightly higher temperature compared to that of pure PBT and becomes broader. This can be explained by the strong interaction between the MWCNT and the matrix. This increase in property diminishes due to the increasing concentration of MWCNT which causes agglomeration in the matrix. The maximum of tan δ is taken as the glass transition temperature (T_g) and is almost the same for all the material.

3.5. Scanning Electron Microscopy. Morphology is a major factor in determining the mechanical properties of the polymer blends and multiwall carbon nanocomposites.

Figures 4(a), 4(b), and 4(c) show the microscopic graphs of the fracture morphologies of PBT/PC at different concentrations of MWCNTs. In MWCNT, filled PBT/PC nanocomposites show a uniformed dispersion PC in PBT matrix in presence of MWCNT in the polymer matrix at lower concentration.

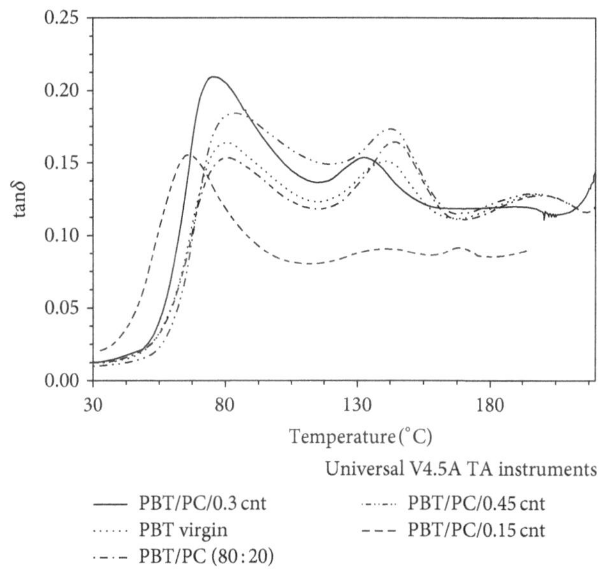

FIGURE 3: Tan delta of virgin PBT, PBT/PC blends, and MWCNT composites.

3.6. X-Ray Diffraction. Wide angle X-ray analysis was conducted on the PBT, PBT/PC nanocomposites to investigate the effect on the structure of the nanocomposites; WAXD patterns of the nanocomposites are shown in Figure 5. The characterization peaks of pure PBT were also observed for the nanocomposites and the position of the peaks remained almost unchanged with the introduction of MWCNT, despite some change in the peak intensity. The intensity of the peak is less in PBT/PC when compared to the nanocomposite due to the semicrystalline nature of PC which reduces the intensity of the peak. This result shows that the incorporation of MWCNT into the PBT and PBT/PC matrix does not change the crystal structure of the nanocomposites. The crystallinity of the nanocomposites was slightly increased with the introduction of MWCNT, which may be explained by the super cooling temperature.

In the nanocomposite, MWCNTs act as a strong nucleating agent in the PBT matrix, and the crystalline temperature shifts to higher temperature, implying that the super cooling of the nanocomposite was increased with the introduction of MWCNT. When the polymer crystallized with more super cooling, it crystallized more perfectly than the less super cooling, and thus the crystallinity of nanocomposite may

Table 4: DSC data of the virgin PBT, PBT/PC blend, and MWCNT.

Sample ID	Melting temperature T_m (°C)	Crystallization temperature T_c (°C)	Heat of fusion (ΔH_m) (J/g)
PBT (100%)	225.00	187.00	26
PBT/PC (80/20)	220.38	185.36	15
PBT/PC (80/20) + 0.15% MWCNT	222.71	197.22	20
PBT/PC (80/20) + 0.3% MWCNT	225.97	202.29	24
PBT/PC (80/20) + 0.45% MWCNT	223.64	199.94	21

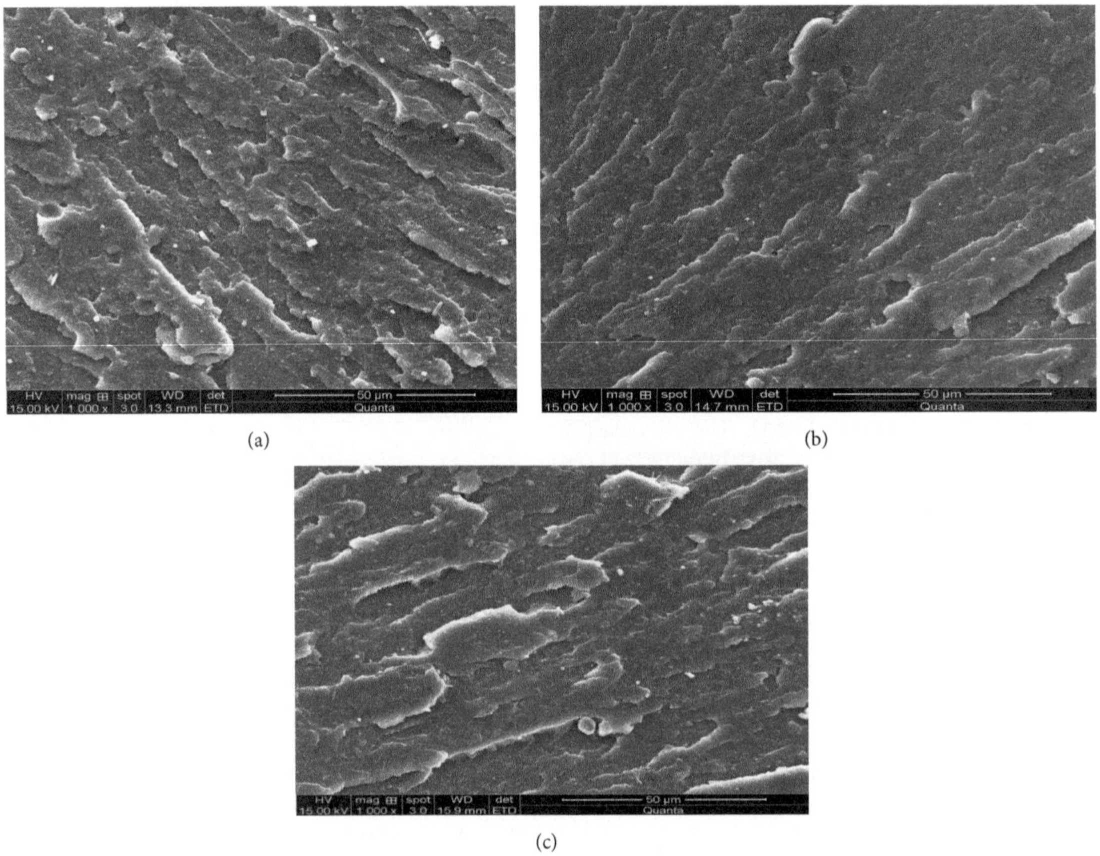

Figure 4: (a) SEM images of PBT/PC/0.15% MWCNT, (b) SEM images of PBT/PC/0.3% MWCNT, (c) and SEM images of PBT/PC/0.45% MWCNT.

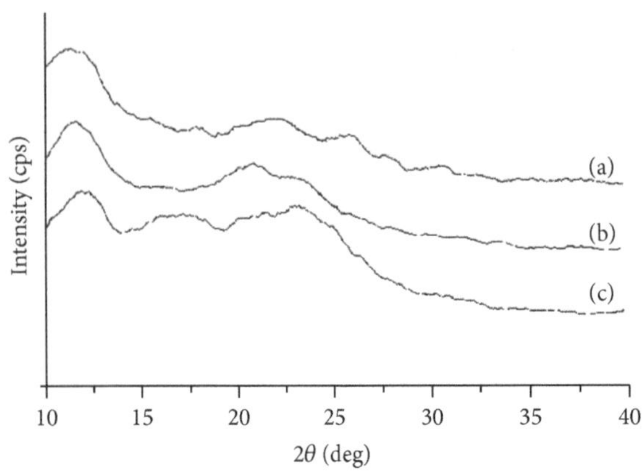

Figure 5: XRD graph of (a) PBT/PC/0.3% MWCNT, (b) virgin PBT, and (c) PBT/PC blend.

be slightly increased with the introduction of MWCNT. Hence, the mechanical properties like tensile strength, flexural strength, and modulus increased and the impact strength was decreased.

In the nanocomposite, MWCNTs act as a strong nucleating agent in the PBT matrix, and the crystalline temperature shifts to higher temperature, implying that the super cooling of the nano composite was increased with the introduction of MWCNT. When the polymer crystallized with more super cooling, it crystallized more perfectly than the less super cooling, and thus the crystallinity of nanocomposite slightly increased with the introduction of MWCNT. Hence, the mechanical properties like tensile strength, flexural strength, and modulus were increased up to 0.30%. The impact strength was increased first and then decreased but was higher than that of PB.

4. Conclusion

The ultimate tensile strength of the nanocomposites increased from 54 MPa to 85 MPa with addition of MWCNT up to 0.3% and then decreased. The tensile modulus values were increased to about 60% and the flexural modulus was more than about 80%. The impact strength was also 60% improved with 20% PC and 50% improved with 0.15% MWCNT. The HDT also improved from 127°C to 205°C. It can be seen from XRD result that the crystallinity of PBT is less affected by incorporating MWCNT. The crystallization temperature for all nanocomposites is little higher than that of the virgin PBT. Faster cooling can be achieved. Hence, the mechanical properties like tensile strength, flexural strength, and modulus were increased up to 0.30%. The impact strength was also improved with 20% PC to about 60% and with 0.15% MWCNT to about 50% and then decreased but was higher than that of PBT. The results suggest that the MWCNts act as strong reinforcement in PBT/PC blends. The crystallizing temperature was increased and the MWCNT may act as strong nucleating agent. The nanotubes are well dispersed in the polymer matrix (up to 0.3% of MWCNT).

The result shows that a small amount of MWCNT is enough to improve the mechanical properties of the PBT/PC matrix. In the present work, it has been shown that PBT/PC/MWCNT thermoplastic nanocomposites can be produced with tensile strength, tensile modulus, flexural strength, and flexural modulus considerably higher than those of PBT or PBT/PC blend. In conclusion, the addition of MWCNT offers a simple and effective means to produce nanocomposites. The MWCNT in PBT/PC blends may be useful as an ultraviolet (UV) stabilizer in the outdoor applications like automotive applications with higher mechanical and thermal properties though the electrical insulation may be little lower.

Conflict of Interests

The authors declare that there is no conflict of interests regarding the publication of this paper.

References

[1] S. Y. Hobbs, V. L. Groshans, M. E. J. Dekkers, and A. R. Shultz, "Erratum," *Polymer Bulletin*, vol. 17, no. 5, p. 496, 1987.

[2] S. Y. Hobs, M. E. J. Dekkers, and V. H. Watkins, "Toughened blends of poly(butylene terephthalate) and BPA polycarbonate," *Journal of Materials Science*, vol. 23, no. 4, pp. 1219–1224, 1988.

[3] M. E. J. Dekkers, S. Y. Hobbs, and V. H. Watkins, "Toughened blends of poly(butylene terephthalate) and BPA polycarbonate —part 2 Toughening mechanisms," *Journal of Materials Science*, vol. 23, no. 4, pp. 1225–1230, 1988.

[4] J. D. Keitz, J. W. Barlow, and D. R. Paul, "Polycarbonate blends with stryene/acrylonitrile copolymers," *Journal of Applied Polymer Science*, vol. 29, no. 10, pp. 3131–3145, 1984.

[5] R. A. Mendelson, "Miscibility and deformation behavior in some thermoplastic polymer blends containing poly(styrene-co-acrylonitrile)," *Journal of Polymer Science A, Polymer Physics*, vol. 23, no. 10, pp. 1975–1995, 1985.

[6] J. A. Brydson, *Plastics Materials*, Butterworth Heinemann, Oxford, UK, 7th edition, 2011.

[7] L. A. Utarcki, *Commerciasl Polymer Blends*, Chapman & Hall, London, UK, 1998.

[8] L. A. Utarcki, *Polymer Blends Handbook*, vol. 1, Kluwer Academic, London, UK, 2002.

[9] O. Lourie and H. D. Wagner, "Transmission electron microscopy observations of fracture of single-wall Carbon nanotubes under axial tension," *Applied Physics Letters*, vol. 73, no. 24, pp. 3527–3529, 1998.

[10] O. Lourie and H. D. Wagner, "Evaluation of Young's modulus of Carbon nanotubes by micro-Raman spectroscopy," *Journal of Materials Research*, vol. 13, no. 9, pp. 2418–2422, 1998.

[11] J. Sandler, M. S. P. Shaffer, T. Prasse, W. Bauhofer, K. Schulte, and A. H. Windle, "Development of a dispersion process for Carbon nanotubes in an epoxy matrix and the resulting electrical properties," *Polymer*, vol. 40, no. 21, pp. 5967–5971, 1999.

[12] O. Lourie, D. M. Cox, and H. D. Wagner, "Buckling and collapse of embedded Carbon nanotubes," *Physical Review Letters*, vol. 81, no. 8, pp. 1638–1641, 1998.

[13] J. Sandler, G. Broza, M. Nolte, K. Schulte, Y.-M. Lam, and M. S. P. Shaffer, "Crystallization of Carbon nanotube and nanofiber polypropylene composites," *Journal of Macromolecular Science—Physics*, vol. 42, no. 3-4, pp. 479–488, 2003.

[14] F. H. Gojny and K. Schulte, "Functionalisation effect on the thermo-mechanical behaviour of multi-wall Carbon nanotube/epoxy-composites," *Composites Science and Technology*, vol. 64, no. 15, pp. 2303–2308, 2004.

[15] C. Zhao, L. Ji, H. Liu et al., "Functionalized Carbon nanotubes containing isocyanate groups," *Journal of Solid State Chemistry*, vol. 177, no. 12, pp. 4394–4398, 2004.

[16] H. Kong, C. Gao, and D. Yan, "Functionalization of multiwalled Carbon nanotubes by atom transfer radical polymerization and defunctionalization of the products," *Macromolecules*, vol. 37, no. 11, pp. 4022–4030, 2004.

[17] S. Barrau, P. Demont, A. Peigney, C. Laurent, and C. Lacabanne, "Dc and ac conductivity of Carbon nanotubes-polyepoxy composites," *Macromolecules*, vol. 36, no. 14, pp. 5187–5194, 2003.

[18] M. Lamy de la Chapelle, C. Stéphan, T. P. Nguyen et al., "Raman characterization of singlewalled Carbon nanotubes and PMMA-nanotubes composites," *Synthetic Metals*, vol. 103, no. 1–3, pp. 2510–2512, 1999.

[19] A. K. Ghosh, "Reduced graphene oxide/ethylene vinyl acetate co-polymer composite with improved thermal stability and flame retardancy," *Journal of Polymer Materials*, vol. 29, pp. 411–421, 2012.

[20] R. Ramlee, M. Mariatti, Z. K. Mohd Ishak, and A. R. Mohamed, "Properties of glass Fiber-Carbon nanotube (CNT)/Epoxy composites," *Journal of Polymer Materials*, vol. 30, no. 1, pp. 103–116, 2013.

[21] K. V. Yeole and S. T. Mhaske, "Novel approach for the preparation of conductive nano composites by using Poly Pyrolle MWCNT," *Journal of Polymer Materials*, vol. 29, pp. 145–151, 2012.

[22] M. Verma, "Applications of nano technology in aviation industry," in *Proceedings of the International Symposium on Macro- and Supramolecular Architectures and Materials (MAM '12)*, V. Rajaendran, P. Paramasivam, and K. E. Geckler, Eds., pp. 253–260, Blooms Bury, New Delhi, India, 2012.

[23] S. Bal and S. Bhattacharjee, "Comparison and analysis of physical properties of Carbon nano material doped polymer composites," in *Proceedings of the International Symposium on Macro- and Supramolecular Architectures and Materials (MAM '12)*, V. Rajaendran, P. Paramasivam, and K. E. Geckler, Eds., pp. 69–74, Blooms Bury, New Delhi, India, 2012.

[24] N. Kumar, S. Vastava, and B. B. Khatua, "Highly electrical conducting polystyrene/MWCNT nanocomposites prepared by in-situ polymerization," in *Advancements in Polymeric Materials*, CIPET, Chennai, India, 2011.

[25] Z.-Y. Xiong, L. Wang, Y. Sun, Z.-X. Guo, and J. Yu, "Migration of MWCNTs during melt preparation of ABS/PC/MWCNT conductive composites via PC/MWCNT masterbatch approach," *Journal of Polymer*, vol. 54, no. 1, pp. 447–455, 2013.

13

Permittivity and Electromagnetic Interference Shielding Investigations of Activated Charcoal Loaded Acrylic Coating Compositions

Sharief ud Din Khan,[1] Manju Arora,[2] M. A. Wahab,[1] and Parveen Saini[2]

[1] *Department of Physics, Jamia Millia Islamia, New Delhi 110025, India*
[2] *CSIR-National Physical Laboratory, Dr. K.S. Krishnan Marg, New Delhi 110012, India*

Correspondence should be addressed to Parveen Saini; pksaini@nplindia.org

Academic Editor: Yves Grohens

Acrylic resin (AR) based electromagnetic interference (EMI) shielding composites have been prepared by incorporation of up to 30 wt% activated charcoal (AC) in AR matrix. These composites have been characterized by XRD, Raman spectroscopy, scanning electron microscopy, dielectric, and EMI shielding measurement techniques. XRD patterns and Raman studies confirm the incorporation of AC particles inside AR matrix and suggest possible interactions between phases. The SEM images show that incorporation of AC particles leads to systematic change in the morphology of composites especially the formation of porous structure. The dielectric measurements show that 30 wt% AC loading composite display higher relative permittivity value (~79) compared to pristine AR (~5). Further, the porous structure, electrical conductivity, and permittivity value contribute towards EMI shielding effectiveness value of −36 dB (attenuation of >99.9% of incident radiation) for these composites, thereby demonstrating their suitability for making efficient EMI shielding coatings.

1. Introduction

Electromagnetic (EM) interference (EMI) is an offshoot of explosive growth of electronics and telecommunication in the modern society [1–3]. The EMI among electronic instruments/appliances may lead to degradation of device performance and may even adversely affect human health [4, 5]. Due to possible hazards of EMI only, the use of EM wave receipting/emitting electronic gadgets is prohibited inside sensitive zones, for example, during flight or inside hospital's ICUs [1, 5, 6]. Therefore, systematic strategies and suitable counter measures are essential to prevent/suppress EMI so as to ensure uninterrupted performance of appliances [1–9]. The primary mechanism of shielding is based on reflection and the material used for shielding by reflection requires mobile charge carriers; that is, shield should have conducting property [9, 10]. Consequently, metals (in the form of filler, coatings, or laminates) are the most common shielding material which uses primarily reflection mechanism for shielding along with minor absorption component. However, metals suffer from problems like poor wear/scratch resistance, corrosion susceptibility, high density, difficult processing, and high cost [1, 6, 9]. The secondary shielding mechanism is absorption for which shield material should have electrical or magnetic dipoles [9–12] along-with finite electrical conductivity. For such purpose materials with high dielectric constant like ZnO, SiO_2, TiO_2, $BaTiO_3$, or high magnetic permeability, for example, carbonyl iron, Ni, Co, or Fe metals, γ-Fe_2O_3, Fe_3O_4, and so forth are used [1, 13]. However, these materials or their composites possess problems like low permittivity or permeability at gigahertz frequencies, weight penalties, narrow-band action, and processing difficulties [1, 3, 13, 14]. Other than reflection and absorption, another mechanism of shielding is multiple reflections, which refer to the reflections at various surfaces or interfaces in the shield [1]. This mechanism requires the presence of a large interfacial area and porous structure. An example of a shield with a large surface area is conducting composite material containing filler, foamed composites, and honeycomb structures [3].

Interestingly, the properties of such composites are highly dependent on nature and concentration constituent phases. Carbon based materials (carbon black (CB), graphite, activated carbon (AC), fullerenes, carbon nanofibers (CNFs), cabon nanotubes (CNTs), and recently graphene) are attractive choice as a filler, due to good electrical/thermal conductivity, low density, excellent corrosion resistance, and reinforcing capability [1, 3, 7, 9, 10, 12, 15–21]. Particularly, AC with high electrical conductivity and large surface area is an economical alternative. Similarly, acrylics are extremely popular matrix materials due to low cost, solution/thermal processability, high filler reception capacity and ability of acrylic resins (AR) to form hard, abrasion resistant and mechanically strong/tough impermeable coatings [22–24].

Herein we have synthesized porous and electrically conducting AC/AR composites by solvent assisted dispersion of 1–30 wt% AC particles within AR matrix. The synthesized composites have been characterized by techniques such as XRD, Raman spectroscopy, and scanning electron microscopy. The dielectric and EMI shielding measurements were also performed. These porous compositions, containing electrically conducting filler, are expected to display good EMI shielding performance.

2. Materials and Method

Acrylic resin (AR was procured from Pidilite Industries Limited, India), xylene (Rankem, India), and activated charcoal (AC, Merck) were used as received. Formation of AC/AR composites involves the preparation of separate dispersion and dissolution of AC and AR in xylene using magnetic stirrer (REMI 5MLH) followed by their mixing and continued stirring till the formation of viscous syrup. This syrup was poured into a Petri dish and allowed to dry at 50–60°C to obtain film/sheet of composite. In a typical synthesis, 14 g of AR was dissolved in 20 mL of xylene and a known amount of AC powder was added into it to obtain composites with different wt%, namely, 1, 3, 5, 10, 20, and 30 wt%.

3. Measurement and Characterization

XRD patterns were recorded using Bruker-D8 advanced diffractometer in the 2θ range of 10–70° and using Cu Kα line as radiation source. The morphological information was recorded using scanning electron microscope (SEM, Leo 440, UK). MicroRaman spectrometer (Lab RAM HR 800, Horiba JY) having 632.8 nm laser source was used for recording Raman spectra of these samples in 3500–400 cm^{-1} region. The laser beam was focused on the sample using confocal microscope with LWD ×50 objective. Shielding effectiveness (SE) measurements were performed by placing the samples (~0.7 μm thick) between a 101 GHz frequency microwave source and a horn type antenna-detector system.

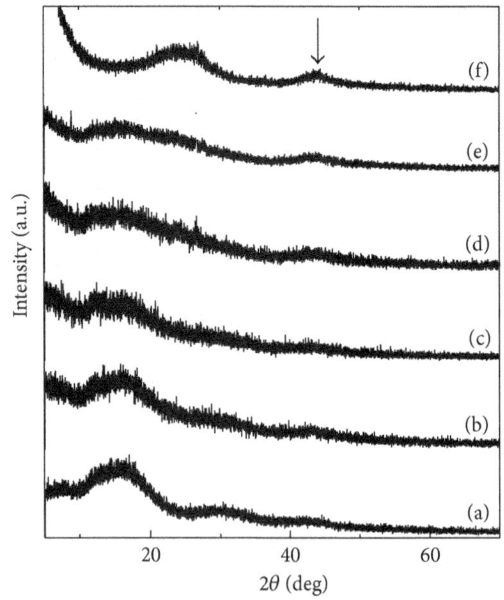

FIGURE 1: X-ray diffraction patterns of (a) pure AR, (b) 5 wt%, (c) 10 wt%, (d) 20 wt%, (e) 30%, and (f) pure AC.

4. Results and Discussion

4.1. Structure Characterization. Figure 1 shows the XRD patterns of pure AR and AC as well as their composites. The characteristic signatures of AR (Figure 1(a)) are located at 2θ values of ~16° and ~31° and 44°, whereas AC's signatures (Figure 1(f)) appear at 25° and 43° [24]. The AC/AR composites give superimposed signatures of both phases (Figures 1(b), (c), (d), and (e)) with systematic shifting of the peak positions and relative change in the peak intensity. Particularly, the shifting of 2θ = 15.95° peak of AR towards lower angle indicates the interaction between the phases.

4.2. Morphological Details. The SEM image of blank AR film (Figure 2(a)) displays smooth and featureless morphology. However, the incorporation of AC within the AC matrix leads to disturbance of AR's morphology and appearance of rough-porous microstructure of AC particles.

As the AC content increases, its signature evolves such that the composite with 30 wt% loading of AC clearly shows the porous structures and channels (bright regions) made up of electrically conducting AC phase coated and infiltrated with nonconducting AR matrix. As these composites display porous structure and electrically conducting in nature, they are expected to display good EMI shielding efficiency, due to reflection, absorption, and multiple reflection phenomena [1–5, 9, 12, 13, 18, 25, 26].

4.3. Raman Spectroscopy. Raman spectra of carbon and its conjugated analogues give two strong bands at ~1560 (G-band) and ~1360 cm^{-1} (D-band) that are dominated by sp^2 (graphitic structure) and sp^3 (defect structures) hybridized carbons, respectively [27, 28]. Figure 3 shows that, for pure AC, G-band appears at 1582 cm^{-1} which shifts to 1598 cm^{-1}

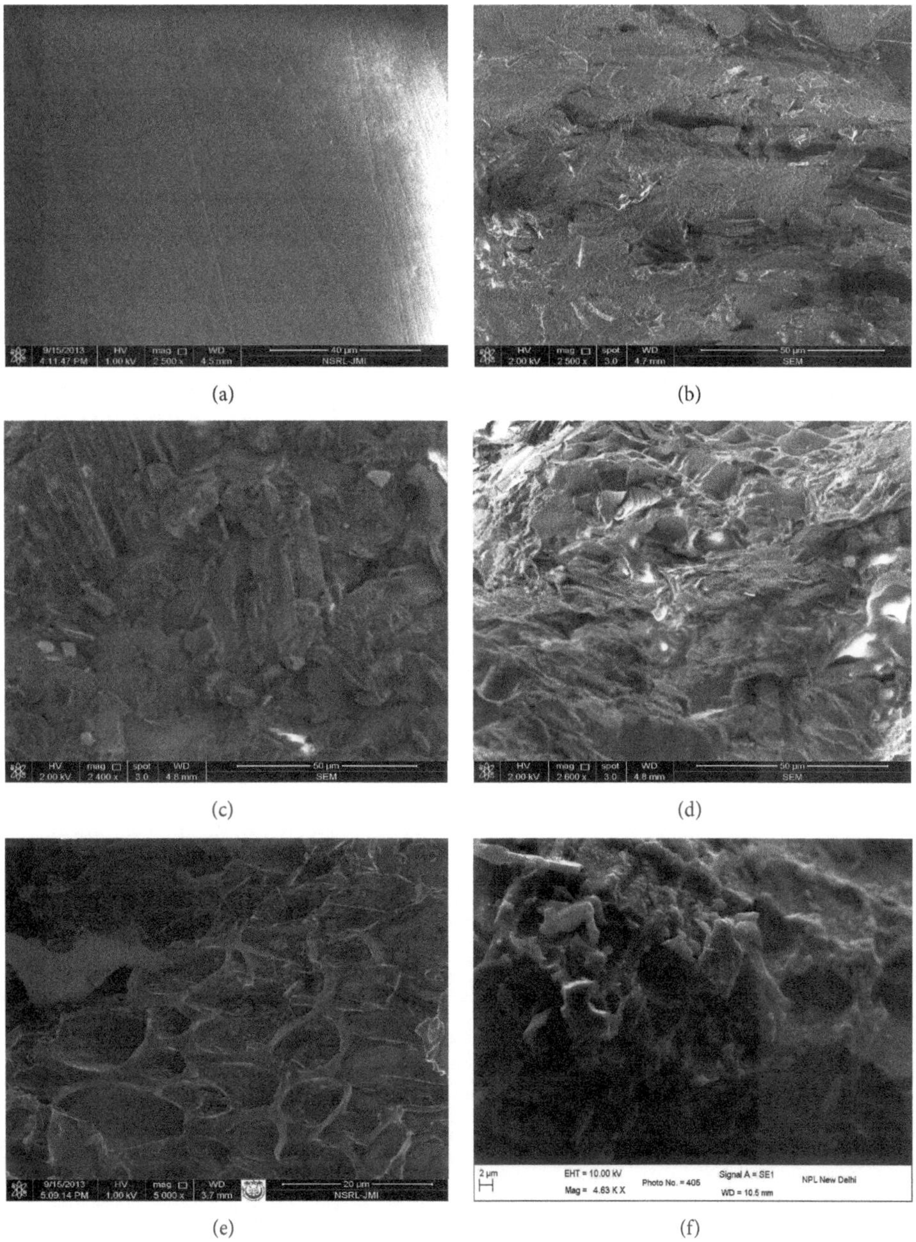

FIGURE 2: SEM images of (a) pure acrylic resin, (b) 1 wt%, (c) 5 wt%, (d) 10 wt%, (e) 20 wt%, and (f) 30 wt% of AC.

on increasing activated carbon percentage in the composites. Similarly, a shift in the disorder band (D-band) from 1304 cm^{-1} (for AC) to 1320 cm^{-1} (for 30 wt% AC containing composite) was observed [27]. Such a shift in the G and D bands on increasing the activated carbon concentration in composites is attributed to the increase in electron-phonon interactions which blueshifts and sharpens the G peak and broadens D band.

4.4. Dielectric Constant or Permittivity Measurements. The study of dielectric constant as a function of temperature or frequency is one of the most convenient and sensitive methods of investigating the polarization and charge localization effects. It is well known that the polarization of a material is controlled by the electronic, ionic, and dipolar polarization. Electronic polarization occurs during a very short interval of time (of the order of 10^{-15} sec) and the process of ionic polarization occurs in the range of 10^{-13}-10^{-12} sec, while the dipole polarization requires relatively longer time [1, 5, 9]. In case of polar polymers, the dielectric constant begins to drop after a certain critical frequency. The dipole molecules cannot orient themselves in the lower temperature region. However, as the temperature increases, orientation of dipoles is facilitated and this increases the dielectric constant. So in this paper room temperature dielectric constant of AC/AR composites has been measured as a function of frequency, as shown in Figure 4, to understand the effect of filler

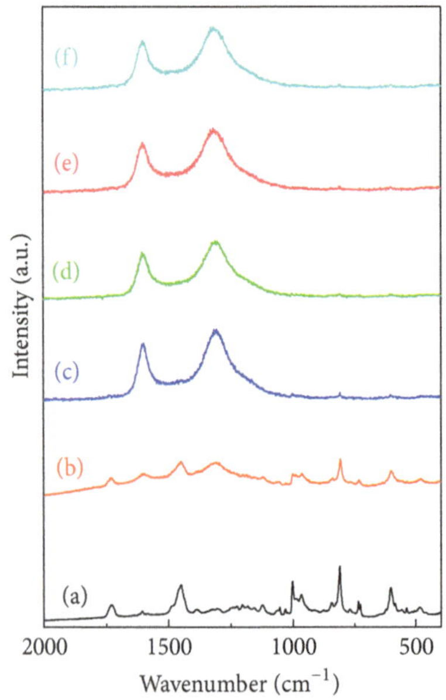

FIGURE 3: Raman spectra of (a) pure acrylic resin, (b) 1 wt%, (c) 5 wt%, (d) 10 wt%, (e) 20 wt%, and (f) 30 wt% of AC.

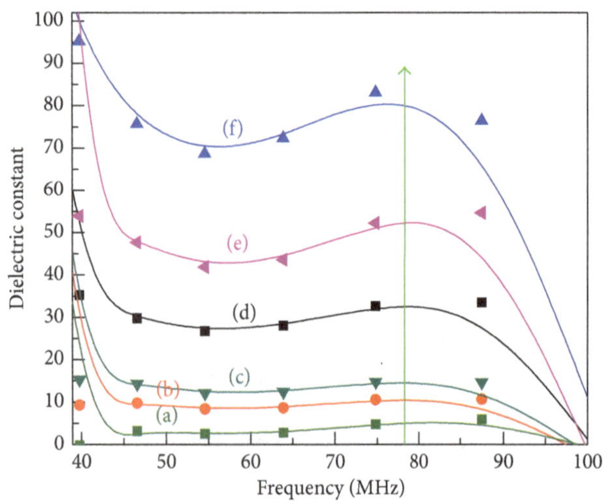

FIGURE 4: Dielectric measurement of (a) pure acrylic resin, (b) 1 wt% (c) 5 wt%, (d) 10 wt%, and (e) 30 wt% of AC.

concentration on polarization behaviour. The results show that, at low frequencies, the electric dipoles have sufficient time to align with the field before the field changes its direction; consequently the dielectric constant is high [1, 29]. However, at higher frequencies, dipoles fail to follow the rapidly changing electric vector; consequently, the dielectric constant value decreases.

Further, it can be seen (Figure 4) that the increase in AC concentration leads to systematic increase in the relative dielectric constant or permittivity (ε'), that is, from ~5 for pure AC to ~79 for 30% AC based composites. This can be attributed to the increase in the space charge buildup due to Maxwell-Wagner interfacial polarization [1, 9, 18, 29] occurring as a consequence of large difference in the electrical conductivity of AC (highly conducting) and AR (insulating matrix). It is important to highlight that good dielectric properties and electrical conductivity values are essential requirements for displaying good EMI shielding performance [1, 8, 9, 30, 31].

4.5. EMI Shielding Performance. The EMI shielding effectiveness (SE) of a material is the ability to attenuate EM radiation that can be expressed in terms of ratio of incoming (incident) and outgoing (transmitted) power [1–3, 6–9, 12, 15, 18, 29–31]. The EMI attenuation offered by a shield may depend on the three mechanisms: reflection of the wave from the front face of shield, absorption of the wave as it passes through the shield's thickness, and multiple reflections of the waves at various interfaces as shown in Figure 5. Therefore, SE of EMI shielding materials is determined by three losses, namely, reflection loss (SE_R), absorption loss (SE_A), and multiple reflection losses (SE_M), and can be expressed as [1–3, 9, 18, 27]

$$\begin{aligned}SE_T\,(\text{dB}) &= (SE_R + SE_A + SE_M) = 10\log_{10}\left(\frac{P_T}{P_I}\right)\\ &= 20\log_{10}\left(\frac{E_T}{E_I}\right) = 20\log_{10}\left(\frac{H_T}{H_I}\right),\end{aligned} \tag{1}$$

where P_I (E_I or H_I) and P_T (E_T or H_T) are the power (electric or magnetic field intensity) of incident and transmitted EM waves, respectively. As, the P_T is always less than P_I, therefore, SE is a negative quantity such that a shift towards more negative value means increase in magnitude of SE_T. It is important to note that the SE_T can be expressed as [1, 6, 32, 33]

$$SE_T\,(\text{dB}) = -10\log_{10}\left(\frac{\sigma_T}{16\omega\varepsilon'\mu'}\right) - 8.68t\left(\frac{\sigma_T\omega\mu'}{2}\right)^{(1/2)}, \tag{2}$$

where $\varepsilon'(\mu')$ and $\varepsilon''(\mu'')$ represent real and imaginary parts of complex permittivity (permeability), respectively, that can be correlated with the total conductivity (σ_T) as

$$\sigma_T = (\sigma_{ac} + \sigma_{dc}) = \omega\varepsilon_o\varepsilon'', \tag{3}$$

where σ_{ac} and σ_{dc} are frequency ($f = \omega/2\pi$) dependent (ac) and independent (dc) components of σ_T. The first and second terms in right-hand side of (2) represent attenuations due to reflection and absorption, respectively. Equations (2) and (3) reveal that, for moderately conducting and nonmagnetic materials ($\mu' \sim 1$ and $\mu'' \sim 0$), both electrical conductivity and permittivity are important for enhancement of total shielding effectiveness. In the present system, good dielectric and electric properties are expected to improve EMI shielding performance [1, 3, 34–36].

The EMI shielding plot (Figure 6) revealed that pure AR displays poor EMI shielding response with SE_T value of only −2.4 dB. However, incorporation of AC phase leads to increase in SE value which scales with AC concentration such that composite containing 30% AC displays SE value of −36 dB. It is important to point out that addition of AC leads to improvement in electrical conductivity as well as

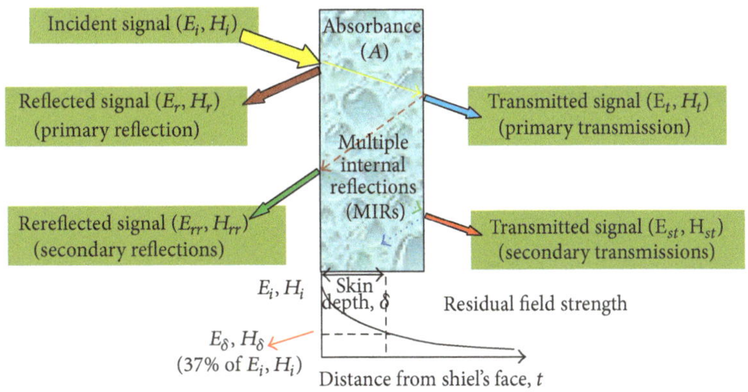

FIGURE 5: Graphical representation of EMI shielding.

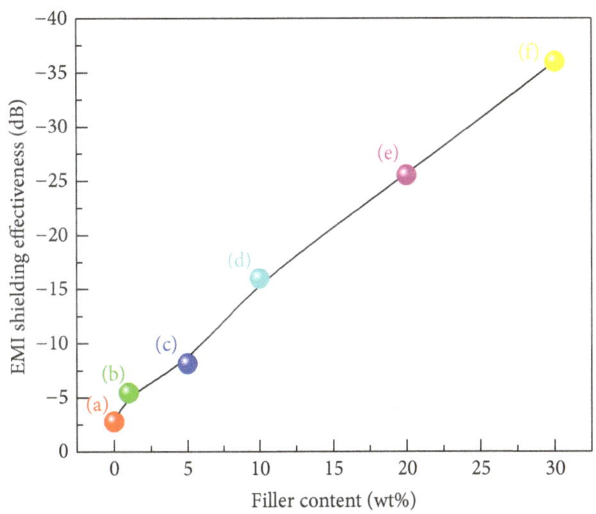

FIGURE 6: EMI shielding effectiveness of (a) pure AR and its composites containing (b) 1 wt%, (c) 5 wt%, (d) 10 wt%, (e) 20 wt%, and (f) 30 wt% of AC.

permittivity, which collectively contributes towards improved antiradiation performance. The above attenuation value is near the total shielding value of −30 dB for phase segregated composites based on 0.66 vol% [37] loading of reduced graphene oxide in thermoplastic matrix. However, the use of activated charcoal (cheap, porous, and electrically conducting filler), compared to other nanofillers (costly, difficult to process due to agglomeration and dispersion issues, and commercially unviable) to obtain comparable EMI shielding performance (at lesser thickness), highlights the technological importance of our work. It should also be noted that the regions where filler is present are electrically conducting (therefore reflecting), whereas the acrylic containing regions are electrically insulating (therefore nonreflecting). These, two regions form a kind of electromagnetically active pseudopore and also contribute towards multiple reflections. Nevertheless, the high attenuation value in our case (i.e., blocking of >99.9% of the incident electromagnetic radiation) demonstrates the technological potential of these composites for making EMI shields.

5. Conclusion

AC/AR composites containing up to 30 wt% AC have been successfully prepared via a facile and scalable solution blending approach. The XRD patterns show superimposed signatures of AC and AR with systematic variation in the peak position and relative intensity. This confirms the incorporation of AC particles inside AR matrix and interactions between phases that were also complemented by Raman spectroscopy results. The morphological investigations revealed that incorporated AC particles regulate the composite's morphology and promote the formation of porous structure. The dielectric measurements revealed that pure AR displays low relative permittivity value (~5) that increases with incorporation of AC and becomes ~79 for 30 wt% AC containing composite, which also displays EMI shielding effectiveness value of −36 dB. This shielding value corresponds to blockage of >99.9% of the incident electromagnetic radiation and suggests that AC/AR composites are promising material for making efficient EMI shields.

Conflict of Interests

The authors declare that there is no conflict of interests regarding the publication of this paper.

References

[1] P. Saini and M. Arora, "Microwave absorption and EMI shielding behavior of nanocomposites based on intrinsically conducting polymers, graphene and carbon nanotubes," in *New Polymers for Special Applications*, A. De Souza Gomes, Ed.A. De Souza Gomes, Ed., InTech, 2012.

[2] H. W. Ott, *Electromagnetic Compatibility Engineering*, John Wiley & Sons, Hoboken, NJ, USA, 2009.

[3] P. Saini, "Electrical properties and electromagnetic interference shielding response of electrically conducting thermosetting nanocomposites," in *Thermoset Nanocomposites*, V. Mittal, Ed., Wiley-VCH Verlag GmbH & Co. KGaA, Weinheim, Germany, 2013.

[4] L. Olmedo, P. Hourquebie, and F. Jousse, *Handbook of Organic Conductive Molecules and Polymers*, vol. 2, John Wiley & Sons, Chichester, UK, 1997.

[5] P. Saini, M. Arora, G. Gupta, B. K. Gupta, V. N. Singh, and V. Choudhary, "High permittivity polyaniline-barium titanate nanocomposites with excellent electro-magnetic interference shielding response," *Nanoscale*, vol. 5, pp. 4330–4336, 2013.

[6] P. Saini, V. Choudhary, B. P. Singh, R. B. Mathur, and S. K. Dhawan, "Enhanced microwave absorption behavior of polyaniline-CNT/polystyrene blend in 12.4-18.0 GHz range," *Synthetic Metals*, vol. 161, no. 15-16, pp. 1522–1526, 2011.

[7] Y. Yang, M. C. Gupta, K. L. Dudley, and R. W. Lawrence, "Conductive carbon nanofiber-polymer foam structures," *Advanced Materials*, vol. 17, no. 16, pp. 1999–2003, 2005.

[8] N. F. Colaneri and L. W. Shacklette, "EMI shielding measurements of conductive polymer blends," *IEEE Transactions on Instrumentation and Measurement*, vol. 41, no. 2, pp. 291–297, 1992.

[9] P. Saini, V. Choudhary, B. P. Singh, R. B. Mathur, and S. K. Dhawan, "Polyaniline-MWCNT nanocomposites for microwave absorption and EMI shielding," *Materials Chemistry and Physics*, vol. 113, no. 2-3, pp. 919–926, 2009.

[10] D. D. L. Chung, "Electromagnetic interference shielding effectiveness of carbon materials," *Carbon*, vol. 39, no. 2, pp. 279–285, 2001.

[11] S. M. Abbas, A. K. Dixit, R. Chatterjee, and T. C. Goel, "Complex permittivity and microwave absorption properties of $BaTiO_3$-polyaniline composite," *Materials Science and Engineering B*, vol. 123, no. 2, pp. 167–171, 2005.

[12] D. D. L. Chung, "Materials for electromagnetic interference shielding," *Journal of Materials Engineering and Performance*, vol. 9, no. 3, pp. 350–354, 2000.

[13] P. Saini, V. Choudhary, N. Vijayan, and R. K. Kotnala, "Improved electromagnetic interference shielding response of poly (aniline)-coated fabrics containing dielectric and magnetic nanoparticles," *Journal of Physical Chemistry C*, vol. 116, pp. 13403–13412, 2012.

[14] P. Singh, V. K. Babbar, A. Razdan, S. L. Srivastava, and R. K. Puri, "Complex permeability and permittivity, and microwave absorption studies of $Ca(CoTi)_xFe_{12-2x}O_{19}$ hexaferrite composites in X-band microwave frequencies," *Materials Science and Engineering B*, vol. 67, no. 3, pp. 132–138, 1999.

[15] N. Li, Y. Huang, F. Du et al., "Electromagnetic Interference (EMI) shielding of single-walled carbon nanotube epoxy composites," *Nano Letters*, vol. 6, no. 6, pp. 1141–1145, 2006.

[16] P. Saini and V. Choudhary, "Enhanced electromagnetic interference shielding effectiveness of polyaniline functionalized carbon nanotubes filled polystyrene," *Journal of Nanoparticle Research*, vol. 15, article 1415, 7 pages, 2013.

[17] S. N. Tripathi, P. Saini, D. Gupta, and V. Choudhary, "Electrical and mechanical properties of PMMA/reduced graphene oxide nanocomposites prepared via in situ polymerization," *Journal of Material Science*, vol. 48, pp. 6223–6232, 2013.

[18] P. Saini, V. Choudhary, K. N. Sood, and S. K. Dhawan, "Electromagnetic interference shielding behavior of polyaniline/graphite composites prepared by in situ emulsion pathway," *Journal of Applied Polymer Science*, vol. 113, no. 5, pp. 3146–3155, 2009.

[19] J. Wu and D. D. L. Chung, "Increasing the electromagnetic interference shielding effectiveness of carbon fiber polymer-matrix composite by using activated carbon fibers," *Carbon*, vol. 40, no. 3, pp. 445–447, 2002.

[20] E. Varrla, S. Venkataraman, and R. Sundara, "Functionalized graphene-PVDF foam composites for EMI shielding," *Macromolecular Materials and Engineering*, vol. 296, no. 10, pp. 894–898, 2011.

[21] Y. Yang, M. C. Gupta, K. L. Dudley, and R. W. Lawrence, "Novel carbon nanotube—polystyrene foam composites for electromagnetic interference shielding," *Nano Letters*, vol. 5, no. 11, pp. 2131–2134, 2005.

[22] P. K. Vallittu, "Flexural properties of acrylic resin polymers reinforced with unidirectional and woven glass fibers," *The Journal of Prosthetic Dentistry*, vol. 81, no. 3, pp. 318–326, 1999.

[23] H.-S. Park, I.-M. Yang, J.-P. Wu et al., "Synthesis of silicone-acrylic resins and their applications to superweatherable coatings," *Journal of Applied Polymer Science*, vol. 81, no. 7, pp. 1614–1623, 2001.

[24] S. D. Khan, M. Arora, C. Puri, M. A. Wahab, and P. Saini, "Synthesis and characterization of acrylic resin/activated carbon composites," *Indian Journal of Pure and Applied Physics*, vol. 52, no. 4, pp. 251–254, 2014.

[25] L. Olmedo, P. Hourquebie, and F. Jousse, "Microwave properties of conductive polymers," *Synthetic Metals*, vol. 69, no. 1-3, pp. 205–208, 1995.

[26] C. K. Das and A. Mandal, "Microwave absorbing properties of DBSA-doped polyaniline/$BaTiO_3$-$Ni_{0.5}Zn_{0.5}Fe_2O_4$," *Journal of Materials Science Research*, vol. 1, no. 1, pp. 45–53, 2012.

[27] P. Saini and V. Choudhary, "Electrostatic charge dissipation and electro-magnetic interference shielding response of polyaniline based conducting fabrics," *Indian Journal of Pure and Applied Physics*, vol. 51, pp. 112–117, 2013.

[28] R. Kumar, S. R. Dhakate, P. Saini, and R. B. Mathur, "Improved electromagnetic interference shielding effectiveness of light weight carbon foam by ferrocene accumulation," *RSC Advances*, vol. 3, no. 13, pp. 4145–4151, 2013.

[29] P. Saini and M. Arora, "Formation mechanism, electronic properties and microwave shielding by nano-structured polyanilines prepared by template free route using surfactant dopants," *Journal of Materials Chemistry A*, vol. 1, no. 31, pp. 8926–8934, 2013.

[30] J. Joo and A. J. Epstein, "Electromagnetic radiation shielding by intrinsically conducting polymers," *Applied Physics Letters*, vol. 65, no. 18, pp. 2278–2280, 1994.

[31] L. W. Shacklette, N. F. Colaneri, V. G. Kulkarni, and B. Wessling, "EMI Shielding of intrinsically conductive polymers," *Journal of Vinyl and Additive Technology*, vol. 14, no. 2, pp. 118–122, 1992.

[32] P. Saini, V. Choudhary, and S. K. Dhawan, "Electrical properties and EMI shielding behavior of highly thermally stable polyaniline/colloidal graphite composites," *Polymers for Advanced Technologies*, vol. 20, no. 4, pp. 355–361, 2009.

[33] P. Saini and V. Choudhary, "Structural details, electrical properties, and electromagnetic interference shielding response of processable copolymers of aniline," *Journal of Materials Science*, vol. 48, no. 2, pp. 797–804, 2013.

[34] Y. K. Hong, C. Y. Lee, C. K. Jeong, D. E. Lee, K. Kim, and J. Joo, "Method and apparatus to measure electromagnetic interference shielding efficiency and its shielding characteristics in broadband frequency ranges," *Review of Scientific Instruments*, vol. 74, no. 2, pp. 1098–1102, 2003.

[35] H.-B. Zhang, Q. Yan, W.-G. Zheng, Z. He, and Z.-Z. Yu, "Tough graphene-polymer microcellular foams for electromagnetic interference shielding," *ACS Applied Materials and Interfaces*, vol. 3, no. 3, pp. 918–924, 2011.

[36] Y. Yang, M. C. Gupta, K. L. Dudley, and R. W. Lawrence, "A comparative study of EMI shielding properties of carbon nanofiber and multi-walled carbon nanotube filled polymer composites," *Journal of Nanoscience and Nanotechnology*, vol. 5, no. 6, pp. 927–931, 2005.

[37] D.-X. Yan, H. Pang, L. Xu et al., "Electromagnetic interference shielding of segregated polymer composite with an ultralow loading of *in situ* thermally reduced graphene oxide," *Nanotechnology*, vol. 25, no. 14, Article ID 145705, 2014.

14

Studies on Structural and Conducting Properties of Goethite Nanoparticles Doped HPMC Polymer Films

Thejas Gopal Krishne Urs, Mahadevaiah Dasaiah, and Rudrappa Somashekar

Department of Physics, University of Mysore, Manasagangotri, Mysore 570006, India

Correspondence should be addressed to Rudrappa Somashekar; rs@physics.uni-mysore.ac.in

Academic Editor: Yeong-Soon Gal

We have carried out systematic studies on Goethite nanoparticles doped (dispersed) HPMC polymer films to quantify the changes in their structural properties, the conductivity, and IR absorbance of films. It is observed that the addition of nanoparticles has significant influence on changing structural parameters and hence enhancing the conductivity of the polymer composite considerably. These polymer composites are water soluble.

1. Introduction

Recent works have focused on the designed synthesis and exploration of applications of binary oxide films, such as TiO_2–CeO_2 [1], TiO_2–ZnO [2], TiO_2–SiO_2, and TiO_2–WO_3 [3]. These were considered to be effective semiconductor, advanced materials, heterogeneous catalysts, and catalyst supports. The nanoparticles doped polymers have wide applications in devices transducers, gas sensors, and as catalysts [4–6]. This is expected as they are inexpensive. The properties of the materials are found to vary with their constituents. There are several methods of preparation of nanoparticles, which we dope to the polymer matrix and among them hydrothermal approach [7, 8] has great advantages. Nanoparticles doped polymers have created a great interest in the field of research and our work is based on the preparation of a new material to the class of polymer nanocomposites. We have used hydroxypropyl methylcellulose (HPMC) as the polymer base in this work, which is a water soluble polymer and is widely used in the field of medicine due to its solubility in water and high inertness, possessing a nontoxic nature. It is mainly used in manufacturing of capsules due to its property of controlled drug release [9, 10]. It is easily accepted and has a variety of viscosities and types of substitution, and the necessary doses are easy to formulate [11]. HPMC is a potential biodegradable cellulose derivative that produces films with good oil barrier and excellent mechanical properties and it is tasteless. The addition of nanoparticles to a polymer changes the structural and other properties like conductivity and IR absorbance considerably from its parent polymer. In this paper we report preparation of materials and characterization studies using X-ray, conductivity, and FT-IR techniques.

2. Preparation

2.1. Preparation of Nanoparticles. The hydrothermal technique has been the best technique to prepare Goethite (FeO(OH)) particles of desired sizes with homogeneity in composition and high degree crystallinity. Goethite (FeO(OH)) nanoparticles were prepared with 1% to 5% weight of Goethite and were tested for their photodegradation efficiency. The synthesis has been carried out using Morey autoclaves (designed and fabricated at the Department of Environmental Science, University of Mysore, Mysore) provided with Teflon liners. The conditions selected for the synthesis of Goethite particles are at temperature of $T = 200^{\circ}C$ and a pressure less than 50 bars. The starting material $FeCl_3$ was taken in a Teflon liner. A required amount of an effective mineralizer was added to the mixture. The mixture is stirred thoroughly to obtain a homogenous and relatively viscous solution. The Teflon liner is placed in an autoclave and then the autoclaves were placed in the oven and kept at a predetermined temperature for a desired period (12–72 hrs). The mineralizer used was 1 M NaOH. After the experimental runs the autoclaves were quenched and the liner is taken out. The contents of the liner are thoroughly washed in double distilled water repeatedly which is later dried in an oven (at 40°C) before using it for the study. The precursors were prepared using double distilled water. It is one of the thermodynamically most stable iron oxides at ambient temperatures and it is, for this reason, often found as an end product of many transformations.

2.2. Preparation of the Polymer Nanocomposites. The nanoparticles doped polymer films are prepared by solution casting method. The HPMC solution of 5% was prepared at room temperature and, for this, the different proportions of 0.01, 0.02, and 0.03 grams of Goethite are added in the separate beakers containing HPMC pure solutions of the same amount and were stirred for around 1 hour each and then it was poured into glass plates and allowed to cast for desired number of days. A pure HPMC film was also prepared in the same procedure for comparative studies.

3. Experimental

3.1. XRD Recording. The casted polymer films were cut into desired pieces of required size (2 cm × 2 cm) and were taken in the sample holder of Rigaku-Denki miniflex II desktop X-Ray diffractometer with the settings of 30 kV and 15 mA, scanning rate of 5° per minute, and for the range of 5° to 40° with step size of 0.02° being recorded. The integral breadth of the diffraction peaks is related to the apparent size of the crystals and to their microstrains. If the size and strain broadening exist simultaneously, then crystallite size and strain can be calculated by Williamson-Hall plot (W-H plot). For relative comparison of the parameters, W-H plot is a reasonably reliable one [12]. The W-H plot considers both limited size of the crystals and the presence of crystallographic distortions which leads to Lorentzian intensity distributions. The slope of the W-H plot represents the average strain in the crystal, whereas intercept with the y-axis gives the crystallite size. The Williamson-Hall relation is given by

$$\beta\frac{\cos\theta}{\lambda} = \frac{1}{D} + 4\varepsilon\frac{\sin\theta}{\lambda}, \quad (1)$$

where "β" is the full width at half maximum (FWHM) of the peak measured in radians, "D" is average crystallite size, and "ε" is average lattice strain. The obtained XRD plots and the values of crystallite size and average strain are given in Figure 1 and Table 1, respectively.

3.2. Conductivity. The conductivity measurements of samples were determined by using HIOKI LCR-Hi tester 3532-50 model available. Our samples were taken in the sample holder provided in the instrument. This instrument gives the transconductance (G) of the sample which is dependent on its thickness and the area. The conductivity (σ) of the sample can be obtained by the relation,

$$\sigma = \frac{t * G}{A}, \quad (2)$$

where "t" is the thickness and "A" is the area of the sample.

This instrument gives the transconductance values for different frequencies from 50 Hz to 5 MHz. The obtained transconductance values for various frequencies were evaluated to get the conductivity results for the same corresponding frequencies and a graph of conductivity for various frequencies was plotted as a function of its concentration.

This plot and tabulation are shown in Figure 2 and Table 2, respectively. At 1 kHz, the variation of conductivity with dopant is given in Figure 3.

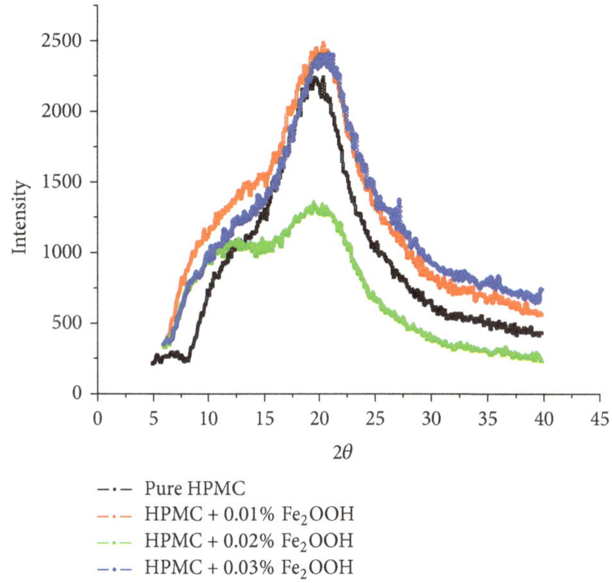

Figure 1: The consolidated XRD plots of pure HMPC and Goethite nanoparticles doped HPMC polymers.

Table 1: W-H plot results from the obtained XRD pattern.

Samples	Crystallite size (Å)	Percentage strain
Pure	11.44	1.90
0.01	10.04	4.48
0.02	11.41	1.51
0.03	15.51	6.16

3.3. FT-IR. The FT-IR spectrum for these composites was recorded using JASCO FT-IR 4100 type A instrument, for the scanning range from 4000 cm^{-1} to 400 cm^{-1}. The infrared spectrum taken for pure HPMC and Goethite doped HPMC polymers is given in Figure 5.

4. Results

It is evident from Table 1 that the average crystallite size of the pure HPMC is found to decrease first with the

Table 2: Conductivity values of the samples at different AC frequencies.

Samples	Conductivity σ in Sm^{-1} ($\times 10^{-6}$)			
	1 kHz	10 kHz	100 kHz	1 MHz
Pure HPMC	0.062	0.175	7.84	4.65
0.01% Goethite: HPMC	9.63	19.2	23.5	28.1
0.02% Goethite: HPMC	18.5	47.6	95.1	139.0
0.03% Goethite: HPMC	18.6	47.8	92.9	137.0

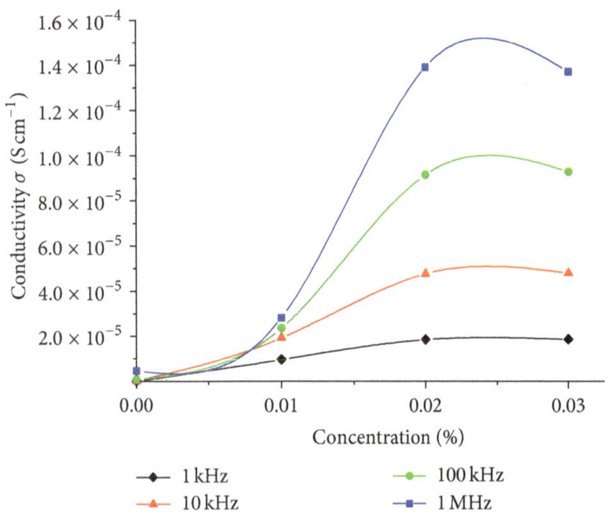

Figure 2: The plot of ionic conductivity of the pure HPMC and Goethite nanoparticles doped HPMC polymers for various frequencies as a function of its concentration.

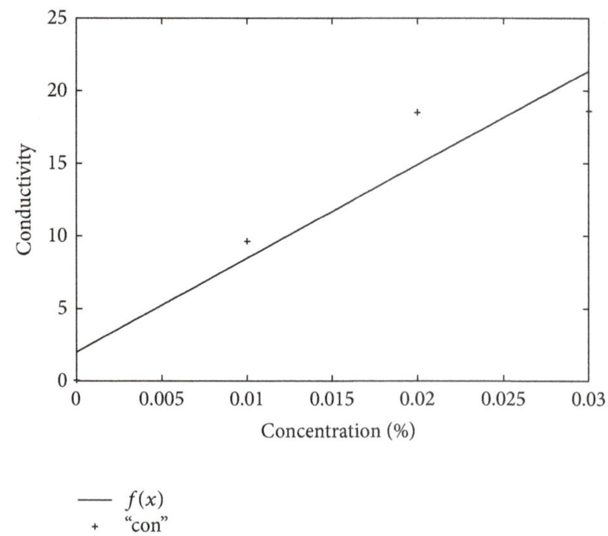

Figure 3: The plot of conductivity as a function of dopant concentration.

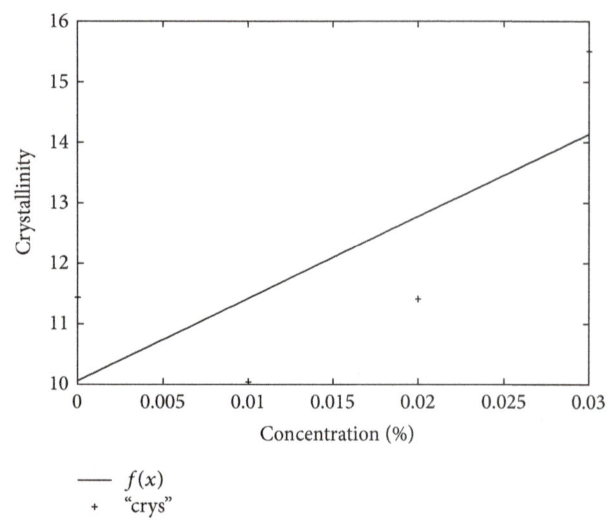

Figure 4: The variation of crystallinity as a function of dopant concentration.

dopant concentration of 0.01% but later it increases with the concentration of Goethite nanoparticles. This is because the addition of nanoparticles to the polymer matrix affects the total crystallite size of the parent polymer, but later the crystallite size of the nanoparticles adds up with the overall crystallite size resulting in the increase of composite crystallite size with the concentration. It is seen that this variation makes an interesting change in conducting property of the polymer composite. The variation of crystallinity with dopant is given in Figure 4. The total strain in the crystal is due to the mismatch between the material and the substrate and other crystallographic defects that may be present in the crystal.

For pure HPMC polymer sample, 3 reflections were considered which were centered at 13.06, 19.67, and 25.47 on the 2θ axis; for these corresponding θ values, d_{hkl} was calculated using Bragg's equation and by the relation for monoclinic crystal lattices; that is,

$$\frac{1}{d_{hkl}^2} = \frac{h^2}{a^2 \sin^2 \beta} - \frac{2hl \cos \beta}{a^2 \sin^2 \beta} + \frac{l^2}{c^2 \sin \beta} + \frac{k^2}{b^2}. \quad (3)$$

The respective h, k, l values were determined. Here λ is the wavelength of X-ray, "a" is 0.787 nm, "b" is 0.817 nm, "c" is 1.034 nm, and "β" is 96.3°, which are the parameters of cellulose [13]. Leonardi et al. have reported that the crystallite size is different along different Bragg angles [14]; considering this and taking all the above stated parameters and Bragg's reflections, an attempt was done to sketch crystallite shape of pure HPMC polymer showing all three reflections along with unit cell crystallite in Figure 6.

The conductivity of HPMC pure film has been enhanced with the addition of Goethite nanoparticles and is significantly high for the concentration of 0.02% but on an average conductivity increases with the increase in dopant percentage. Plot of conductivity versus concentration for various frequencies is shown in Figure 2, which shows the respective increase in the conducting property of HPMC polymer with increase in its concentration.

From the FT-IR spectrum we have studied the nature of chemical bonds in polymers that can be studied through

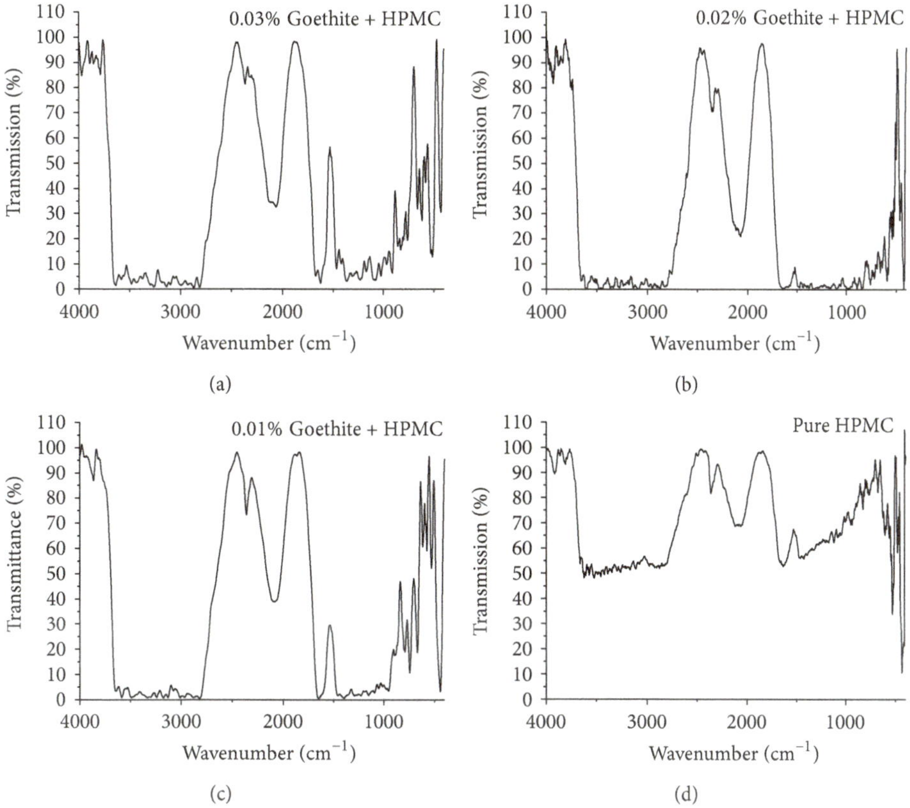

FIGURE 5: IR spectrum obtained for the pure HPMC and Goethite nanoparticles doped HPMC polymers and Goethite nanoparticles doped HPMC polymers.

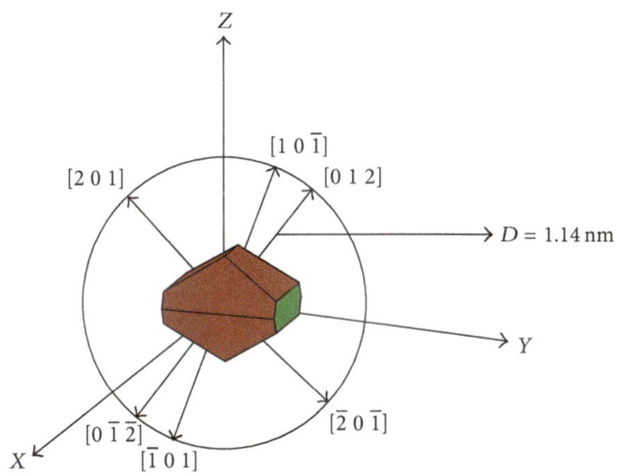

FIGURE 6: Unit cell crystallite shape with crystallite size along different Bragg angles.

the characterization of the vibration modes by infrared spectroscopy [15]. We have calculated few parameters such as equilibrium vibration frequency ($\bar{\upsilon}_e$), anharmonicity constant, fundamental and first overtone transition frequencies, zero point energy, and force constant of O–H stretch. For pure HPMC the fundamental and first overtone transitions of O–H bonds are centered at 3620 cm^{-1} and 2117.5 cm^{-1}, respectively. The equilibrium vibration frequency of the chain about this bond is calculated by solving the equations given below [16].

For the pure HPMC polymer:

Frequency of the fundamental transition

$$\bar{\upsilon}_e (1 - 2\chi_e) = 2117.5. \quad (4)$$

Frequency of the first overtone

$$2\bar{\upsilon}_e (1 - 3\chi_e) = 3620.7. \quad (5)$$

The obtained equilibrium vibration "$\bar{\upsilon}_e$" for pure HPMC is 2731.68 cm^{-1}. In the same manner the equilibrium vibration for all the other samples was calculated.

The zero-point energy is given by the following relation:

$$\varepsilon_0 = \frac{\bar{\upsilon}_e}{2(1 - \chi_e/2)}, \quad (6)$$

where $\bar{\upsilon}_e$ is the equilibrium vibration and χ_e is the anharmonicity constant.

The force constant is calculated using the relation

$$k = 4\pi^2 \mu c^2 \bar{\upsilon}_e^2, \quad (7)$$

where "μ" is the reduced mass of the system, "c" is the light velocity, and "$\bar{\upsilon}_e$" is equilibrium vibration frequency.

TABLE 3: IR parameters calculated.

Samples	Fundamental transition frequency (cm^{-1})	First overtone frequency (cm^{-1})	Equilibrium vibration frequency $\bar{\upsilon}_e$ (cm^{-1})	Anharmonicity constant χ_e	Zero-point energy ε_0 (cm^{-1})	Force constant "K" (N/m)
Pure	2117.46	3620.7	2731.68	11.2×10^{-2}	1289.35	412.65
0.01%	2096.24	3587.91	2700.81	11.2×10^{-2}	1274.7	403.37
0.02%	2071.17	3604.30	2609.21	10.3×10^{-2}	1236.76	376.42
0.03%	2062.5	3618.76	2568.74	9.85×10^{-2}	1220.15	364.89

Using all these equations the IR spectrum parameters which were quoted before were calculated and tabulated in Table 3. From these results it is evident that the force constant between O–H bond is found to be decreased with increase in dopant concentration and also the zero-point energy ε_0 is also found shifted towards the lower frequency with the increased dopant concentration. The anharmonicity constant χ_e and the equilibrium vibration frequency "$\bar{\upsilon}_e$" do change in the same manner with respect to the dopant concentration. From this we can say that there are no such changes encountered in the structural properties of Goethite nanoparticles doped HPMC polymer as that found in the microstructural parameters such as crystallite size and strain calculated using WAXS data [15].

5. Conclusions

The conductivity of Goethite doped HPMC polymers has been increased with the increase in its concentration. But this increase in the conducting property of these doped polymers is not linear with the increase in the concentration of dopants. This nonlinearity in conducting property of Goethite: HPMC polymers is due to its crystallite size variation. The variation of crystallite sizes has been tabulated in Table 2 and shown in Figure 4. We have obtained the conductivity of pure HPMC at room temperature around 6.28×10^{-6} Sm^{-1} and the conductivity for doped ones 9.63×10^{-4} to 1.86×10^{-3} Sm^{-1} which shows that the conductance of these doped HPMC polymers is more than that of silicon and conducting polymers like polythiophene and polyacetylene which possess a conductance in the range 4.3×10^{-4} and 10^{-10} to 10^{-8} Sm^{-1}, respectively.

Conflict of Interests

The authors declare that there is no conflict of interests regarding the publication of this paper.

Acknowledgments

The authors thank UGC (University Grants Commission), India, for the grants under UPE (University with potential of Excellence) and CPEPA (Centre with potential for Excellence in particular area) research program for the University of Mysore, India.

References

[1] K. Nakagawa, Y. Murata, M. Kishida, M. Adachi, M. Hiro, and K. Susa, "Formation and reaction activity of CeO_2 nanoparticles of cubic structure and various shaped CeO_2-TiO_2 composite nanostructures," *Materials Chemistry and Physics*, vol. 104, no. 1, pp. 30–39, 2007.

[2] K. H. Yoon, J. Cho, and D. H. Kang, "Synthesis and characterization of TiO2-ZnO nanocomposite by a two-step chemical method," *Materials Research Bulletin*, vol. 34, pp. 1451–1461, 1999.

[3] I. Shiyanovskaya and M. Hepel, "Bicomponent WO_3/TiO_3 films as photoelectrodes," *Journal of the Electrochemical Society*, vol. 146, no. 1, pp. 243–249, 1999.

[4] P. Q. Mantas and J. L. Baptista, "The barrier height formation in ZnO varistors," *Journal of the European Ceramic Society*, vol. 15, no. 7, pp. 605–615, 1995.

[5] F. C. M. van de Pol, F. R. Blom, and T. J. A. Popma, "R.f. planar magnetron sputtered ZnO films I: structural properties," *Thin Solid Films*, vol. 204, no. 2, pp. 349–364, 1991.

[6] M. J. Mayo, "Processing of nanocrystalline ceramics from ultrafine particles," *International Materials Reviews*, vol. 41, no. 3, pp. 85–115, 1996.

[7] L. Vayssieres, "Growth of arrayed nanorods and nanowires of ZnO from aqueous solutions," *Advanced Materials*, vol. 15, no. 5, pp. 464–466, 2003.

[8] M. Guo, P. Diao, and S. Cai, "Hydrothermal growth of well-aligned ZnO nanorod arrays: dependence of morphology and alignment ordering upon preparing conditions," *Journal of Solid State Chemistry*, vol. 178, no. 6, pp. 1864–1873, 2005.

[9] H. Konno, T. Handa, D. Alonzo, and L. Yaylor, "Effect of polymer type on the dissolution profile of amorphous solid dispersions containing felodipini," *European Journal of Pharmaceutics and Biopharmaceutics*, vol. 70, no. 2, pp. 493–499, 2008.

[10] E. Karavas, E. Georgarakis, and D. Bikiaris, "Application of PVP/HPMC miscible blends with enhanced mucoadhesive properties for adjusting drug release in predictable pulsatile chronotherapeutics," *European Journal of Pharmaceutics and Biopharmaceutics*, vol. 64, no. 1, pp. 115–126, 2006.

[11] H. McPhillips, D. Q. M. Craig, P. G. Royall, and V. L. Hill, "Characterisation of the glass transition of HPMC using modulated temperature differential scanning calorimetry," *International Journal of Pharmaceutics*, vol. 180, no. 1, pp. 83–90, 1999.

[12] J. I. Langford and D. Louër, "Powder diffraction," *Reports on Progress in Physics*, vol. 59, no. 2, article 131, 1996.

[13] O. M. Samir and R. Somashekar, "Intrinsic strain effect on crystal and molecular structure of (dch32) cotton fiber," *Powder Diffraction*, vol. 22, no. 1, pp. 20–26, 2007.

[14] A. Leonardi, M. Leoni, S. Siboni, and P. Scardi, "Common volume functions and diffraction line profiles of polyhedral domains," *Journal of Applied Crystallography*, vol. 45, pp. 1162–1172, 2012.

[15] S. Sangappa, S. Asha, T. Demappa, G. Sanjeev, P. Parameswara, and R. Somashekar, "Spectroscopic and thermal studies of 8 MeV electron beam irradiated HPMC films," *Nuclear Instruments and Methods in Physics Research. Section B: Beam Interactions with Materials and Atoms*, vol. 267, no. 14, pp. 2385–2389, 2009.

[16] J. Davenas, X. L. Xu, G. Boiteux, and D. Sage, "Relation between structure and electronic properties of ion irradiated polymers," *Nuclear Instruments and Methods in Physics Research B*, vol. 39, no. 1–4, pp. 754–763, 1989.

15

The Effects of Antioxidants Content on Mechanical Properties and Water Absorption Behaviour of Biocomposites Prepared by Single Screw Extrusion Process

Mohd Hafizuddin Ab Ghani, Mohd Nazry Salleh, Ruey Shan Chen, Sahrim Ahmad, Mohd Rashid Yusof Hamid, Ismail Hanafi, and Nishata Royan Rajendran Royan

Material Science Programme, School of Applied Physics, Faculty of Science and Technology, The National University of Malaysia, 43600 Bangi, Malaysia

Correspondence should be addressed to Mohd Hafizuddin Ab Ghani; mr_merce7@yahoo.com

Academic Editor: Yves Grohens

The performance of hybrid fillers between rice husk and sawdust filled recycled high density polyethylene (rHDPE) with the presence of antioxidants (IRGANOX 1010 and IRGAFOS 169, with the ratio of 1 : 1) was investigated. The biocomposites with 30 wt% of matrix and around 70 wt% of hybrid fillers (rice husk and sawdust) and different antioxidants' contents (0 to 0.7 wt%) were prepared with single screw extruder. Increasing the amount of antioxidants in biocomposites reduced the modulus of elasticity and modulus of rupture on flexural testing. The addition of antioxidants increased the tensile and impact strength of biocomposites. From the study, samples with 0.5 wt% of antioxidants produce the most reasonable strength and elasticity of biocomposites. Furthermore, the effect of antioxidants content on water uptake was minimal. This might be caused by the enhanced interfacial bonding between the polymer matrix and hybrid fillers, as shown from the morphology by using scanning electron microscopy (SEM).

1. Introduction

The large amount of disposable bottles presently produced makes the search for alternative procedures for recycling of these materials imperative. As chemical processing is most often costly and sometimes aggressive to the environment, a possible solution is the recycling of such material by thermomechanical techniques [1]. Great attention has been focused on the addition of natural fibres as reinforcement for plastics, replacing glass fibres and other synthetic materials. This is due to environmental concerns and also for providing a good combination of high performance, great versatility, and processing advantages at lower cost. Some drawbacks reduce the application of natural fibers as reinforcement in polymers, such as the incompatibility between fibres and polymer matrices, the tendency to form aggregates during processing, and the poor resistance to moisture. Incompatibility of components is responsible for poor thermal and mechanical properties. The stress transfer from the matrix to the fibre depends on fibre-fibre and fibre-matrix interactions [2].

To overcome this problem, it is necessary to promote polymer modification with polar groups (such as maleic anhydride or glycidyl methacrylate stearic acid) to enhance the adhesion between the composite components. The coupling agent more often used for this application is a polyethylene graft maleic anhydride (MAPE). Interactions between the anhydride groups of maleated coupling agents and the hydroxyl groups of natural fibres can overcome the incompatibility and increase tensile and flexural strengths of natural fibre thermoplastic composites [2].

In order to protect the biocomposites against thermooxidative degradation, antioxidants are used. In this study, two types of antioxidants, namely, IRGANOX 1010 (Pentaerythritol Tetrakis(3-(3,5-di-tert-butyl-4-hydroxyphenyl)propionate) and IRGAFOS 168 (Tris(2,4-di-tert-butylphenyl)phosphate), are used. IRGANOX 1010 is a sterically hindered phenolic antioxidant that is highly effective, nondiscoloring stabilizer for organic materials such as plastics, synthetic fibers, elastomers, adhesives, waxes, oils, and fats. This type of antioxidant can be applied in polyolefins, such as polyethylene, polypropylene, polybutene, and olefin copolymers such as ethylene-vinyl acetate copolymers. Also, its use is recommended in other polymers such as polyacetals, polyamides, and polyurethanes [3]. IRGAFOS 168 antioxidant is also known as a trisarylphosphite processing stabiliser. It is a highly effective, low-volatile, and hydrolysis resistant antioxidant for coating resins. It protects the resin polymer against oxidation during resin synthesis, manufacturing of the paint, processing (thermal curing and over baking), and the designed life time of the final coating. IRGAFOS 168 provides excellent protection against discoloration and change of physical properties caused by excessive heat exposure. In application, it provides outstanding thermal and color stability to a variety of coating resins upon exposure to high temperature during synthesis and subsequent processing [4].

Regarding economic and ecological factors, wood fibre plastic was established as a standard material, but unfortunately the price of raw wood fibre has increased by approximately 25–30% for the last two years and the sources are being limited. Therefore, new sources could be suitable alternatives for wood fibre. Grain byproducts, that is, rice husk and wheat husk, have sufficient fibre values, meaning that the proper utilization of waste materials will provide cheap engineering materials [5]. Recent interest in the environmental impact of polymer-based materials has led to the development of new products prepared with recycled polymers containing biodegradable materials. Lignocellulosic plastic composites constitute an important set within this kind of materials showing several advantages over traditional mineral-filled plastic composites such as low density, low production costs, biodegradability, and renewability. Hardness, stiffness, and dimensional stability of plastics have also been improved by incorporation of lignocellulosic fillers [6].

The combination of wood and plastics (thermoplastic) can produced an extrudable, injectable, or thermoformable composite. That new composite can be used in many applications especially in nonstructural industry. The wood industry can expand into new sustainable markets with the formation of a new class of composites with the combination of the wood industry and the plastics industry. The wood component, usually a flour or fiber, is combined with a thermoplastic to form an extrudable, injectable, or thermoformable composite that can be used in many nonstructural applications [7].

2. Experimental

2.1. Materials. The raw materials such as matrix and fillers were obtained from RePlas Sdn. Bhd. The matrix is recycled high density polyethylene (rHDPE). The fillers are mixed between sawdust (40 mesh) and rice husk (60 mesh) to create hybrid filler. And 3% of MAPE was added as a compatibilizer to improve the compatibility of hydrophobic polymer and hydrophilic hybrid fillers and thus improves the interfacial bonding between matrix and fillers as discussed by Rosa et al. [6]. The antioxidants (IRGANOX 1010 and IRGAFOS 168 as a ratio of 1 : 1 from Ciba) used in this study had been varied to examine their presence on mechanical properties and water absorption behavior. Five samples were produced according to the formulation in Table 1.

Table 1: Formulation of biocomposite.

rHDPE (wt%)	Additives (wt%)		Fillers (wt%)	
	Antioxidants	MAPE	Rice husk	Sawdust
30	0.0	3	45	22.0
30	0.1	3	45	21.9
30	0.3	3	45	21.7
30	0.5	3	45	21.5
30	0.7	3	45	21.3

2.2. Processing. Rice husk was oven dried at 80°C for 24 h in order to reduce the moisture content to approximately 1–5% and then stored in sealed plastic bag before compounding. Biocomposites were processed in a single screw extruder (ULTRASLIT OHP 500) with an L/D ratio of 28. In all cases polymer matrix, fillers, antioxidants, and compatibilizing agent were fed together into the extruder hopper. The extrusion was conducted at temperature (from feed to die) of 160°C, 170°C, 170°C, and 180°C with a screw speed of 70 rpm. The extruded strand was cooled at room temperature and pelletized.

WPC test specimens were produced using hot and cold press processes (LP50, LABTECH Engineering Company, LTD). Before hot and cold press processes, WPC pellets were ground using grinder to make it fine to suit with the mold. The ground pellets were then put in the mold size 14 mm × 14 mm × 3 mm. The temperatures of the hot press were set 140°C for upper platen and lower platen, respectively. The period of preheating, venting, and full pressing were set at 3, 2, and 9 minutes, respectively. The cold press was set at 5 minutes to cool the specimens and the temperature used was 25°C. The pressure used to press the specimens was set at 1000 psi for both hot and cool press. After finishing the molding processes, the specimens had dried in oven for 24 hours with the temperature 70°C in order to remove any moisture left. The specimens were then cut according to ASTM specifications.

2.3. Testing

2.3.1. Tensile. The tensile tests were carried out in a universal testing machine (Testometric Universal) with cross head speed of 5 mm/min, according to ASTM D638-03 standard for Young's modulus and tensile strength values. Ten samples for each formulation are tested.

2.3.2. Flexural. The flexural tests were carried out following the ASTM D256-05 standard at room conditions using a universal testing machine (Testometric Universal) with a loading rate of 2.8 mm/min for the modulus of rupture

(MOR) and modulus of elasticity (MOE) values. Ten samples for each formulation are tested.

2.3.3. Impact. The impact test was carried out according to ASTM D 256-05 standard at room conditions using an Izod-Charpy (Ray-Ran) testing machine. The weight and velocity of the pendulum used were 0.452 kg and 3.46 ms^{-1}, respectively. All the samples were notched on the center of one longitudinal side according to the ASTM D 256-05. For each treatment level, ten replications were tested. Ten samples for each formulation are tested.

2.3.4. Water Absorption. The water absorption test was performed following the ASTM D 570-98 standard. Water absorption was determined by the weight gain relative to the dry weight of the samples. Water absorption of the sample was recorded after 2 h, 12 h, and 24 h until 2 months of immersion in distilled water at room temperature, 24°C. Five specimens of each formulation were dried in an oven for 24 h at 80 ± 2°C. The dried specimens were weighed with precision of 0.001 g. All specimens were immersed in distilled water. After reaching the time, specimens were removed from the distilled water, the surfaces of specimens were wiped off using tissue, and wet weight values were determined with the same precision (0.001 g). Water absorption (M) percent was calculated using the following formula:

$$M = \left[\frac{m_t - m_o}{m_o}\right] \times 100, \quad (1)$$

where m_o and m_t represent the oven-dry weight and weight after time t, respectively. The same method was applied to study water absorption of biocomposites by using sea water.

2.3.5. Morphology. The scanning electron microscopy (SEM) was used to evaluate the fractured surface of the biocomposites. The microscope was operated at a voltage of 15 kV. The surface was sputtered with a thin gold layer.

3. Results and Discussion

3.1. Tensile Test. Figures 1 and 2 represent graph of tensile strength and Young's modulus, respectively. From the results of Figure 1, it is shown that samples with antioxidants content have higher result compared to control sample (without antioxidants content). As can be seen from the graph, sample with 0.5 wt% of antioxidants gives the highest value which is 8.1 MPa, 11% higher than control sample. The tensile strength markedly increased with increasing antioxidants content by creating good interfacial bonding between the hydrophobic matrix polymer hydrophilic filler and coupling agent. The sufficient amount of antioxidants content can increase the tensile strength of biocomposites [7], while Chong et al. [8] explain that the decrement of tensile strength on biocomposites is due to the weak bonding between the matrix and filler. In addition, filler agglomerations also lead to uneven dispersion in matrix, thus weakening the tensile strength of the material.

Graph of Young's modulus shows decrement trend from control sample to 0.7 wt% antioxidants content. The highest

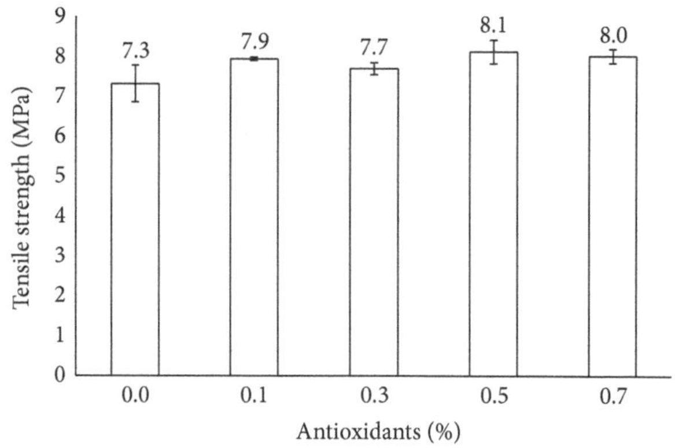

FIGURE 1: Tensile strength.

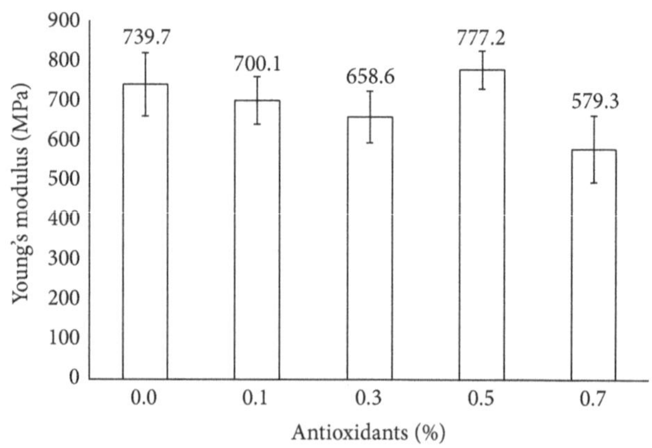

FIGURE 2: Young's modulus.

and lowest values are 777.2 MPa and 579.3 MPa recorded from 0.5 wt% and 0.7 wt% antioxidants' contents, respectively. Sample with 0.5 wt% antioxidants content gives unexpected increment maybe due to contaminants in the sample. The composites produced by using recycled plastics should expect contamination effect in matrix that can be one of the drawbacks in this field [7].

3.2. Flexural Test. Figures 3 and 4 illustrate the modulus of elasticity (MOE) and modulus of rupture (MOR), respectively. The results show the trend that addition of antioxidants reduces the values of MOE and MOR. These results were in contradiction with our previous study [9]. Our last study showed that the values of MOR and MOE were increased by increment amount of antioxidants content. These differentiations can be explained by using different types of extruder machine that will give the different flexural testing values.

Graph modulus of elasticity shows the decrement trend from control sample to sample with content of 0.5 wt% of antioxidants. There is slight increment value shown by sample content of 0.7 wt% of antioxidants. The control sample gives strength of 2548.5 MPa, while sample with 0.1 wt% antioxidants content has 2302.8 MPa. This value shows the highest reading among all the five samples tested. The lowest

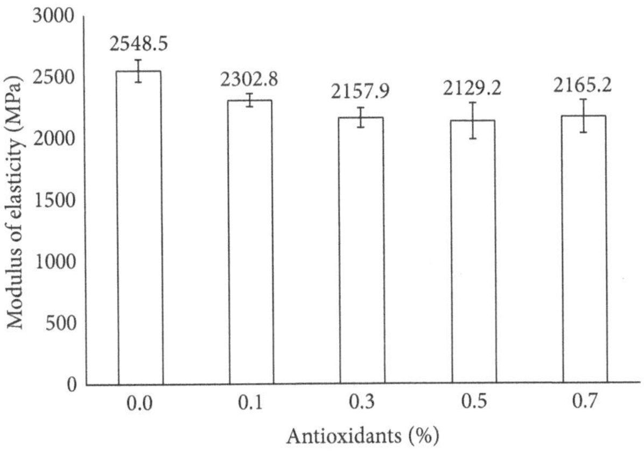

FIGURE 3: Modulus of elasticity.

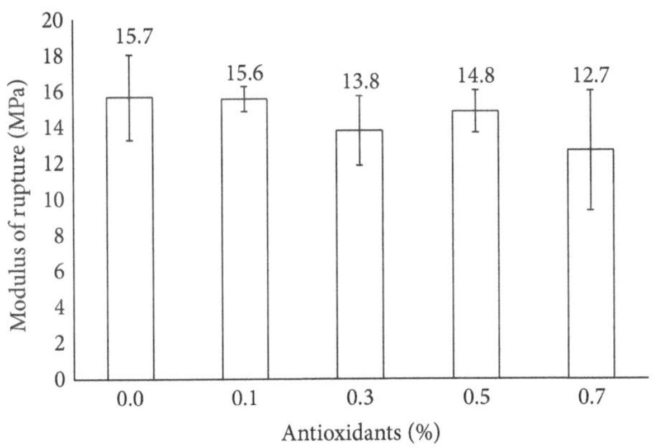

FIGURE 4: Modulus of rupture.

MOE reading was recorded by sample content of 0.5 wt% of antioxidants which is 2129.2 MPa. Percentage differentiation between the highest and the lowest values of MOE is approximately 17%.

Mixing recycled high density polyethylene as a matrix, wood powder as filler, and coupling agent with a mixture ratio of 47 : 50 : 3 gives MOE value equal to or higher than biocomposites produced by using virgin plastic. In addition, this type of mixture can produce strong, stable composites and can be used as building materials [10]. According to Chen et al. [11], MOE of composites increased with increasing amount of wood particles content up to 60 wt%. However, if more than 60 wt% of wood particles mix in sample this will weaken the mechanical properties of the composite.

The modulus of rupture also shows the decrement trend on plotted graph. The control sample recorded the highest value for MOR which is 15.7 MPa. Sample containing 0.1 wt% of antioxidants decreased 0.6% of control sample value and recorded 15.6 MPa for MOR. Sample with 0.3 wt% antioxidants content reduces by 12% compared to control sample and gives the MOR value of 13.8 MPa. Another two samples, namely, 0.5 wt% and 0.7 wt%, of antioxidants content give 14.8 MPa and 12.7 MPa, respectively, for MOR. These two values show the decreasing trend as much as 5% and 19% from control sample, respectively.

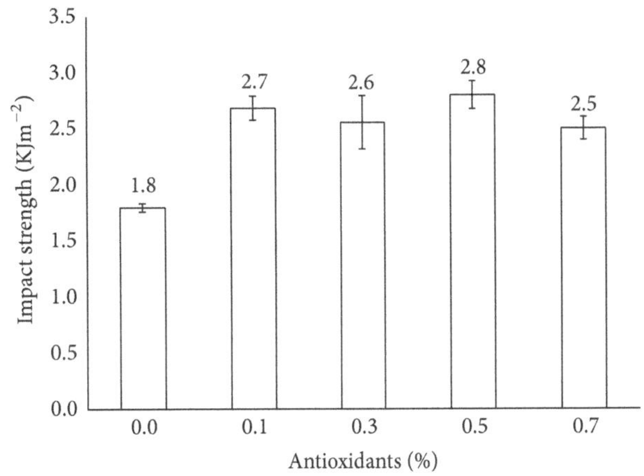

FIGURE 5: Impact strength.

The presence of antioxidants prevents free radical atom from diffusing in biocomposites, while the density of material acts as a gatekeeper to control the number of oxygen molecules penetrating into the matrix. Thus, production of high density biocomposites can reduce the existence of excess oxygen molecules in composite material. A small amount of oxygen molecules in biocomposites will reduce the oxidative degradation process and consequently decrease the flexural values [7].

3.3. Impact Test. Figure 5 shows the impact strength graph. The graph has an increment trend from control sample to the fifth sample with the content of 0.7 wt% of antioxidants. Although there is a slight decrement shown from sample with the content of 0.3 wt% of antioxidants, this result is still higher compared to the control sample.

From the graph, the control sample gives the value 1.8 KJm^{-2} which is 36% less than the highest value recorded from sample with the content of 0.5 wt% antioxidants. The highest value for impact testing is 2.8 KJm^{-2} as plotted in the graph above. This result could be explained by chemical composition of fiber that had been used. Wood fiber contains high cellulose and hemicellulose; these two types of cell are hydrophilic. Any moisture penetrated into samples while processing or assembling will affect the values of testing. Besides that, poor adhesion between matrix and hybrid fibers generates void spaces at interfacial of polymer-filler. These lead to lower impact strength of the samples [7].

3.4. Water Absorption Test. Figures 6 and 7 show graph of the water absorption for samples immersed in two types of water, namely, distilled and sea water, respectively. From the graphs, both show the same increment trend until reaching the optimum water absorption of samples. In this study, the amounts of antioxidants added could affect the bonding between the matrix and fibres. Addition of antioxidants helped in decreasing the water uptake of the biocomposites. This is because the antioxidants as mineral fillers have distributed

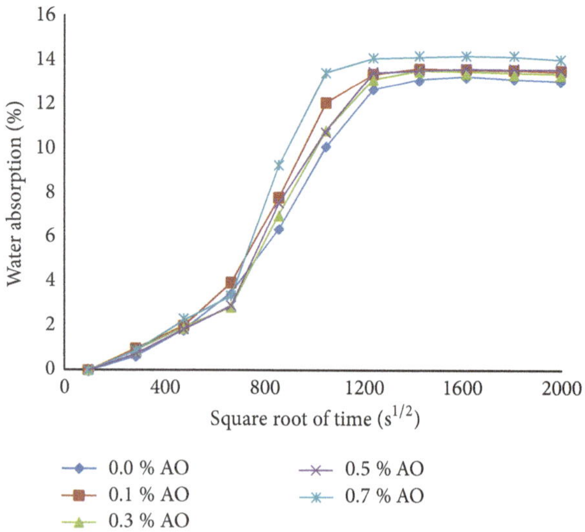

FIGURE 6: Distilled water.

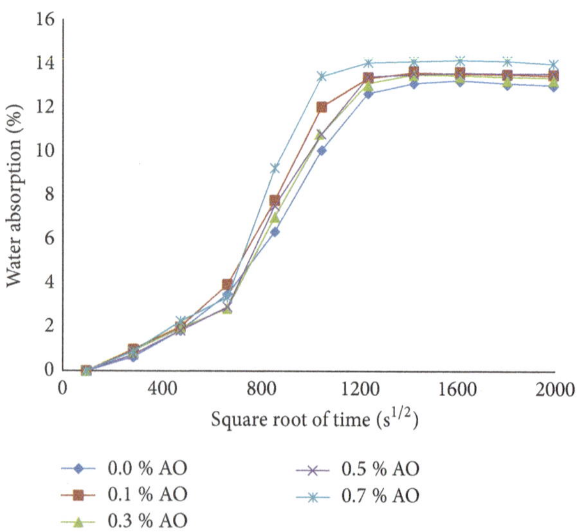

FIGURE 7: Sea water.

and penetrated well into the fibres, which prevents water from entering the samples. The water uptake increases substantially with increasing fibre content. According to Klyosov [7], the water absorption for biocomposites is typically 0.7–2% after 24 hours and 1–5% after a week and increases to 18–22% after several months of submersion in water.

This result could be explained by the wood chemical composition. Organic fibre especially sawdust contains high hydrophilic content. Celluloses and hemicelluloses are the main cause for the high water absorption of natural fibres. This is because they contain hydroxyl groups. Besides that, poor wettability between fibre particles and the polymer matrix generates void spaces around the fibre particles. Because of poor wetting, the filler particles create hollow spaces, introducing porosity to the internal structure of the biocomposites as indicated in Figure 8. Dewetting of the polymer at the interphase creates stress concentration points.

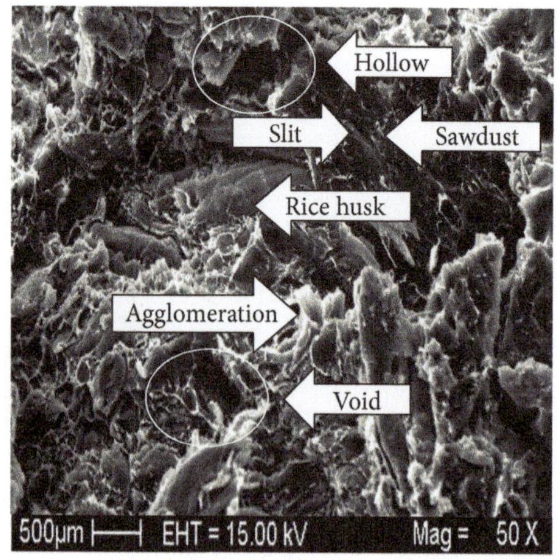

FIGURE 8: With 0.0 wt% antioxidants content.

These lead to higher water uptake in the composites natural fibre base.

According to Panthapulakkal and Sain study [12], the absorption of water on thermoplastic composite reinforced natural fibre mainly caused by microcracks that form on the surface and also in the material. Besides that, swelling fiber caused by water or moisture while producing the samples can also affect the water absorption rate. They also stressed on disconnection bond between matrix and filler due to high ambient temperature contributing to the absorption of water in tested samples.

The absorption of water in composite increased proportionally with the time of immersion until reaching the saturation point. Water absorption rate is higher in sea water compared to distilled water. This is caused by the presence of metal ions such as Na^+, K^+, Ca^{++}, and Mg^{++} in sea water [13]. The finding was supported by Silva et al. [14]; the sedimentation of metal ions was formed on the filler in reinforced thermoplastic composite. Therefore, it can be concluded that high amounts of metal ions of sea water will form sediment in the container. The sediment formed due to chemical reaction between metal ions and biocomposites. As a result, it will affect the rate of water absorption and thus deteriorate the mechanical properties of the material.

3.5. Morphology Test. Figures 8 and 9 show the results of scanning electron microscopy (SEM) morphology for samples without antioxidants and with 0.5 wt% antioxidants content, respectively. Sample without antioxidants content shows the agglomeration of fibers. The agglomeration is formed due to uneven distribution of fibers during extrusion and compression processes. The agglomeration of materials can affect the mechanical and physical properties of biocomposites. A slit is seen through the sawdust fiber; this could be explained by stress transfer from the matrix to the filler. The stress-strain process occurred from weaker part to stronger part. In this biocomposites, the recycled high density polyethylene matrix is weaker than hybrid fillers.

In Figure 9 image, there is no clear cleft that can be seen between matrix and fibers; this situation indicates

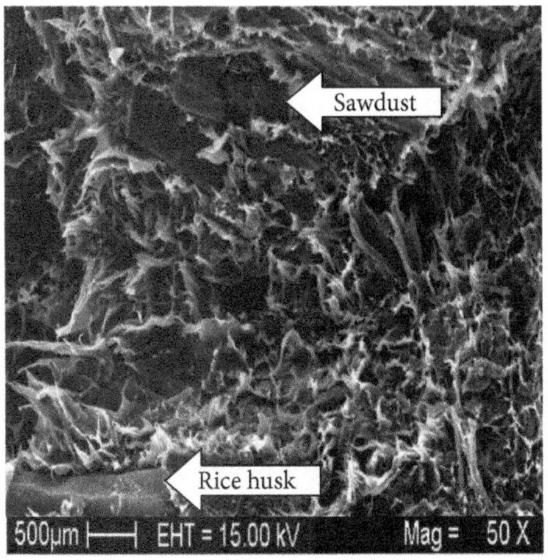

FIGURE 9: With 0.5 wt% antioxidants content.

the good interface bonding between them. In addition, the morphology shows that the fibres were utterly attached to the matrix and it is also evident that the fibres were strongly imbedded in the matrix, indicating the efficiency of the mixing, which was attributed to a good interfacial interaction between hybrid fibres and matrix [15]. The interfacial bonding was improved due to the esterification mechanism, and the fracture occurred at the fibres themselves. This means that the stress is well propagated between the fibres/fillers and the matrix polymer, resulting in enhanced tensile strength and modulus in response to stress [10].

4. Conclusion

Antioxidants were proven to be effective in increasing the mechanical properties and water absorption of the biocomposites. This indicates that the addition of antioxidants as additive improves the properties and the durability performance of recycled high density polyethylene reinforced with a hybrid of rice husks and sawdust. Besides the addition of antioxidants, there are various parameters influencing the mechanical properties of fibre-reinforced composites including the stress transfer at the interface, fibre aspect ratio, mixing temperatures, addition of a coupling agent, and the fibre-matrix adhesion. In conclusion, the mechanical properties of the biocomposites significantly increased with the addition of antioxidants content except for flexural test. The increment of antioxidants content reduced the modulus of elasticity and modulus of rupture.

Conflict of Interests

The authors declare that there is no conflict of interests regarding the publication of this paper.

References

[1] A. F. Ávila and M. V. Duarte, "A mechanical analysis on recycled PET/HDPE composites," *Polymer Degradation and Stability*, vol. 80, no. 2, pp. 373–382, 2003.

[2] J. R. Araújo, W. R. Waldman, and M. A. de Paoli, "Thermal properties of high density polyethylene composites with natural fibres: coupling agent effect," *Polymer Degradation and Stability*, vol. 93, no. 10, pp. 1770–1775, 2008.

[3] S. C. Ciba, *Irganox 1010 Phenolic Primary Antioxidant for Processing and Long-Term Thermal Stabilization*, Ciba Specialty Chemicals, 1998.

[4] S. C. Ciba, *Irgafos 168, Processing Stabiliser*, Ciba Specialty Chemicals, 2005.

[5] A. K. Bledzki, A. A. Mamun, N. N. Bonnia, and S. Ahmad, "Basic properties of grain by-products and their viability in polypropylene composites," *Industrial Crops and Products*, vol. 37, no. 1, pp. 427–434, 2012.

[6] S. M. L. Rosa, E. F. Santos, C. A. Ferreira, and S. M. B. Nachtigalt, "Studies on the properties of rice-husk-filled-PP composites—effect of maleated PP," *Materials Research*, vol. 12, no. 3, pp. 333–338, 2009.

[7] A. A. Klyosov, *Plastic Composites*, John Wiley & Sons, New York, NY, USA, 2007.

[8] E. L. Chong, I. Ahmad, H. M. Dahlan, and I. Abdullah, "Reinforcement of natural rubber/high density polyethylene blends with electron beam irradiated liquid natural rubber-coated rice husk," *Radiation Physics and Chemistry*, vol. 79, no. 8, pp. 906–911, 2010.

[9] M. H. Ab Ghani and S. Ahmad, "Effects of antioxidants content on the physical and mechanical properties of wood plastic composites," *Key Engineering Materials*, vol. 471-472, pp. 151–156, 2011.

[10] K. B. Adhikary, S. Pang, and M. P. Staiger, "Dimensional stability and mechanical behaviour of wood-plastic composites based on recycled and virgin high-density polyethylene (HDPE)," *Composites B: Engineering*, vol. 39, no. 5, pp. 807–815, 2008.

[11] H. C. Chen, T. Y. Chen, and C. H. Hsu, "Effects of wood particle size and mixing ratios of HDPE on the properties of the composites," *European Journal of Wood and Wood Products*, vol. 64, no. 3, pp. 172–177, 2006.

[12] S. Panthapulakkal and M. Sain, "Agro-residue reinforced high-density polyethylene composites: fiber characterization and analysis of composite properties," *Composites A: Applied Science and Manufacturing*, vol. 38, no. 6, pp. 1445–1454, 2007.

[13] S. K. Najafi and H. Y. Kordkheili, "Effect of sea water on water absorption and flexural properties of wood-polypropylene composites," *European Journal of Wood and Wood Products*, vol. 69, no. 4, pp. 553–556, 2011.

[14] R. V. Silva, E. M. F. Aquino, L. P. S. Rodrigues, and A. R. F. Barros, "Curaua/glass hybrid composite: the effect of water aging on the mechanical properties," *Journal of Reinforced Plastics and Composites*, vol. 28, no. 15, pp. 1857–1868, 2009.

[15] M. R. Y. Hamid, M. H. Ab Ghani, and S. Ahmad, "Effect of antioxidants and fire retardants as mineral fillers on the physical and mechanical properties of high loading hybrid biocomposites reinforced with rice husks and sawdust," *Industrial Crops and Products*, vol. 40, no. 1, pp. 96–102, 2012.

16

Preparation and Characterization of Talc Filled Thermoplastic Polyurethane/Polypropylene Blends

Emi Govorčin Bajsić,[1] Vesna Rek,[1] and Ivana Ćosić[2]

[1] *Department of Polymer Engineering and Organic Chemical Technology, Faculty of Chemical Engineering and Technology, University of Zagreb, Marulićev trg 19, 10 000 Zagreb, Croatia*
[2] *Department of Industrial Ecology, Faculty of Chemical Engineering and Technology, University of Zagreb, Marulićev trg 19, 10 000 Zagreb, Croatia*

Correspondence should be addressed to Emi Govorčin Bajsić; egovor@fkit.hr

Academic Editor: Yves Grohens

The effect of the addition of talc on the morphology and thermal properties of blends of thermoplastic polyurethane (TPU) and polypropylene (PP) was investigated. The blends of TPU and PP are incompatible because of large differences in polarities between the nonpolar crystalline PP and polar TPU and high interfacial tensions. The interaction between TPU and PP can be improved by using talc as reinforcing filler. The morphology was observed by means of scanning electron microscopy (SEM). The thermal properties of the neat polymers and unfilled and talc filled TPU/PP blends were studied by using dynamic mechanical analysis (DMA), differential scanning calorimetry (DSC), and thermogravimetric analysis (TGA). The addition of talc in TPU/PP blends improved miscibility in all investigated TPU/T/PP blends. The DSC results for talc filled TPU/PP blends show that the degree of crystallinity increased, which is due to the nucleating effect induced by talc particles. The reason for the increased storage modulus of blends with the incorporation of talc is due to the improved interface between polymers and filler. According to TGA results, the addition of talc enhanced thermal stability. The homogeneity of the talc filled TPU/PP blends is better than unfilled TPU/PP blends.

1. Introduction

Blending has been widely and effectively used to modify or control the properties of polymer by appropriately compounding miscible polymers. Miscible polymer blends can create new materials with completely different properties, and fabricated articles may possess good mechanical properties. Thermoplastic polyurethanes (TPUs) are linear, segmented copolymers consisting of alternating hard segments (HSs) and soft segments (SSs). HSs, comprised of diisocyanate and short chain extender (CE) molecules such as diols or diamines, are rigid and highly polar. HSs have high interchain interaction due to hydrogen bonding between the urethane/urea groups. The hydrogen bonding associations within the HSs in TPUs act as a reinforcing filler for the soft matrix. On the other hand, SSs, formed from linear long-chain diols or polyols, are flexible and weakly polar [1]. Phase separation in TPUs occurs because of the thermodynamic immiscibility or incompatibility between the hard and soft phase. The degree of phase separation in Pus depends on the types of diisocyanate and polyol and polyol employed to produce the PU, type of polyfunctional compound used in the crosslinking process, NCO : OH ratio, size of HSs and SSs, and method of synthesis. Polypropylene (PP), a semicrystalline polyolefin polymer with very interesting mechanical properties (ductility and stiffness), good thermal properties, and excellent chemical and moisture resistance is also easy to process. Limitation of PP is its poor adhesion to the surface of other phases such as rubber or polar materials. This is primarily a result of the nonpolar nature of PP [2, 3]. The incorporation of proper filler increase interfacial adhesion between matrix and disperse phase refines blends morphology and therefore leads to an improvement of processing and application properties of the final material. Traditional fillers for PP are calcium carbonate, talc, glass fiber, wollastonite, mica, glass beads, and wood flour. Filled PP containing talc is used extensively because of a combination of stiffness, dimensional stability, and, importantly, low cost. It is well known that mineral fillers increase the rigidity of the polymers, but they also decrease ductility and toughness [4]. In studying the complex structure and morphology of polymers modified by mineral fillers, some problems may arise concerning the character and extent of interaction at the polymer-filler interface, the homogeneity of filler distribution, the filler orientation in the case of filler anisometric particles, and the polymer-filler adhesion [5]. The talc filled TPU/PP blends can be processed by injection molding, which is a common method to make products from thermoplastics. The properties of the molded product depend on the nature of the filler, its concentration, and the molding conditions. The most important characteristics of the filler are its size, shape and its ability to act as nucleating agent for crystallization [5].

2. Materials and Methods

2.1. Materials. Thermoplastic polyurethane (TPU, Desmopan 588E a polyester-polyether type, Bayer A.G., Germany), isotactic polypropylene (i-PP, Moplen HP 500N, Basel Polyolefines, Basel, Switzerland with MFI = 12 g/10 min at 190°C), talc (Luzenac 10 MO, Solvay, France), with platelet diameter (400 nm), specific surface (6.50 $m^2 g^{-1}$), and density (2.8 $g cm^{-3}$).

2.2. Preparation of Blends. The TPU/PP blends without talc (TPU/PP: 80/20, 50/50, and 20/80 blends) and with 5 wt% of talc (TPU/T/PP: 80/5/20, 50/5/50, and 20/5/80 blends) were prepared by melt blending using a Rondol 21 mm LAB TWIN twin-screw extruder. Before processing, the talc was added to the TPU and PP pellets and the components were mixed and charged into the twin-screw extruder. The temperatures in the extruder were set to 170°C, 180°C, 180°C, 200°C, and 200°C, the temperature of the die was 200°C, the screw speed was 60 rpm, and the torque was between 60 and 95%. Extrudates were cooled in a water bath, pulled, and pelletized. Injection moulding was done using a Rondol, High Force 5 injection press to produce test specimens at 180°C.

2.3. Characterization. The viscoelastic properties of the neat TPU, PP; TPU/T, and PP/T composites as well as TPU/PP blends, also the miscibility of the TPU/PP blends, were carried out with a Dynamic Mechanical Analyser DMA 983 (TA Instruments, USA) at a frequency of 1 Hz.

The changes of storage (E') and loss (E'') moduli were measured over the temperature range of −100 to 250°C at a heating rate of 3°C/min. The sample length between the clamps was approximately 25 mm.

The crystallization behavior of neat polymers (TPU, PP) and TPU/PP and TPU/T/PP blends were studied by differential scanning calorimetry (Mettler Toledo DSC 822^e, Switzerland) under nitrogen flow (40 mL/min). DSC was calibrated with indium standard prior to use. The thermal cycling applied was that samples were heated to 190°C at a heating rate of 10°C/min, annealed at 190°C for 10 min in order to eliminate any previous thermal history, and then quenched with liquid nitrogen to −100°C and heated to 190°C. From the second heating run, the melting enthalpy (ΔH_m) and crystallization enthalpy (ΔH_c) were obtained. The TPU/PP and TPU/T/PP blend samples were sliced from injection-moulded specimens, a sample weight of approximately 10 mg was used.

The degree of crystallinity, χ_c, was calculated by comparing the measured melting heat to melting heat of 100% crystalline iPP (1):

$$\chi_c = \frac{\Delta H_m}{\Delta H^0_{\mathrm{PP}}} \times 100, \qquad (1)$$

where ΔH_m is the enthalpy of malting and ΔH^0_{PP} enthalpy of malting for completely crystalline PP, equal to 165 J/g [6, 7].

In order to assess the thermal stability of the neat TPU, PP; TPU/T and PP/T composites and TPU/PP blends were characterized by thermogravimetric analysis (TGA) (Q500, TA Instruments, USA). The sample (about 10 mg) was loaded on the pan and heated from room temperature to 700°C at a heating rate of 10°C/min under nitrogen flow (60 mL/min).

A Vega 3 Tescan scanning electron microscope (Brno, Czech Republic) was used to study the morphology. Samples were fractured in liquid nitrogen and the fracture surface was coated with gold before being examined with the microscope.

3. Results and Discussion

The viscoelastic properties of the neat TPU, PP: TPU/T, PP/T composites and TPU/PP blends were characterized using the dynamic mechanical analysis (DMA). The temperature dependence of the storage (E') and loss moduli (E'') for neat TPU, PP and TPU/T, PP/T composites are presented in Figures 1(a) and 1(b), respectively. The experimental results are listed in Table 1. The temperature associated with the peak maximum of E'' corresponds to the glass transition temperature (T_g) of TPU and PP. The E''/T curve for the neat TPU shows the T_g of the soft segments of the polyurethane at −27.3°C [8, 9] and this value increases by adding talc (from −27.3 to −21.7°C) (Figure 1(b), Table 1). The value of T_g of a polymer depends on the macromolecule chain segment mobility in the polymer matrix. The presence of talc restricted the mobility of the molecular chain of TPU, and so T_g increases as a consequence of interactions between talc and the TPU chains. In the case of PP and PP/T composite, E'' showed three distinct relaxations (Figure 1(b), Table 1). The maxima on relaxation peak at 24.3°C could be ascribed to the glass transition (T_g) of the PP amorphous phase [10], at −44.3°C the γ-relaxation of the PP appears associated with the motions of small-chain groups like methyl and methylene [11], and another relaxation at higher temperatures appears, α_C relaxation shoulder (at approximately 57.2°C), and could be related to a more rigid chain with an amorphous character related to the PP crystallites (tie molecules, etc.).

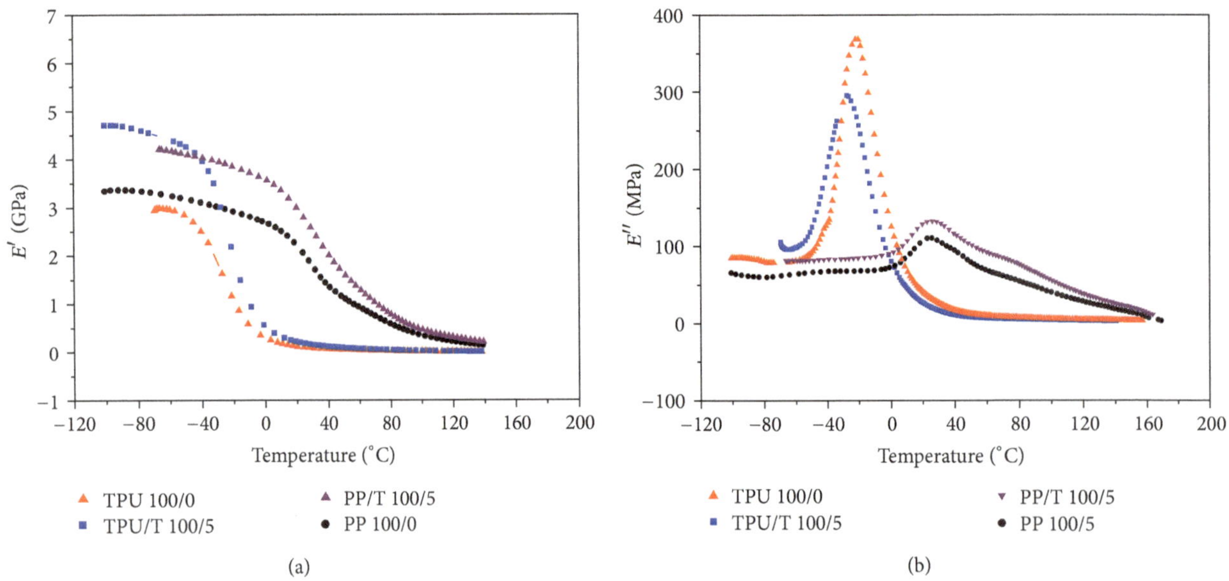

FIGURE 1: Plot of (a) storage modulus (E') and (b) loss modulus (E'') for the neat TPU, PP and TPU/T and PP/T composites.

TABLE 1: DMA results of the neat TPU, PP, TPU/T and PP/T composites, and unfilled and talc filled TPU/PP blends.

Sample	$T_{g,\mathrm{TPU}}$ (°C, E'')	$T_{g,\mathrm{PP}}$ (°C, E'')	E' (GPa) at 25°C	α_c (°C)
TPU 100/0	−27.3	—	0.1	—
TPU/T 100/5	−21.7	—	0.2	—
PP 100/0	—	24.3	2.3	57.2
PP/T 100/5	—	26.4	2.8	69.4
TPU/PP 80/20	−22.2	—	0.2	—
TPU/T/PP 80/5/20	−21.1	—	0.5	—
TPU/PP 50/50	−21.2	16.6	1.0	72.3
TPU/T/PP 50/5/50	−18.0	18.2	1.2	77.0
TPU/PP 20/80	−26.7	20.7	1.9	75.3
TPU/T/PP 20/5/80	−20.3	22.5	2.0	76.7

The transition temperatures obtained in the DMA measurements of the investigated unfilled and talc filled TPU/PP blends with different TPU/PP content are shown in Figure 2 and summarized in Table 1. Figure 2(a) shows the temperature dependency of E'' for unfilled and talc filled TPU/PP blends, in which the two distinct relaxations observed corresponded to the T_g of TPU and T_g of PP at all compositions except for the TPU/PP 80/20 blends. Compared with the neat TPU and PP, the T_g of TPU shifted to higher temperatures and the T_g of PP shifted to lower temperatures for the TPU/PP and TPU/T/PP blends and the maximum peaks broader.

This might be explained with a lower segmental mobility which is suppressed by the crystalline phase of PP and with filler.

In the case of talc filled blends (TPU/T/PP blends), the addition of talc in the blends the T_g maxima of the TPU and PP shift towards to each other. This result may be related to the same interaction in the blends due to the presence of talc, which caused better interphase for both polymers [11, 12].

Figure 2(b) shows the temperature dependence of the storage modulus (E') for neat polymers, composites, unfilled, and talc filled TPU/PP blends. The results for all investigated samples at 25°C are listed in Table 1. As shown in Figure 2(b) the E' of the neat TPU and TPU/T composite decreased sharply in the T_g region due to the mobility of the soft segments. The results shown in Table 1 indicate that in the measured region, the E' of the neat PP was higher than that of the neat TPU and at 25°C a value of TPU (0.08 GPa) is much lower than that of the neat PP (2.87 GPa). In all prepared blends, the E' have intermediate value of the neat TPU and PP. The values of E' of blends increased with increasing PP content in blends and with the addition of talc. These results may be caused by an interaction between the TPU and PP, as well as interactions with talc particles.

Figure 3 shows DSC thermograms for unfilled and talc filled TPU/PP blends and the corresponding values are summarized in Table 2.

It can be observed from Table 2 that the melting temperatures (T_m) of the TPU/PP blends do not change with

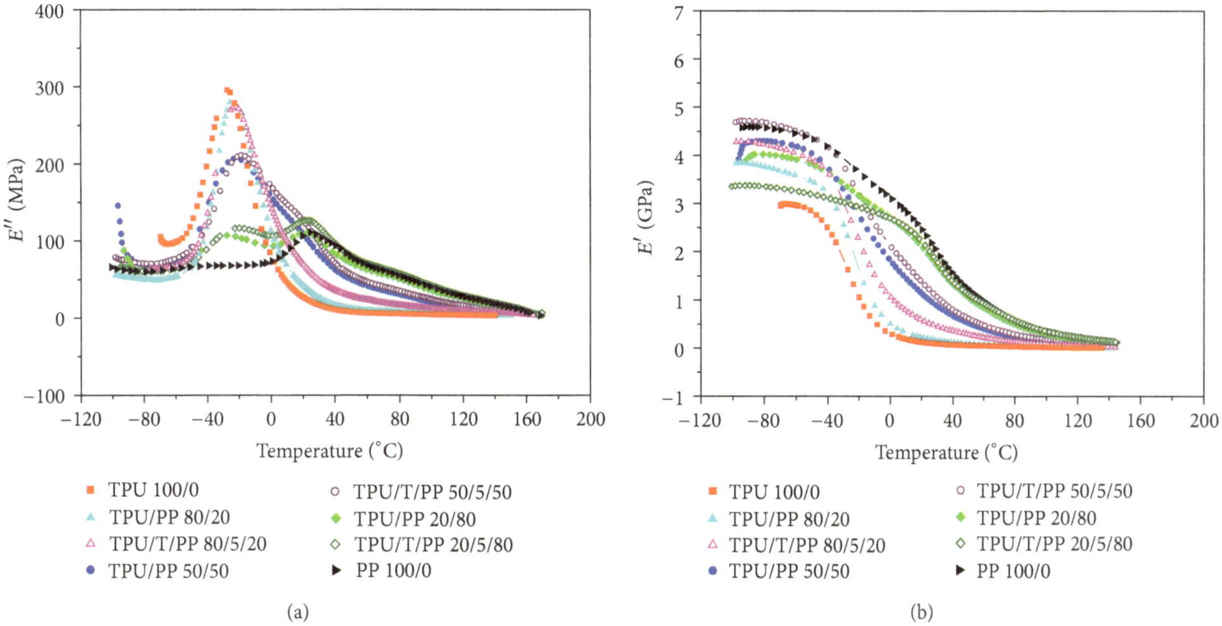

FIGURE 2: Plot of (a) loss modulus (E'') and (b) storage modulus (E') of unfilled and talc filled TPU/PP blends.

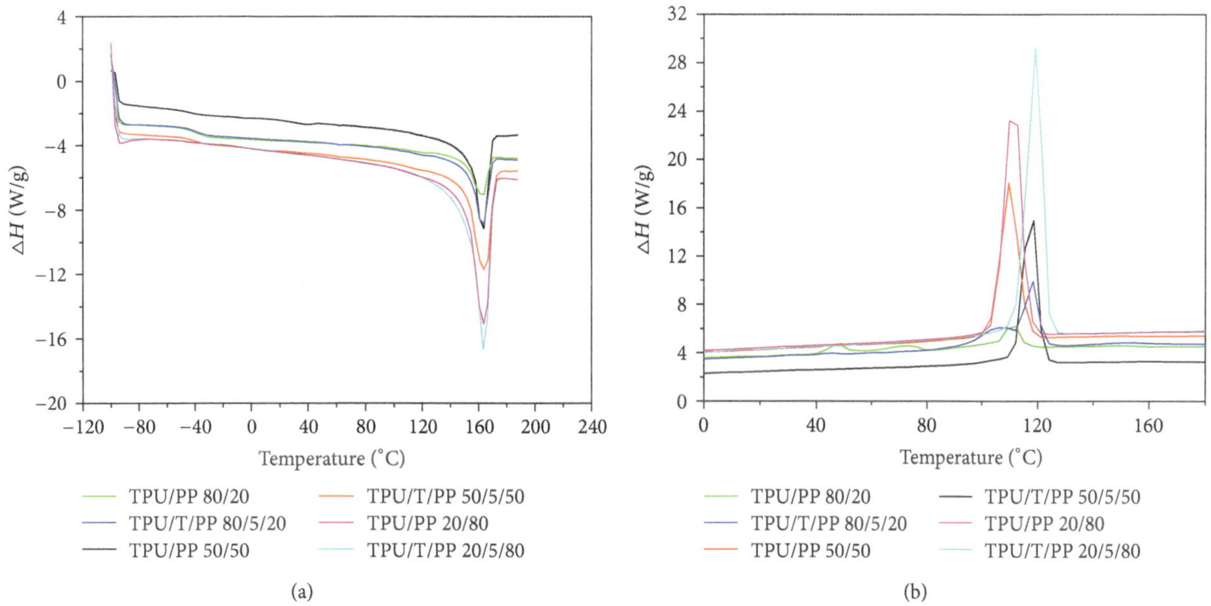

FIGURE 3: DSC heating (a) and crystallization (b) curves of unfilled and talc filled TPU/PP blends.

TABLE 2: Thermal properties of the unfilled and talc filled TPU/PP blends.

Sample	T_m (°C)	T_{cp} (°C)	ΔH_m (J/g)	ΔH_c (J/g)	χ_c (%)
TPU/PP 80/20	163.4	111.4	15.0	9.3	9.1
TPU/T/PP 80/5/20	163.0	118.1	24.0	28.0	14.5
TPU/PP 50/50	164.3	110.0	42.1	45.8	25.5
TPU/T/PP 50/5/50	163.0	117.5	46.6	53.8	28.2
TPU/PP 20/80	164.3	111.3	74.1	77.4	45.0
TPU/T/PP 20/5/80	164.0	120.0	77.9	79.8	47.2

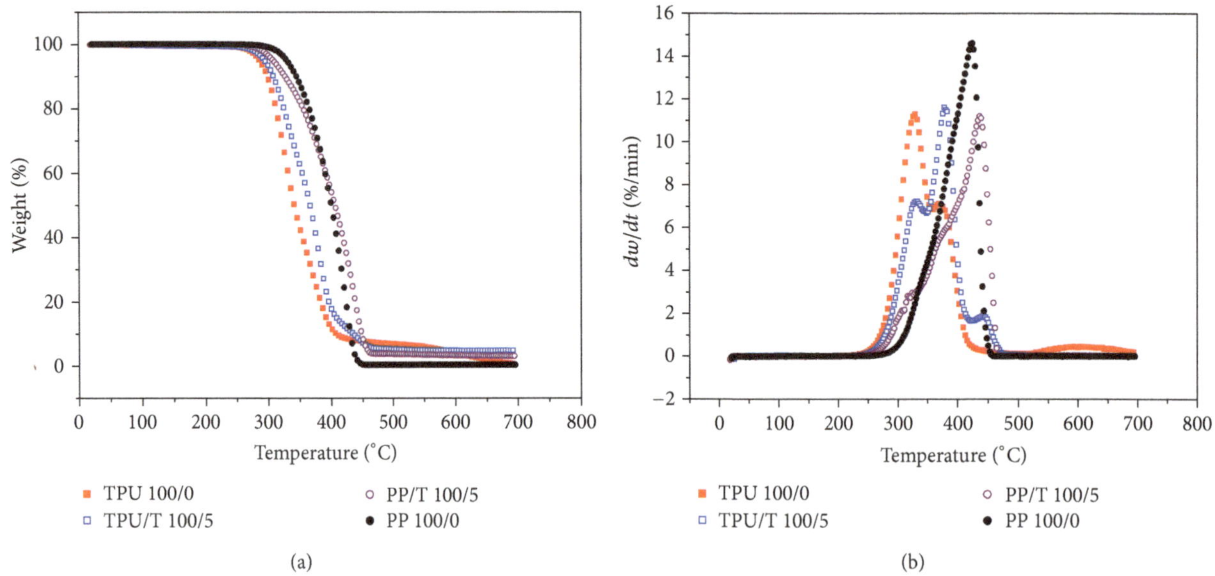

FIGURE 4: TG curves (a) and DTG curves (b) of the neat TPU, PP and TPU/T and PP/T composites.

TABLE 3: Results of TGA measurement for the neat TPU, PP, TPU/T and PP/T composites, and unfilled and talc filled TPU/PP blends.

Sample	T_5 (°C)	T_1^{max} (°C)	Δ_{m1} (%)	T_2^i (°C)	T_2^{max} (°C)	Δ_{m2} (%)	T_3^i (°C)	T_3^{max} (°C)	Δ_{m3} (%)	T_{end} (°C)	Residue at 700°C (%)
TPU 100/0	288.1	328.8	62.3	357.6	369.8	30.2	—	—	—	400.4	0.7
TPU/T 100/5	295.6	331.1	34.3	348.2	380.0	53.4	425.4	441.3	6.5	458.4	4.9
PP 100/0	312.3	423.9	99.7	—	—	—	—	—	—	437.7	0.3
PP 100/5	330.7	435.5	96.4	—	—	—	—	—	—	455.5	3.2
80/20	299.6	337.2	44.1	352.2	368.9	51.7	—	—	10.4	—	2.8
80/5/20	300.0	337.8	33.0	347.2	378.3	45.9	421.2	430.4	15.2	451.5	5.0
50/50	301.7	318.7	18.3	347.5	388.7	39.4	413.2	439.5	40.4	454.6	1.4
50/5/50	305.1	319.3	18.9	349.6	398.0	34.1	411.0	442.5	43.4	458.9	3.3
20/80	318.2	319.09	8.6	—	—	—	346.8	440.6	90.6	458.9	0.7
20/5/80	319.5	319.08	7.7	—	—	—	335.2	442.1	89.7	457.5	2.4

T_5: temperature at which 5% weight loss occurs from the TG curve; $T_{1,2,3}^{max}$, temperature of maximum rate of decomposition for the 1st, 2nd, and 3rd steps of degradation, respectively, from the DTG curve; $\Delta_{m1,2,3}$: mass loss on the end of the 1st, 2nd, and 3rd steps of degradation; T^i: temperature of initial decomposition for the 2nd and 3rd steps of degradation; T_{end}: final degradation temperature.

the addition of talc. Figure 3(b) shows that the TPU/PP blends with talc (TPU/T/PP blends) exhibited a much greater exothermic crystallization peak. The peak temperature of the crystallization (T_{cp}), enthalpy of melting (ΔH_m) and enthalpy of crystallization (ΔH_c) was higher for talc filled TPU/PP blends than that of unfilled blends. The higher values of the degree of crystallinity (χ_c) for talc filled TPU/PP blends indicated that the talc acts as a nucleating agent [13, 14].

In order to determine the thermal stability of the neat polymers, composites and unfilled and talc filled TPU/PP blends thermogravimetric analysis (TGA) was used. The thermal stability of the investigated samples was determined in terms of T_5, temperature at which 5% weight loss occur. Figure 4 shows the TG curves (Figure 4(a)) and DTG curves (Figure 4(b)) of the unfilled and talc filled TPU and PP; the values of characteristic temperatures are shown in Table 3.

From the DTG curves obtained for the unfilled and talc filled TPU it was evident that the thermal degradation occurred in two steps of degradation. The first step of degradation is assigned to the degradation of the hard segment, while the second step is related to the degradation of the soft segments in TPU [15]. The results in Table 3 show that the first step of decomposition starts at 288.1°C for unfilled TPU (TPU 100/0) and 295.0°C for talc filled TPU (TPU/T 100/5). In the first step of degradation a weight loss of 62.3% (TPU) and 33.4% (TPU/T) was observed with a maximum rate of weight lost (T_1^{max}) at 328.8 and 331.1°C, respectively, while in the second step it lost 30.2% (TPU) and 53.4% (TPU/T) with T_2^{max} (368.8 and 380.0°C) and the first step of degradation finished at T_{end} (400.4 (TPU) and 458.4°C (TPU/T)). From the obtained results it can be concluded that the degradation of TPU takes place at temperatures from 288.1°C to 400.4°C

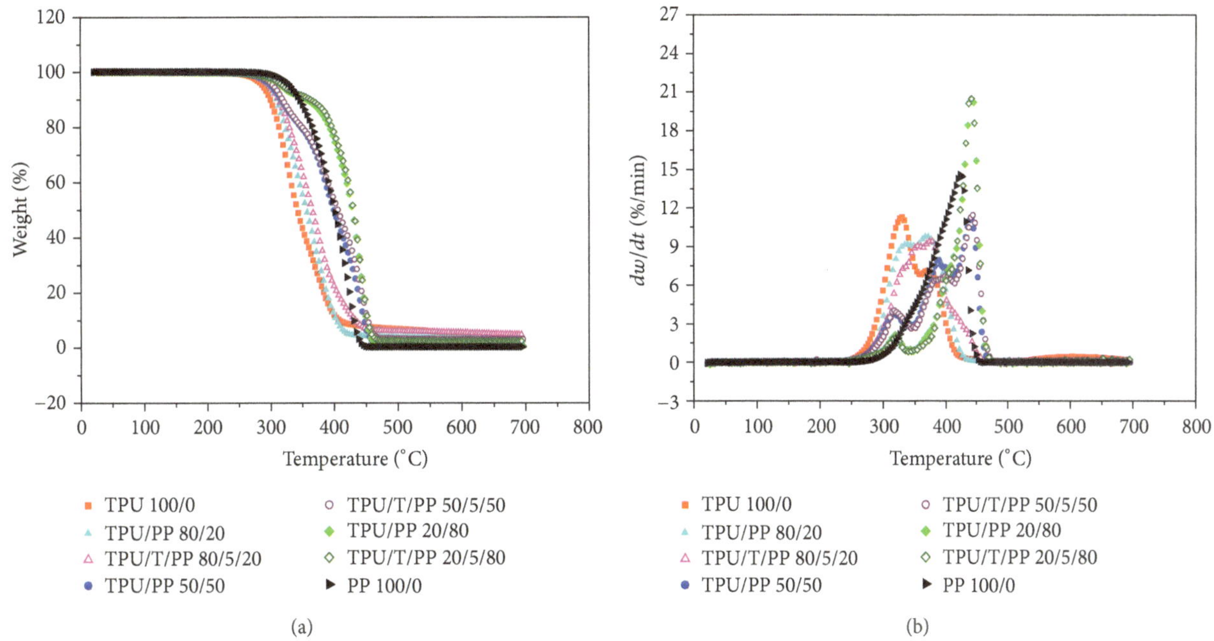

FIGURE 5: TG curves (a) and DTG curves (b) of the neat TPU, PP and unfilled and talc filled TPU/PP blends.

and for TPU/T from 295.0°C to 458.4°C. The weight residues at the temperature of 700°C are 0.7% and 4.9%, respectively.

One step of degradation was observed for the neat PP which started at 330.7°C with a maximum rate of weight lost (T_1^{max}) at 423.9°C and end set decomposition temperature (T_{end}) at 437.7°C. The PP lost 99.7% of its weight and the weight residue at $T = 700$°C is 0.3%. Based on the obtained results it can be concluded that PP degrades faster than TPU, but decomposition starts at higher temperatures, which confirms its higher thermal stability. The thermal stability of talc filled PP increased compared with the neat PP.

Figures 5(a) and 5(b) represent TG and DTG curves of all investigated TPU/PP blends; the characteristic values related to the decomposition temperatures are shown in Table 3.

All blends display three degradation steps, the first and the second related to the degradation of the hard and soft segments in TPU, while the third to the degradation of PP. The T_5 for the TPU/T/PP: 80/5/20, 50/5/50, and 20/5/80 are at 300.0, 305.1, and 319.5°C respectively, which are all greater than 288.1°C for the neat TPU and 299.6, 301.7, and 318.2°C, respectively, for the unfilled TPU/PP blends. The results of T_5 suggest that the incorporation of talc into TPU and TPU/PP blends could improve the thermal stability of the blends. This is probably due to the good polymer-filler interaction. The residues of all talc filled TPU/PP blends were all larger than those of unfilled TPU/PP blends and decreased with the increase of PP content.

SEM micrographs of the cryogenically fractured samples indicated a two-phase morphology with no sign of interfacial adhesion. In the SEM micrograph of the 80/20 TPU/PP blend more black hollows were shown to have arisen from dispersed PP particles that were pulled out (Figure 6(a)). The smaller hollows can be observed in the SEM micrograph of the 20/80 TPU/PP blend (Figure 7(a)). The observed two-phase morphology of the TPU/PP blends indicated the immiscibility and low compatibility of the components. Since the immiscible blends are thermodynamically unstable, they must be stabilized to prevent coalescence during melt processing. The stability of polymer blends is enhanced by the addition of a useful block copolymer or filler. The influence of the talc on the TPU/PP blends is shown in Figures 6(b) and 7(b), respectively. With the addition of talc, the relatively finer dispersity of the phases was found and the smaller hollows as well. So, in this case, the homogeneity of the blend is better than that of the blend without talc.

4. Conclusion

The thermal and morphology properties of neat components (TPU, PP) TPU/T and PP/T composites and TPU/PP blends were investigated by differential scanning calorimetry (DSC), dynamic mechanical analysis (DMA), thermogravimetric analysis (TGA), and scanning electron microscopy (SEM). In DMA results the presence of two T_g's in the blends which were shifted toward each other suggests that TPU/PP blends are partially compatible. The intensity of the glass transition peak of the amorphous phase of TPU decreased upon the addition of PP. This might be explained with a lower segmental mobility which is suppressed by the crystalline phase of PP. The DSC results indicated that a higher degree of crystallinity (χ_c) was obtained for TPU/PP blends when talc is added, which can be attributed to the nucleating effect induced by talc particles. This finding implies that the introduction of talc into TPU/PP blends improved compatibility of TPU and PP. According to the TGA results the thermal stability of talc filled TPU/PP blends was higher compared to the unfilled

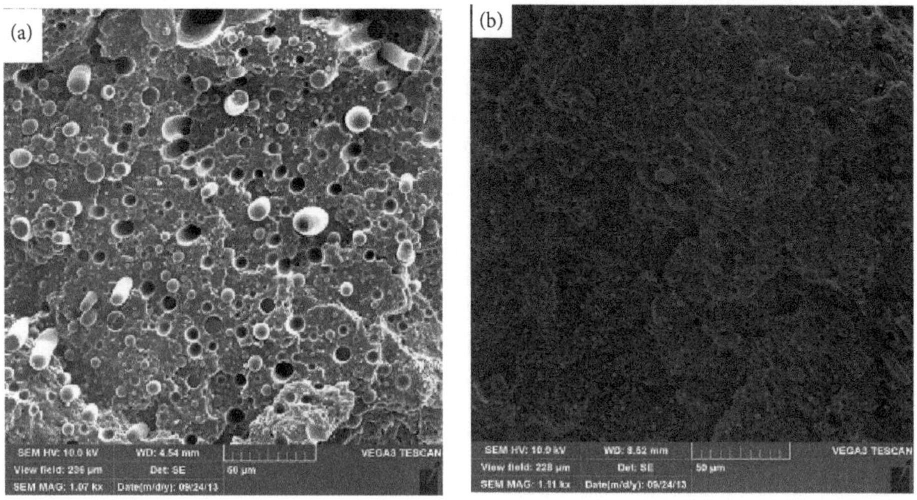

(a) (b)

FIGURE 6: SEM micrographs of the TPU/PP 80/20 blend (a) and 80/5/20 blend (b).

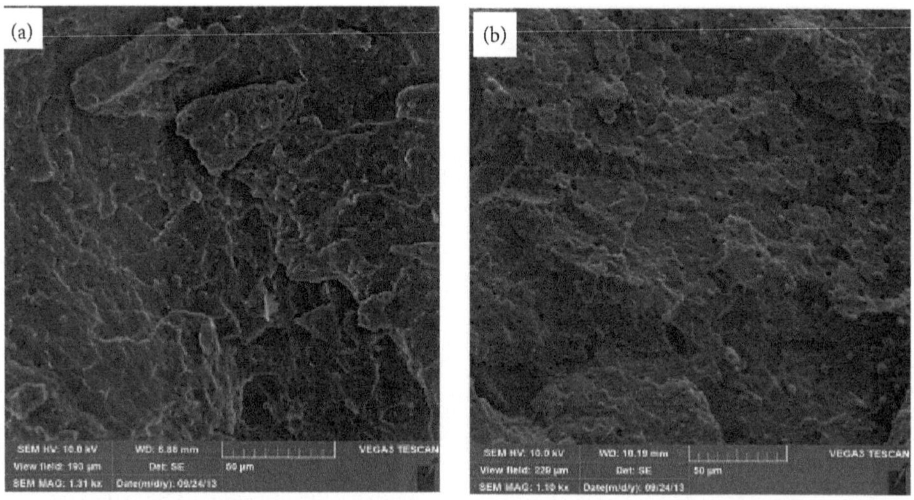

(a) (b)

FIGURE 7: SEM micrographs of the TPU/PP 20/80 blend (a) and 20/5/80 blend (b).

TPU/PP blends. With the addition of talc, the relatively finer dispersity of the phases was found and the smaller hollows as well. The homogeneity of the talc filled TPU/PP blends is better than unfilled TPU/PP blends.

Conflict of Interests

The authors declare that there is no conflict of interests regarding the publication of this paper.

Acknowledgment

The authors gratefully acknowledge the financial support of this work by the Ministry of Science, Education and Sport of the Republic of Croatia (Project no. 125-1252971-2578).

References

[1] S. L. Cooper and A. V. Tobolsky, "Properties of linear elastomeric polyurethanes," *Journal of Applied Polymer Science*, vol. 10, pp. 1837–1844, 1966.

[2] B. Lotz, J. C. Wittmann, and A. J. Lovinger, "Structure and morphology of poly(propylenes): a molecular analysis," *Polymer*, vol. 37, no. 22, pp. 4979–4992, 1996.

[3] J. Varga, "Supermolecular structure of isotactic polypropylene," *Journal of Materials Science*, vol. 27, no. 10, pp. 2557–2579, 1992.

[4] B. Pukanszky, "Composites," in *Polypropylene: Structure, Blends and Composites*, J. Karger-Kocsis, Ed., vol. 3, pp. 1–70, Chapman & Hall, London, UK, 1995.

[5] M. Fujiyama and T. Wakino, "Crystal orientation in injection molding of talc-filled polypropylene," *Journal of Applied Polymer Science*, vol. 42, no. 1, pp. 9–20, 1991.

[6] B. Wunderlich, *Crystal Structure, Morphology, Defects, Macromolecular Physics*, Academic Press, New York, NY, USA, 1973.

[7] B. Wunderlich, *Crystal Melting, Macromolecular Physics*, Academic Press, New York, NY, USA, 1980.

[8] Y. Kim, W. J. Cho, and C. S. Ha, "Dynamic mechanical and morphological studies on the compatibility of plasticized PVC/thermoplastic polyurethane blends," *Journal of Applied Polymer Science*, vol. 71, no. 3, pp. 415–422, 1999.

[9] P. Potschke, K. Wallheinke, H. Fritsche, and H. Stutz, "Rheology modification of PVC plastisols with poly(butylene terephthalate)-*b*-poly(tetramethylene glycol)," *Journal of Applied Polymer Science*, vol. 66, no. 4, pp. 749–753, 1997.

[10] R. H. Boyd, "Relaxation processes in crystalline polymers: experimental behaviour—a review," *Polymer*, vol. 26, no. 3, pp. 323–347, 1985.

[11] B. E. Read, G. D. Dean, and P. E. Tomlins, "Effects of physical ageing on creep in polypropylene," *Polymer*, vol. 29, no. 12, pp. 2159–2169, 1988.

[12] D. Sudhin and J. L. David, *Polymer Compatibilizers*, Hanser Publishers, Munich, Germany, 1996.

[13] J. Menczel and J. Varga, "Influence of nucleating agents on crystallization of polypropylene. I: talc as a nucleating agent," *Journal of Thermal Analysis*, vol. 28, no. 1, pp. 161–174, 1983.

[14] J. Varga and F. Schulek-Tóth, "Filled compounds of the β-modification of polypropylene," *Die Angewandte Makromolekulare Chemie*, vol. 188, no. 1, pp. 11–25, 1991.

[15] H. K. Lee and S. W. Ko, "Structure and thermal properties of polyether polyurethaneurea elastomers," *Journal of Applied Polymer Science*, vol. 50, no. 7, pp. 1269–1280, 1993.

17

Thermal Degradation Studies of Terpolymer Derived from 2-Aminothiophenol, Hexamethylenediamine, and Formaldehyde

P. U. Belsare,[1] A. B. Zade,[1] P. P. Kalbende,[2] and M. S. Dhore[1]

[1] *Department of Chemistry, Laxminarayan Institute of Technology, Rashtrasant Tukadoji Maharaj Nagpur University, Nagpur, Maharashtra 440010, India*

[2] *Department of Applied Chemistry, Datta Meghe Institute of Engineering, Technology and Research, Sawangi (Meghe), Wardha 442001, India*

Correspondence should be addressed to A. B. Zade; ab_zade18@yahoo.com

Academic Editor: Yeong-Soon Gal

Terpolymer (2-ATPHMDAF-I) has been synthesized by the condensation of 2-aminothiophenol and hexamethylenediamine with formaldehyde in the presence of 2 M hydrochloric acid as a catalyst with 1 : 1 : 2 molar proportion of reacting monomers. The structure of newly synthesized terpolymer has been elucidated and confirmed on the basis of elemental analysis and various spectral techniques, that is, UV-visible, FT-IR, and ^{1}H-NMR spectroscopy. Number average molecular weight ($\overline{Mn}$) has been determined by conductometric titration in nonaqueous medium. The viscosity measurements in dimethyl sulfoxide (DMSO) have been carried out to ascertain the characteristic functions and constants. The studies have been further extended to nonisothermal thermogravimetric analysis for determination of their mode of decomposition and relative thermal stability. Activation energy (E_a), order of reaction (n), and frequency factor (z) were calculated by Friedman, Chang, Sharp-Wentworth and Freeman-Carroll methods. Activation energy calculated by Friedman and Chang methods are in close agreement with each other while the results obtained from Freeman-Carroll and Sharp-Wentworth's methods are found to be in a similar order.

1. Introduction

Study of thermal analysis comprises a group of techniques in which a physical property of substance is measured as a function of temperature when the substance is subjected to a controlled temperature program. Thermally stable terpolymers have recently become boon to polymer chemist due to their superior and high performance utility. Since the modern history of thermogravimetry, thermal degradation of polymers and the study of their kinetics have been at the center of thermal analysis. Many researchers tried to improve the thermal stability at elevated temperature by changing the monomer composition in polymer synthesis.

The thermal degradation study of terpolymer has become a subject of recent interest, being an important property which primarily decides thermal stability and processability. A wide variety of thermally stable polymers have been synthesized and studied their thermogravimetric property and finds many applications such as ion-exchangers [1–5], semiconductors [6], high dielectric constant for energy storage capacitors [7], packaging, adhesives and coatings in electrical sensors, activators, catalysts and thermally stable materials [8–10]. Phenolic resins are known for their wide applications in various areas because of their thermal stability, easy availability, cost effectiveness, and some of their excellent properties [11].

The thermal stability of terpolymers has been extensively studied by employing the method of thermogravimetric analysis (TGA) by several authors [12–21]. Thermoanalytical and kinetic studies of terpolymer resins derived from 8-hydroxyquinoline-5-sulphonic acid/p-cresol, oxamide/melamine with formaldehyde have been reported by Singru et al. [12, 13]. Thermal and metal ion bonding properties of terpolymer resin synthesized from resorcinol-thiourea-formaldehyde have been reported by Karunakaran et al. [14]. The thermal degradation kinetics of some new terpolymers

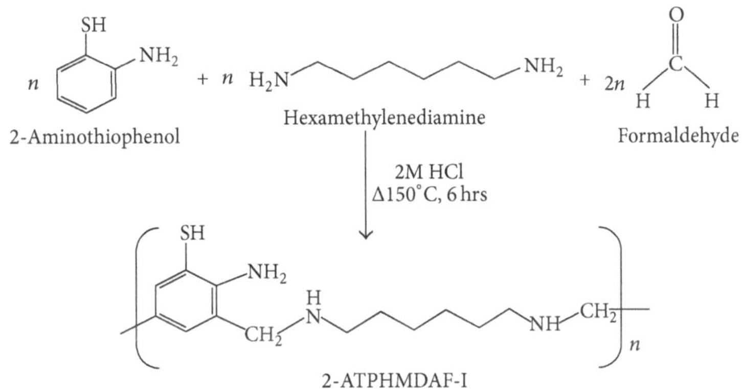

Figure 1: Chemical reaction of 2-ATPHMDAF-I terpolymer.

derived from 2,4-dihydroxypropiophenone, oxamide, and formaldehyde have been studied by Tarase et al. [15]. Nonisothermal decomposition and kinetic analysis of copolymer derived from 2,4-dihydroxybenzoic acid, melamine, and formaldehyde have been reported by Butoliya et al. [16]. Michael et al. carried out thermal degradation of terpolymers synthesized from salicylic acid/8-hydroxyquinoline and guanidine with formaldehyde [17, 18]. Thermogravimetric analysis of terpolymer resins derived from 2,4-dihydroxyacetophenone, dithiooxamide, and formaldehyde by Rahangdale et al. [19], 8-hydroxyquinoline, dithiooxamide, and formaldehyde by Katkamwar et al. [20], and 8-hydroxyquinoline-5-sulphonic acid, catechol, and formaldehyde by Mandavgade et al. [21] has been studied in detail. Methods for the estimation of kinetic parameters from thermogravimetric studies are generally based on the assumption that the Arrhenius equation is valid with thermal and diffusion barriers being negligible.

In our earlier communications [3, 12, 13, 15–17, 19, 20, 22], various studies on synthesis, characterization, and thermogravimetric analysis of some new polymeric resins have been reported. Hence, in the present investigation, it has been planned to study the nonisothermal thermogravimetric analysis of terpolymer derived from 2-aminothiophenol, hexamethylenediamine, and formaldehyde which has not been reported so far in literature.

2. Materials and Methods

2.1. Starting Materials. 2-Aminothiophenol and hexamethylenediamine used in the present investigation of analytical grade purity were purchased from Acros Chemicals, Belgium, and Aldrich, USA, respectively. Formaldehyde (37%) was purchased from S.D. Fine Chemicals, India. All the used solvents like N,N-dimethylformamide, dimethyl sulfoxide, tetrahydrofuran, acetone, and diethyl ether were procured from Merck, India.

2.2. Synthesis. 2-ATPHMDAF-I terpolymer was prepared by condensing 2-aminothiophenol and hexamethylenediamine with formaldehyde in the presence of 2 M hydrochloric acid as a catalyst in 1 : 1 : 2 molar proportions at temperature 150°C in an oil bath for about 6 hrs. The yellowish white colored solid product was obtained. The product obtained was extracted with diethyl ether to remove the excess of 2-aminothiophenol-formaldehyde copolymer which might be present along with 2-ATPHMDAF-I terpolymer. The chemical reaction of above synthesis has been depicted in Figure 1.

2.3. Characterization of Terpolymer. Newly synthesized and purified terpolymer was subjected to elemental analysis for carbon, hydrogen, nitrogen, and sulphur on Elementar Vario EL-III Elemental Analyzer and ultraviolet-visible spectra of terpolymer in dimethyl sulfoxide (DMSO) solvent recorded on Varian Carry 5000 UV-Vis spectrophotometer in the range from 200 to 800 nm at Sophisticated Test and Instrumentation Center Cochin, University of Science and Technology, Cochin. ^{1}H-NMR study was performed in DMSO as solvent on Bruker Advance-II 400 NMR spectrophotometer. Infrared spectrum was recorded in nujol mull on Perkin Elmer Spectrum RX-I spectrophotometer in the range from 4000 to 400 cm^{-1} at Sophisticated Analytical Instrumentation Facility (SAIF), Punjab University, Chandigarh. The nonisothermal thermogravimetric analysis of newly prepared terpolymer has been carried out using Perkin Elmer Diamond 3-II thermogravimetric analyzer, in air atmosphere with a heating rate of 10°C min^{-1} in the temperature range from 40 to 1000°C at Vishveshwarya National Institute of Technology (VNIT), Nagpur.

2.4. Analytical and Physicochemical Studies. The number average molecular weight ($\overline{Mn}$) was determined by conductometric titration in nonaqueous medium such as dimethyl sulfoxide (DMSO) using ethanolic KOH as a titrant. Conductometric titration in nonaqueous media has been proved to be a simple yet effective method used by earlier research workers [13, 23, 24] for the determination of number average molecular weight of phenol-formaldehyde resins [25].

The intrinsic viscosity of newly synthesized terpolymer has been evaluated using Ubbelohde viscometer [26, 27] fabricated in our research laboratory at different concentrations ranging from 0.3 to 0.05% of terpolymer in DMSO at 30°C.

TABLE 1: Elemental analysis data of 2-ATPHMDAF-I terpolymer.

Terpolymer	Carbon (%)		Hydrogen (%)		Nitrogen (%)		Sulphur (%)		Yield (%)
	Expt.	Calc.	Expt.	Calc.	Expt.	Calc.	Expt.	Calc.	
2-ATPHMDAF-I	63.15	63.40	8.43	8.68	15.74	15.85	11.87	12.08	80.55

Intrinsic viscosity $[\eta]$ was calculated from relevant plots of Huggins equation (1) [28] and Kraemer's equation (2) [29]. Consider

$$\frac{\eta_{\text{sp}}}{\text{C}} = [\eta] + K_1[\eta]^2 \cdot C \tag{1}$$

$$\ln\frac{\eta_{\text{rel}}}{\text{C}} = [\eta] + K_2[\eta]^2 \cdot C, \tag{2}$$

where η_r is the relative viscosity, $[\eta] = \lim_{C\to 0}(\eta_{\text{sp}}/\text{C})$, K_1 is Huggins' constant, and K_2 is Kraemer's constant.

2.5. Thermal Studies

2.5.1. Theoretical Consideration. Thermogram was interpreted and analyzed to obtain information about the % weight loss at different temperatures which gives information about sample composition, product formed after heating, and kinetic parameters. Kinetics parameters have been determined using Friedman [30], Chang [31], Sharp-Wentworth [32], and Freeman-Carroll [33] methods as follows.

Friedman's technique is as follows:

$$\ln\left(\frac{d\alpha}{dt}\right) = \ln(z) + n\ln(1-\alpha) - \frac{E_a}{RT}, \tag{3}$$

where α is the conversion at time t, R is the gas constant (8.314 J/mol/K), and T is the absolute temperature (K). From the slope of the linear plot of $\ln(1-\alpha)$ versus $1/T$, n can be obtained. The plot of $\ln(d\alpha/dt)$ versus $1/T$ should be linear with the slope E_a/R, from which E_a can be obtained.

Chang's technique is as follows:

$$\ln\frac{(d\alpha/dt)}{(1-\alpha)^n} = \ln(z) - \frac{E_a}{RT}. \tag{4}$$

A plot of $[\ln(d\alpha/dt)/(1-\alpha)^n]$ versus $1/T$ will yield a straight line if the order of decomposition reaction, n, is selected correctly. The slope and intercept of this line will provide the $(-E_a/R)$ and $\ln(z)$ values, respectively.

Sharp-Wentworth's technique is as follows:

$$\log\frac{dc/dt}{1-c} = \log\left(\frac{A}{\beta}\right) - \frac{E_a}{2.303R}\cdot\frac{1}{T}, \tag{5}$$

where dc/dt is the rate of change of fraction of weight with change in temperature; β is linear heating rate, dT/dt; c is the fraction of polymer decomposed at time t. Thus, a linear plot of $\log((dc/dt)/(1-c))$ versus $1/T$ is obtained whose slope gives the value of E_a and A may be evaluated from the intercept. The linear relationship confirmed that the assumed order is correct.

Freeman-Carroll's technique is as follows:

$$\frac{\Delta\log(dw/dt)}{\Delta\log W_r} = \left(-\frac{E_a}{2.303R}\right)\cdot\frac{\Delta(1/T)}{\Delta\log W_r} + n, \tag{6}$$

where dw/dt is the rate of change of weight with time, $W_r = W_c - W$, W_c is the weight loss at the completion of reaction; W is the total weight loss up to time, E_a is the energy of activation, and n is the order of reaction.

The $\Delta\log(dw/dt)$ and $\Delta\log W_r$ values were taken at regular intervals of $1/T$. In this case, $\Delta\log(dw/dt)/\Delta\log W_r$ versus $\Delta(1/T)/\Delta\log W_r$ gives a straight line. The slope and intercept are equal to $-(E_a/R)$ and n, respectively.

3. Results and Discussion

3.1. Elemental Analysis. The results of elemental analysis are shown in Table 1 used to assign empirical formula and empirical weight for 2-ATPHMDAF-I terpolymer. Composition of terpolymer was assigned on the basis of elemental analysis and was found to be in good agreement with that of calculated values.

The number average molecular weight $(\overline{\text{Mn}})$ of this terpolymer has been determined by conductometric titration method in nonaqueous medium using standard potassium hydroxide (0.05 M) in absolute ethanol as a titrant. The specific conductance was plotted against milliequivalents of ethanolic KOH required for neutralization of 100 g of each terpolymer. There are several breaks before the complete neutralization of all phenolic thiol groups. The first break in the plot was the smallest break and it is assumed that this corresponds to a stage in titration when an average of one phenolic thiol group of each chain was neutralized. From the plot, the first and final breaks were noted. The average degree of polymerization $(\overline{\text{Dp}})$ and hence the number average molecular weight $(\overline{\text{Mn}})$ of terpolymer have been determined using the following formula:

$$(\overline{\text{Dp}}) = \frac{\text{Total milliequivalents of base required for complete neutralisation}}{\text{Milliequivalents of base required for smallest interval}},$$

$$(\overline{\text{Mn}}) = (\overline{\text{Dp}}) \times \text{Repeat unit weight}. \tag{7}$$

It is observed that the molecular weight of terpolymer increases with the increase in hexamethylenediamine content. This observation is in agreement with the trend observed by earlier researchers [17]. Conductometric titration curve of terpolymer was shown in Figure 2 and the results are presented in Table 2.

The intrinsic viscosity $[\eta]$ was determined by the corresponding linear plots (Figure 3). Huggins and Kraemer's

TABLE 2: Molecular weight determination and viscometric data of 2-ATPHMDAF-I terpolymer.

Terpolymer	Empirical formula of repeat unit	Empirical weight of repeat unit	Average degree of polymerization ($\overline{Dp}$)	Average molecular weight ($\overline{Mn}$)	Intrinsic viscosity $[\eta]$ dL g^{-1}	Huggins constant (K_1)	Kraemer's constant (K_2)	$K_1 + K_2$
2-ATPHMDAF-I	$C_{14}H_{23}N_3S_1$	265	32	8480	0.82	0.286	0.272	0.558

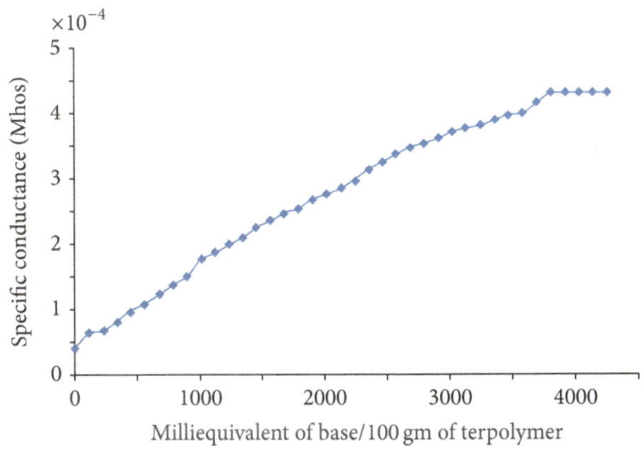

FIGURE 2: Conductometric titration curve of 2-ATPHMDAF-I terpolymer.

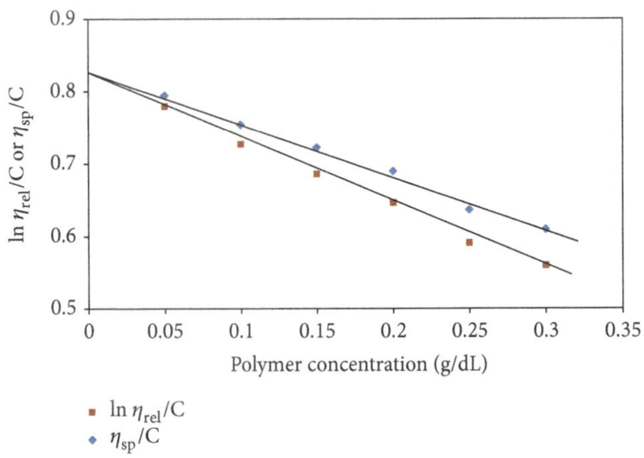

FIGURE 3: Viscometric plot of 2-ATPHMDAF-I terpolymer.

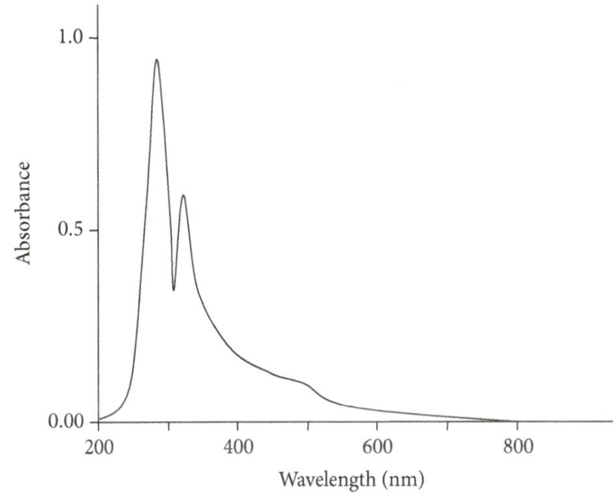

FIGURE 4: UV-visible spectrum of 2-ATPHMDAF-I terpolymer.

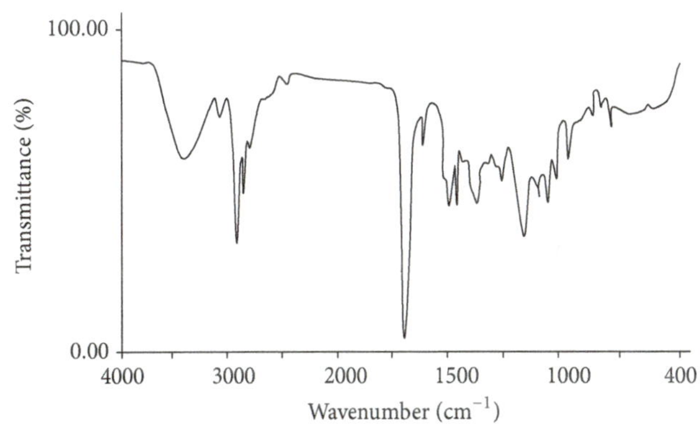

FIGURE 5: FT-IR spectrum of 2-ATPHMDAF-I terpolymer.

constants were determined by (1) and (2). According to the above relations, the plots of η_{sp}/C and $\ln \eta_{rel}/C$ against C were linear with slopes of K_1 and K_2, respectively. By extrapolating linear plot to zero concentration, intercepts on the viscosity function axis give $[\eta]$ value in both plots. The calculated values of the constants K_1 and K_2 (Table 2) in most cases satisfy the relation $K_1 + K_2 = 0.5$ favorably. It was observed that terpolymer having higher ($\overline{Mn}$) shows higher value of $[\eta]$.

3.2. UV-Visible Spectra. The UV-visible spectrum is depicted in Figure 4 for 2-ATPHMDAF-I terpolymer sample in pure DMSO recorded in the region from 200 to 800 nm. The spectrum exhibits two absorption maxima in the regions 280–300 and 300–330 nm. These observed positions for the absorption bands have different intensities. The more intense band is due to $\pi \rightarrow \pi^*$ allowed transition of conjugation in aromatic ring which is shifted from the basic value 269 nm to 280–300 nm and the medium band at 300–330 nm is due to $n \rightarrow \pi^*$ forbidden transition of –SH group.

The more intense band is due to the fact that bathochromic shift may be due to the combining effect of conjugation and the phenolic –SH and $-NH_2$ groups. Both bathochromic and hyperchromic effects of conjugation are shown by chromophore (aromatic ring) due to auxochromic substituents –SH and $-NH_2$ groups in terpolymer. This observation is in good agreement with the proposed most probable structure of above terpolymer.

3.3. FT-IR Spectra. The FT-IR spectrum of 2-ATPHMDAF-I terpolymer depicted in Figure 5 and spectral data are

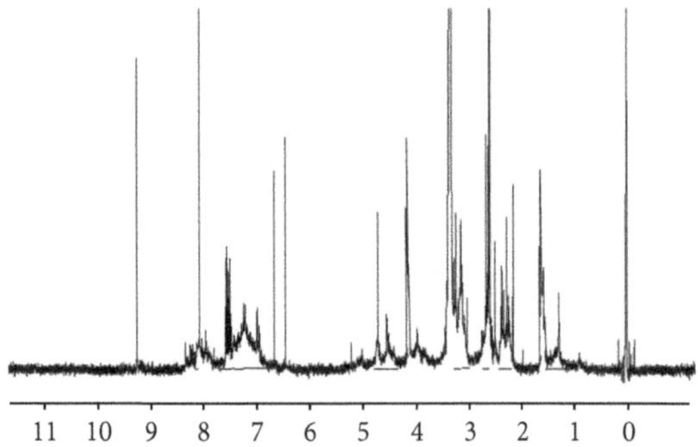

FIGURE 6: ^{1}H-NMR spectrum of 2-ATPHMDAF-I terpolymer.

TABLE 3: IR frequencies of 2-ATPHMDAF-I terpolymer.

Observed wavenumber (cm^{-1})	Assignments
3434 b, st	>NH stretching (sec. amine)
3024 m	>C–H stretching (aromatics)
2600 w	–SH (phenolic thiol)
1650–1700 sh	>NH bending (sec. amine)
1482 m	>C=C< (aromatic ring)
1236 m	C–N stretching (aromatic amine)
807 w 1013 m	Tetrasubstitution in benzene skeleton
	Methylene bridge
1467 m	$-CH_2$ bending
1345 b, m	$-CH_2$ wagging
729 m	$-CH_2$ rocking

b: broad; sh: sharp; st: strong; m: medium; w: weak.

TABLE 4: ^{1}H-NMR spectral data of 2-ATPHMDAF-I terpolymer.

^{1}H-NMR chemical shift (δ) ppm of terpolymer	Nature of protons assigned
8.05 s	Meta proton of Ar–H (aromatic –H)
7.2 s	Proton of Ar–SH (thiophenol)
6.7 s	Amino proton of $-CH_2-NH-CH_2-$ (NH-bridge)
6.5 s	Proton of $Ar-NH_2$ (phenolic $-NH_2$)
4.1 t	Methylene proton of $-NH-CH_2-CH_2-$ linkage
2.5 s	Methylene proton of $Ar-CH_2-NH-$ linkage
1.5 q	Methylene proton of $-CH_2-CH_2-CH_2-$ linkage

tabulated in Table 3. A broad and strong band appeared at 3434 cm^{-1} which may be assigned due to the >NH stretching (sec. amine). A medium and sharp band displayed at 3024 cm^{-1} may be assigned due to the stretching vibration of aromatic C–H group. A weak band at 2600 cm^{-1} indicates the presence of a –SH (phenolic thiol) group. A sharp band appearing in the region of 1650–1700 cm^{-1} may be due to bending vibration of >NH (sec. amine) group. A medium band appearing at 1482 cm^{-1} indicates the presence of >C=C< (aromatic) group. Aromatic C–N stretch appeared in the region of 1236 cm^{-1} which shows medium band. The presence of tetrasubstituted benzene skeleton shows weak and medium peaks at 807 cm^{-1} and 1013 cm^{-1}. The methylenic bridge shows broad and medium peaks for bending, wagging, and rocking vibration appearing at 1467, 1345, and 729 cm^{-1}.

3.4. ^{1}H-NMR Spectra. ^{1}H-NMR spectral data are incorporated in Table 4 and spectrum is presented in Figure 6. Spectrum reveals different patterns of peaks, since each of them possesses a set of protons having different proton environment. The significant singlet signal appearing at the region of δ 8.05 ppm is due to metaproton of Ar–H. A singlet observed at δ 7.2 ppm is due to proton of Ar–SH (thiophenol). Amino proton of $-CH_2-NH-CH_2-$ linkage gives singlet that is observed at δ 6.7 ppm. Amino proton of aromatic amine gives singlet at δ 6.5 ppm. Methylenic proton of $-NH-CH_2-CH_2-$ linkage gives triplet at δ 4.1 ppm. Singlet is observed at δ 2.5 ppm due to methylenic proton of $Ar-CH_2-NH-$ linkage. The quintet at δ 1.5 ppm may be attributed to methylenic proton of $-CH_2-CH_2-CH_2-$ linkage.

3.5. Thermogravimetric Analysis. 2-ATPHMDAF-I terpolymer was subjected to thermogravimetric analysis and the data was used to assess the degradation pattern. Thermal degradation behavior of synthesized terpolymer has been incorporated in Table 5 and decomposition pattern is shown in Figure 7 in temperature range from 40 to 1000°C.

Decomposition pattern of terpolymer thermogram shows four decomposition steps in which the loss of water molecule (6.12% expected and 6.36% calculated) has been observed in the first step up to 130°C. Second decomposition step starts from 130 to 230°C which represents the degradation of –SH and $-NH_2$ groups attached to aromatic nucleus (23.52% expected and 23.67% calculated). The third decomposition step starts from 230 to 320°C, corresponding to 59.66% mass loss of aromatic nucleus with two methylene groups against calculated 59.71%. The fourth decomposition step starts from 320 to 990°C which corresponds to loss of complete hexamethylenediamine moiety (98.67% expected and 100% calculated). Consequently, no residue may be assigned after complete degradation.

3.6. Kinetics of Thermal Decomposition by the Friedman, Chang, Sharp and Wentworth, and Freeman-Carroll Methods. The observed thermal stability of 2-ATPHMDAF-I terpolymer may be due to the stronger intermolecular hydrogen bonding present in polymer which may be attributed due to the presence of water of crystallization resulting in the resistance to higher temperature. By applying the thermogravimetric data to four thermal degradation kinetic methods, that is, the Friedman, Chang, Sharp and Wentworth, and Freeman-Carroll methods, it shows four different degradation steps corresponding to loss of respective groups. The thermoanalytical data has been determined for different stages as given in Table 6. This kinetic analysis should be

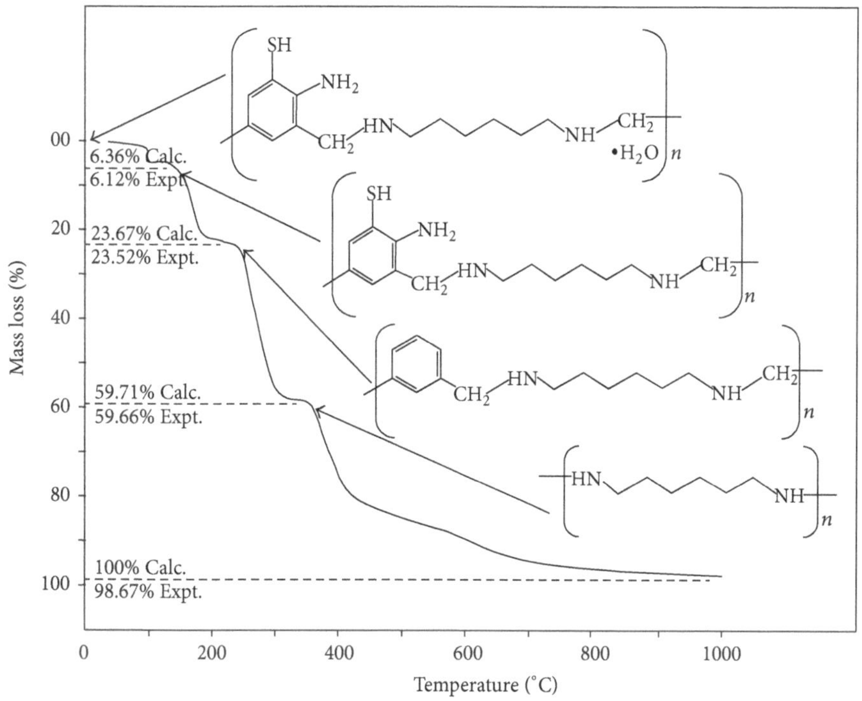

FIGURE 7: Decomposition pattern of 2-ATPHMDF-I terpolymer.

TABLE 5: Thermal degradation behavior of 2-ATPHMDAF-I terpolymer.

Terpolymer	Stages of decomposition	Temp. range (°C)	Species degraded	Wt. loss (%) Expt.	Wt. loss (%) Calc.
2-ATPHMDAF-I	First	40–130	One water molecule	6.12	6.36
	Second	130–230	One –SH and one –NH_2 group	23.52	23.67
	Third	230–320	Benzene ring along with side chain of hexamethylenediamine moiety	59.66	59.71
	Forth	320–990	Complete loss of hexamethylenediamine moiety	98.67	100

TABLE 6: Thermoanalytical data for each degradation step of 2-ATPHMDF-I terpolymer.

Methods	I E_a	I n	I z	II E_a	II n	II z	III E_a	III n	III z
Friedman	32.11	1.14	5.34	42.01	1.05	12.23	28.67	1.15	12.01
Chang	31.93	1.0	5.38	43.90	1.0	12.44	28.34	1.0	12.03
Sharp-Wentworth	13.95	1.0	7.04	18.19	1.0	7.66	12.42	1.0	8.47
Freeman-Carroll	15.56	0.9	6.12	25.19	1.33	5.30	8.12	1.32	5.21

(Decomposition steps I, II, III.)

a starting point to obtain the useful information on the behavior of sample.

To obtain the relative thermal stability of terpolymer, the methods described by Friedman, Chang, Sharp and Wentworth, and Freeman-Carroll were adopted. From the results, it is concluded that the values of kinetic parameters show good agreement with each other in Friedman and Chang's methods and some different results are obtained in Sharp and Wentworth and Freeman-Carroll's methods as shown in Table 6.

Fairly comparable results of kinetic parameters, namely, E_a, n, and z, are obtained for each degradation step by Friedman and Chang method and may be due to analogy in mathematical model. Also, results obtained by Sharp and Wentworth and Freeman-Carroll methods are in good agreement with each other with slight variations between the results; it is concluded that the values of kinetic parameters depend on kinetic methods used as well as degrading species at a particular step. Total calculations obtained from different kinetic methods demonstrated that the numerical value of

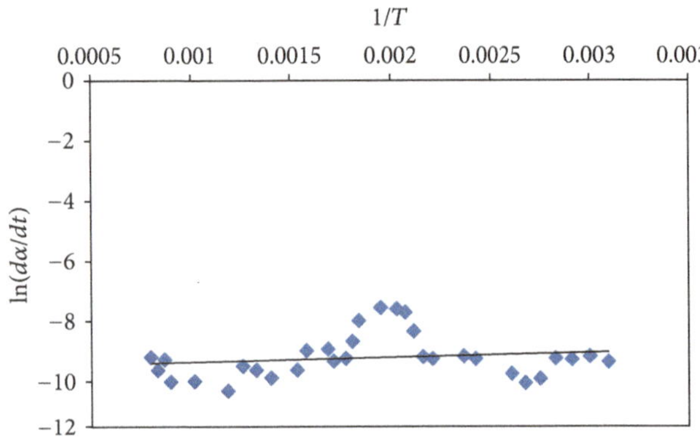

FIGURE 8: Friedman's plot of 2-ATPHMDAF-I terpolymer for energy of activation.

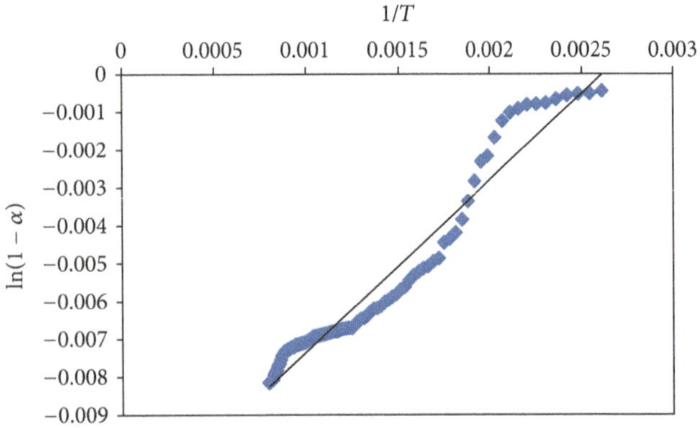

FIGURE 9: Friedman's plot of 2-ATPHMDAF-I terpolymer for order of reaction.

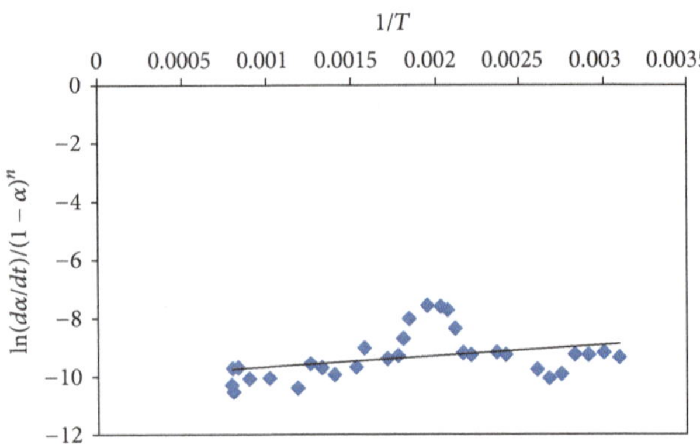

FIGURE 10: Chang's plot of 2-ATPHMDAF-I terpolymer.

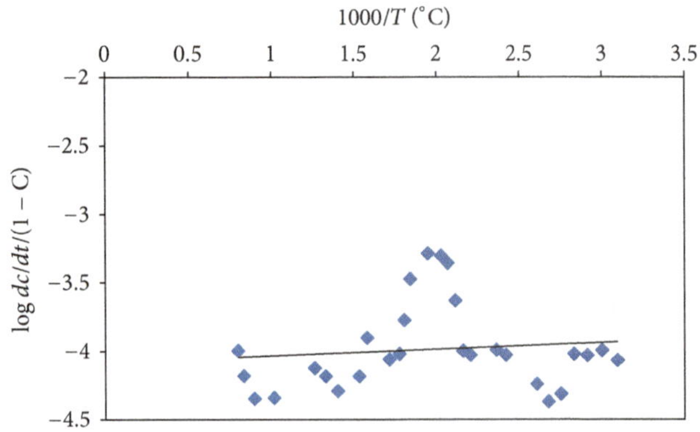

FIGURE 11: Sharp and Wentworth's plot of 2-ATPHMDAF-I terpolymer.

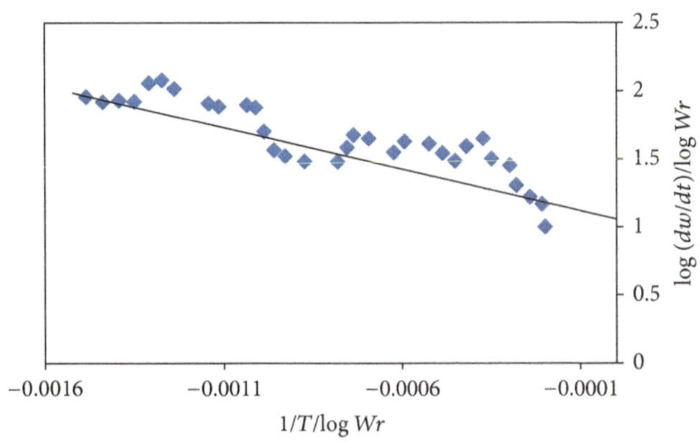

FIGURE 12: Freeman-Carroll's plot of 2-ATPHMDAF-I terpolymer for order of reaction.

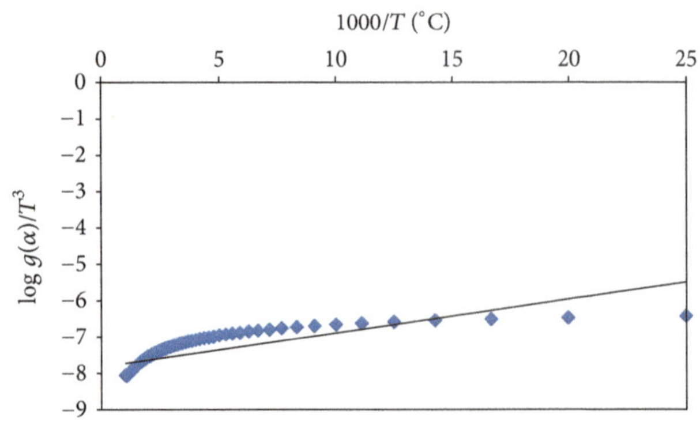

FIGURE 13: Freeman-Carroll's plot of 2-ATPHMDAF-I terpolymer for activation energy.

kinetic parameters depends on the mathematical models used to analyze the experimental data and level of degradation. By using the above mentioned techniques, variations in the results are obtained which represents versatility and great utility of thermal degradation of mathematical kinetics equations in thermogravimetry. However, it is difficult to draw any unique conclusion regarding the decomposition mechanism.

From the thermogravimetric analysis, kinetic plots of the terpolymer have been shown in Figures 8, 9, 10, 11, 12, and 13 to calculate energy of activation, order reaction, and frequency factor.

By using the above mentioned methods, fairly similar results are obtained in Friedman and Chang's methods and slight variations are obtained in between Sharp and Wentworth and Freeman-Carroll method which are found to be in good agreement with each other. From the point of view of chemical kinetics, 2-ATPHMDAF-I is thermally stable.

Low values of frequency factor revealed that decomposition reaction of terpolymer may be slow and no other possible reason can be given. However, in Friedman, Chang, Freeman-Carroll, and Sharp and Wentworth's plots, all points did not fall on straight line, which indicates that the decomposition of terpolymer is not obeying first order of reaction perfectly [13, 34].

4. Conclusion

Synthesis of targeted terpolymer (2-ATPHMDAF-I) has been confirmed which is supported by the results obtained by elemental analysis and spectral data. From the elemental analysis, electronic, IR, and ^{1}H-NMR spectral studies, the proposed structure of the synthesized terpolymer is confirmed. The values of kinetic parameter calculated from the Friedman and Chang methods are in good agreement with each other and values obtained from the Sharp and Wentworth and Freeman-Carroll methods are also in similar order. Thermogram of targeted terpolymer shows four degradation steps and hexamethylenediamine molecule almost degrades completely up to 990°C. From the results obtained, the values of kinetic parameters are significantly controlled by the level of degradation and calculation methods used to analyze the experimental data.

Conflict of Interests

The authors declare that there is no conflict of interests regarding the publication of this paper.

Acknowledgments

The authors wish to express their sincere thanks to University Grand Commission for financial assistance to their laboratory, to SAIF, Punjab University, Chandigarh, for carrying out spectral analysis, and also to VNIT, Nagpur, for providing thermogravimetric facility.

References

[1] W. B. Gurnule, P. K. Rahangdale, L. J. Paliwal, and R. B. Kharat, "Chelation ion-exchange properties of copolymer resin derived from 4-hydroxyacetophenone, oxamide, and formaldehyde," *Journal of Applied Polymer Science*, vol. 89, no. 3, pp. 787–790, 2003.

[2] V. D. Mane, N. J. Wahane, and W. B. Gurnule, "Copolymer resin. VII. 8-hydroxyquinoline-5-sulfonic acid-thiourea- formaldehyde copolymer resins and their ion-exchange properties," *Journal of Applied Polymer Science*, vol. 111, no. 6, pp. 3039–3049, 2009.

[3] M. V. Tarase, A. B. Zade, and W. B. Gurnule, "Resin I: synthesis, characterization, and ion-exchange properties of terpolymer resins derived from 2,4-dihydroxypropiophenone, biuret, and formaldehyde," *Journal of Applied Polymer Science*, vol. 108, no. 2, pp. 738–746, 2008.

[4] A. D. Kushwaha, A. B. Kalambe, V. V. Hiwase, and D. N. Urade, "Structural and antibacterial study of resin-II derived from p-nitrophenol, resorcinol and formaldehyde," *Journal of Chemical and Pharmaceutical Research*, vol. 4, no. 2, pp. 1111–1116, 2012.

[5] W. Hackenberger, R. Alberta, W. Paul, D. Jeong, and Q. Zhang, "High dielectric constant terpolymers for energy storage capacitors," in *Proceedings of the 25th Symposium for Passive Components*, pp. 239–243, Palm Springs, 2005.

[6] S. K. Kapse, V. V. Hiwase, and A. B. Kalambe, "Synthesis and semi conducting behaviour of the ter-polyligand derived from p-hydroxyacetophenone, quinhydrone and melamine," *Journal of Chemical and Pharmaceutical Research*, vol. 4, no. 3, pp. 1734–1739, 2012.

[7] D. N. Urade, V. V. Hiwase, and A. B. Kalambe, "Structural and thermokinetic parameters of terpolymeric resin derived from p-hydroxyacetophenone, bis (2-amino- 1, 3, 4 -thiadiazole) and glycerol," *Journal of Chemical and Pharmaceutical Research*, vol. 4, no. 1, pp. 732–740, 2012.

[8] S. Pashaei, S. Siddaramaiah, M. M. Avval, and A. A. Syed, "Thermal degradation kinetics of nylon6/gf/crysnano nanoclay nanocomposites by TGA," *Chemical Industry and Chemical Engineering Quarterly*, vol. 17, no. 2, pp. 141–151, 2011.

[9] A. Gardziella, L. A. Pilato, and A. Knop, *Phenolic Resins, Chemistry, Applications, Standardization, Safely and Ecology*, Springer, 2000.

[10] G. E. Zaikov, *Degradation and Stabilization of Polymers*, Theory and Practice, Nova Science Publishers, New Delhi, India, 1995.

[11] N. G. Grassie, *Scott Polymer Degradation and Stabilization*, Cambridge Press, Cambridge, Mass, USA, 1985.

[12] R. N. Singru, A. B. Zade, and W. B. Gurnule, "Thermoanalytical study and kinetics of new 8-hydroxyquinoline 5-sulphonic acid-oxamide-formaldehyde terpolymer resins," *E-Journal of Chemistry*, vol. 6, no. 1, pp. S171–S182, 2009.

[13] R. N. Singru, A. B. Zade, and W. B. Gurnule, "Synthesis, characterization, and thermal degradation studies of copolymer resin derived from p-cresol, melamine, and formaldehyde," *Journal of Applied Polymer Science*, vol. 109, no. 2, pp. 859–868, 2008.

[14] M. Karunakaran, C. T. Vijayakumar, C. Magesh, and T. Amudha, "Terpolymer resin-ii -thermal and metal ion binding properties of resorcinolthiourea- formaldehyde terpolymer resin," *International Journal of Engineering Science and Technology*, vol. 3, no. 1, pp. 162–176, 2011.

[15] M. V. Tarase, A. B. Zade, and W. B. Gurnule, "Kinetics of thermal degradation studies of some new terpolymers derived from 2, 4-dihydroxypropiophenone, oxamide, and formaldehyde," *Journal of Applied Polymer Science*, vol. 116, no. 2, pp. 619–627, 2010.

[16] S. S. Butoliya, W. B. Gurnule, and A. B. Zade, "Study of non-isothermal decomposition and kinetic analysis of 2,4-dihydroxybenzoic acid-melamine-formaldehyde copolymer," *E-Journal of Chemistry*, vol. 7, no. 3, pp. 1101–1107, 2010.

[17] P. E. P. Michael, P. S. Lingala, H. D. Juneja, and L. J. Paliwal, "Synthetic, structural, and thermal degradation of a tercopolymer derived from salicylic acid, guanidine, and formaldehyde," *Journal of Applied Polymer Science*, vol. 92, no. 4, pp. 2278–2283, 2004.

[18] P. E. P. Michael, J. M. Barbe, H. D. Juneja, and L. J. Paliwal, "Synthesis, characterization and thermal degradation of 8-hydroxyquinoline-guanidine-formaldehyde terpolymer," *European Polymer Journal*, vol. 43, no. 12, pp. 4995–5000, 2007.

[19] S. S. Rahangdale, A. B. Zade, and W. B. Gurnule, "Thermal degradation studies of terpolymer derived from 2, 4-dihydroxyacetophenone, dithiooxamide and formaldehyde," *Ultra Scienceno*, vol. 19, pp. 213–218, 2007.

[20] S. S. Katkamwar, A. B. Zade, and W. B. Gurnule, *Ultra Science*, vol. 73, no. 2, pp. 103–108, 2007.

[21] S. K. Mandavgade, J. R. Dontulwar, and W. B. Gurnule, "Thermal degradation study of new polymer derived fr om 8-hydroxyquinoline 5-sulphonic acid and catechol," *Der Pharma Chemicano*, vol. 4, no. 4, pp. 1695–1703, 2012.
[22] M. V. Tarase, A. B. Zade, and W. B. Gurnule, "Kinetics of thermal degradation studies of some new terpolymers derived from 2, 4-dihydroxypropiophenone, oxamide, and formaldehyde," *Journal of Applied Polymer Science*, vol. 116, no. 2, pp. 619–627, 2010.
[23] R. N. Singru, W. B. Gurnule, V. A. Khati, A. B. Zade, and J. R. Dontulwar, "Eco-friendly application of p-cresol-melamine-formaldehyde polymer resin as an ion-exchanger and its electrical and thermal study," *Desalination*, vol. 263, no. 1-3, pp. 200–210, 2010.
[24] R. H. Gupta, A. B. Zade, and W. B. Gurnule, "Resin-IV: synthesis and characterization of terpolymers derived from 2-hydroxyacetophenone, melamine, and formaldehyde," *Journal of Applied Polymer Science*, vol. 109, no. 5, pp. 3315–3320, 2008.
[25] G. E. Dunn and R. C. Donald, "Infrared spectra of aqueous sodium benzoates and salicylates in the carboxyl-stretching region: chelation in aqueous sodium salicylates," *Canadian Journal of Chemistry*, vol. 47, no. 24, pp. 4577–4588, 1969.
[26] R. J. Young and P. A. Lovell, *Introduction To Polymers*, Nelson Thornes, 2nd edition, 1991.
[27] N. P. S. Chauhan, "Terpolymerization of p-acetylpyridine oxime, p-methylacetophenone and formaldehyde, and its thermal studies," *Journal of Thermal Analysis and Calorimetry*, vol. 110, no. 3, pp. 1377–1388, 2012.
[28] M. L. Huggins, "The viscosity of dilute solutions of long-chain molecules. IV. Dependence on concentration," *Journal of the American Chemical Society*, vol. 64, no. 11, pp. 2716–2718, 1942.
[29] E. O. Kraemer, "Molecular weights of celluloses and cellulose derivates," *Industrial & Engineering Chemistry Research*, vol. 30, no. 10, pp. 1200–1203, 1938.
[30] H. L. Freidman, "Kinetics of thermal degradation of char-forming plastics from thermogravimetry. Application to a phenolic plastic," *Journal of Polymer Science C: Polymer Symposia*, vol. 6, no. 1, pp. 183–195, 1964.
[31] W. L. Chang, "Decomposition behavior of polyurethanes via mathematical simulation," *Journal of Applied Polymer Science*, vol. 53, no. 13, pp. 1759–1769, 1994.
[32] J. H. Sharp and S. A. Wentworth, "Kinetic analysis of thermogravimetric data," *Analytical Chemistry*, vol. 41, no. 14, pp. 2060–2062, 1969.
[33] E. S. Freeman and B. Carroll, "The application of thermoanalytical techniques to reaction kinetics. The thermogravimetric evaluation of the kinetics of the decomposition of calcium oxalate monohydrate," *The Journal of Physical Chemistry*, vol. 62, no. 4, pp. 394–397, 1958.
[34] J. J. Zsago, "Kinetic analysis of thermogravimetric data," *The Journal of Physical Chemistry*, vol. 72, no. 7, pp. 2406–2411, 1968.

18

New Semi-IPN Hydrogels Based on Cellulose for Biomedical Application

S. K. Bajpai[1] and M. P. Swarnkar[2]

[1] *Polymer Research Laboratory, Department of Chemistry, Government Model Science College, Jabalpur, Madhya Pradesh 482001, India*
[2] *Department of Chemistry, Kymore Science College, Kymore, Katni, Madhya Pradesh 483880, India*

Correspondence should be addressed to S. K. Bajpai; sunil.mnlbpi@gmail.com

Academic Editor: Dirk Kuckling

Cellulose pulp, obtained from a paper industry, has been dissolved in PEG/NaOH system and the resulting solution has been polymerized in the presence of monomer acrylic acid (AA) and crosslinker N,N′ methylene bisacrylamide via free radical polymerization. The Cell/PEG/poly (SA) ternary semi-IPN hydrogel, so prepared, was characterized by FTIR and TG analysis. The dynamic water uptake of various hydrogels, having different compositions, was investigated in the physiological buffer of pH 7.4 at 37°C. The various hydrogels exhibited chain-relaxation controlled swelling behavior. The uptake data was best interpreted by Schott kinetic model. The various diffusion coefficients, that is, initial (D_i), average (D_{ave}), and late time (D_L), were also calculated using the dynamic water uptake data. The hydrogels showed fair pH and salt-dependent swelling behavior.

1. Introduction

Cellulose is one of the most abundant renewable organic materials with annual production of about 5×10^{11} metric tons [1]. Cellulose is a linear syndiotactic homopolymer composed of D-anhydroglucopyranose units, which are linked by β-(1 $\rightarrow$ 4)-glycosidic bonds. This biopolymer exhibits some excellent properties such as mechanical robustness, biodegradability, hydrophilicity, and biocompatibility [2]. These properties are mainly responsible for extensive use of cellulose in a wide spectrum of applications that include pharmacy, agriculture, medical science, industries, and so many other related branches. Some of the major applications include water reservoirs in agriculture [3], body water retainers [4], ocular bandages [5], artificial cartilage [6], controlled drug delivery [7, 8], scaffolds for regenerative medicine [9], stomach bulking agents [10], and wound dressings [11]. Most of the applications of cellulose-based hydrogels, mentioned above, exploit water absorption property of cellulose and so it is essential to have full control on the swellability of this biopolymer in these applications. But insolubility of cellulose in water (due to extensive inter- and intramolecular H-bonding between O and H atoms of cellulose) and lack of ionizable functional groups in cellulose backbone are the major barriers that must be crossed to make this polymer useful in various fields. In recent past, several solvent systems have been used for dissolution of cellulose in effective manners. These include NaOH-urea/thiourea system [12] and ionic liquids [13], which are limited to laboratory use only due to volatility, toxicity, and high cost. In addition, the introduction of desired functionality into cellulose backbone is usually achieved by its chemical modification, which is carried out on alcoholic hydroxyls in each of its AGUs under practical relevance. This sometimes compromises with the basic properties of cellulose.

Recently, a new solvent system consisting of aqueous solution of poly (ethylene glycol) (PEG) and NaOH has been reported for dissolution of cotton linters to prepare cellulose fibers [14]. This newly developed solvent system enabled us to think that addition of a suitable monomer into dissolved cellulose in PEG/NaOH system, followed by free radical induced polymerization, could yield a ternary polymeric hydrogel system with desirable properties like pH and salt sensitivity. Poly (ethylene glycol) is a water soluble biocompatible polymer and has been used extensively in biomedical fields [15]. The Food and Drug Administration (FDA), USA, has approved PEGylated agents for parenteral or topical administration and as a component of various foods, cosmetics, and drug delivery systems, such as liposomes and nasal sprays [16].

In the light of the above-mentioned facts, we hereby report synthesis and water absorption behavior of semi-IPN hydrogel system, comprising of Cell, PEG, and sodium acrylate (SA). The system is not only biocompatible but it also exhibits fair pH and salt sensitive water absorption behavior. Moreover, the presence of three polymers could make a better control over the swelling responses of this hydrogel system.

2. Experimental

2.1. Materials. Cellulose pulp was obtained from a local paper industry (Orient Paper Mill, Amlai, Shahdol, Madhya Pradesh). The monomer acrylic acid (AAc), crosslinker N,N′ methylene bisacrylamide (MB), and initiator potassium per sulphate (KPS) were obtained from Merck Chemical Industry, Mumbai, India, and were analytical grade. Poly ethylene glycol l (PEG) with molar masses of 200, 4000, and 6000 were supplied from Research Lab (Pune, India). The other chemicals such as H_2SO_4, NaOH, sodium dihydrogen phosphate, and acetic acid were obtained from Hi Media Chemicals, Mumbai, India.

2.2. Methods

2.2.1. Preparation of Microcrystalline Cellulose (MCC) from Pulp. The cellulose pulp (CP) was acid hydrolyzed using H_2SO_4 solution (64% wt/vol) to yield microcrystalline cellulose (MCC) powder following the method given elsewhere [17].

2.2.2. Dissolution of MCC in PEG/NaOH Solvent System. A precalculated quantity of MCC powder was mixed into 100 mL of aqueous solution, containing PEG (6% wt/vol) and NaOH (8% wt/vol). The dispersion was allowed to be put in Deep Freezer (Voltas, India) at −5°C for a period of 24 h. The solid frozen mass, so obtained, was kept at room temperature under vigorous stirring for a period of 2 h. The resulting transparent solution was filtered with ordinary filter paper and was used to prepare hydrogels.

2.2.3. Preparation of Cell/PEG/Poly (SA) Semi-IPN Hydrogels. The solution, obtained in the above step, was used to prepare ternary hydrogel system. In brief, preparation of a representative sample may be described as follows: to 5 mL of cellulose solution, 10.63 m mol of monomer AAc, 260 μmol of crosslinker MB, and 200 μmol of initiator KPS were added and the mixture was stirred vigorously to ensure complete dissolution of all the components. The reaction solution was now transferred into test tubes (id 12.5 mm) and kept in electric oven (Tempstar, India) at 60°C for 2 h. After the polymerization was over, the resulting semitransparent hydrogels were taken out of test tubes, cut into slices (thickness 2.6 mm), and then placed in distilled water for 4 h to remove unreacted salts left within the polymers. Finally, the hydrogel discs were taken out and dried in a dust-free chamber until they showed constant weights. In all eight hydrogels, samples were prepared with varying contents of cellulose, sodium acrylate, and crosslinker MB. Moreover, hydrogels with PEG of varying molecular masses were also prepared. The composition of various semi-IPN hydrogels, prepared, is given in Table 1.

2.2.4. Characterization. The MCC obtained by acid hydrolysis of cellulose pulpand the ternary hydrogel synthesized were characterized by thermogravimetric (TG) analysis. In addition, surface morphology of MCC was also investigated by FE-SEM analysis.

The thermogravimetric analysis was performed using a thermogravimetric analyzer (Mettler, Toledo TGA/SDTA 851, Switzerland). A definite quantity of grinded sample was placed in a ceramic crucible and analyzed over the temperature range of 30 to 800°C, at the heating rate of 10°C min^{-1} under the constant flow of N_2 at the rate of 30 mL min^{-1}.

The morphology of the obtained MCC was investigated using a JEOL 6400F microscope operated with an accelerating voltage of 2 kV and a working distance of 4.4 mm. 50 μL sediment suspensions (0.01 wt%) was dropped onto clean silicon wafers followed by air drying for 24 h and then sputtered with an approximately 6 nm layer of gold/palladium. The FTIR spectrum of synthesized polymer was recorded on FTIR-spectrophotometer (Shimadzu, 8400S) using KBr. For this, dry polymer was grinded and mixed with KBr. The scans recorded were the average of 100 scans and spectral range selected was 400 to 4000 cm^{-1}.

2.2.5. Swelling Studies. The water absorption behavior of hydrogels was studied gravimetrically [18]. The preweighed hydrogel sample was placed in phosphate buffer solution of pH 7.4 under sink conditions at 37°C. The sample was taken out at regular time intervals, wiped superficially to remove loosely bound surface water, weighed accurately, and then again put in the swelling medium. The mass measurement process continued until the swollen hydrogel attained constant weight. The percent swelling (PS) was determined using the following expression:

$$\mathrm{PS} = \frac{(M_t - M_0)}{M_0} \times 100, \quad (1)$$

where M_0 and M_t are the initial dry mass and mass at time "t," respectively. The equilibrium percent swelling (EPS) was determined by replacing M_t in the above expression by M_∞, the mass of swollen hydrogel at equilibrium. All the experiments were carried out in triplicate and average values are reported in the data.

TABLE 1: Composition of various semi-IPN hydrogel synthesized[a].

Sample code	Cell (g)	PEG* (g)	SA (mili mol)	MB (μmol)	KPS (μmol)
A	0.2	0.4	10.63	260	200
B	0.2	0.4	10.63	390	200
C	0.2	0.4	10.63	520	200
D	0.2	0.4	6.36	390	200
E	0.2	0.4	14.88	390	200
F	0.4	0.4	10.63	390	200
G	0.6	0.4	10.63	390	200
H	0.2	0.4	10.63	390	200
I	0.2	0.4	10.63	390	200

[a]Volume of feed mixture = 5.0 mL.
*Samples A, B, C, D, E, F, and G were prepared with PEG of molar mass 6000.
**Samples H and I were prepared with PEG of molar masses 4000 and 200, respectively.

3. Results and Discussion

3.1. Rationale for the Semi-IPN Hydrogel System. The basic idea, lying behind the selection of appropriate polymers to prepare hydrogel, was based on the following considerations.

(1) The cellulose pulp, after acid hydrolysis, yields cellulose microparticles (CMP) that can be conveniently dissolved in PEG/NaOH system.

(2) The aqueous solution of cellulose, so obtained, contains another biocompatible polymer PEG in the dissolved state. Therefore, the solution has now two biopolymers in the dissolved state.

(3) The addition of ionizable groups containing monomer, that is, acrylic acid, into the above aqueous solution and subsequent polymerization could render the salt and pH sensitivity to the resulting ternary polymeric hydrogel system. In addition, the presence of NaOH in the solvent system would also transform AAc into sodium acrylate (SA) which imparts greater water absorption property to the resulting hydrogel system. In this way, the system Cell/PEG/poly(SA) could come out with a fair overall biocompatibility, pH and salt sensitivity, and, finally, excellent swelling property. The swelling of this system could be better controlled by variation in contents of these components.

3.2. Synthesis of Cell/PEG/Poly (SA) Semi-IPN Hydrogels. The hydrogels were prepared by free radical induced aqueous polymerization of monomer sodium acrylate (SA) in the presence of crosslinker MB and dissolved polymers cellulose and PEG. The interpenetrating polymer network, so formed, consists of covalently crosslinked poly(sodium acrylate) chains that are physically entangled with cellulose and PEG macromolecular segments. The possibility of H-bonding interactions between O and H atoms associated with cellulose and PEG molecules cannot be ruled out. The overall scheme of formation of ternary hydrogel system is shown in Figure 1.

3.3. Thermogravimetric Analysis (TGA). The thermograms of cellulose microparticles (CMP) and Cell/PEG/poly (SA) ternary hydrogel sample B are shown in Figures 2(a) and 2(b), respectively. It is clear that initial weight loss of nearly 5% is observed from 20 to 120°C, probably due to moisture evaporation (phase I). Later on, the CMP undergo a sharp thermal degradation in the temperature range of 210°C to 440°C, where all cellulose is pyrolyzed. The drastic weight loss is attributable to the liberation of volatile hydrocarbon from rapid thermal decomposition of cellulose. The total weight of CMP is now reduced to 30%. Therefore, in phase II, the weight loss is approximately 65%. Finally, in phase III, there is and observed gradual weight loss, probably due to the steady decomposition of the remaining heavy components mainly from lignin [19]. In this phase there is almost 8% weight loss.

The thermogram of ternary hydrogel system as depicted in Figure 2(B) shows a different behavior due to presence of other polymers, namely, poly (SA) and PEG. Out of the three polymers, namely, CMP, poly (SA), and PEG, present in the hydrogel, the highest content is of poly (SA) and so its thermal stability is expected to govern the overall stability of the thermogram. It can be seen that below 350°C there is weight loss of almost 31%, which is attributable to the loss of water molecules bound to the polar groups of polymeric constituents, and decarboxylation of –COO groups associated with poly (SA) chains [20]. Beyond 350°C, the weight loss of polymer continues up to 433°C due to presence of small amount of PEG in the sample. It has been reported by Manson et al. [21] that in the pure PEG the weight loss occurs in the range of 300 to 450°C. Finally, the second sharp weight loss occurs in the temperature range of 415°C to 510°C which is attributable to decomposition of both the polymer constituents with the elimination of CO and CO_2 by the way of abundant backbone scission and formation of small concentration of unsaturation. The slow and gradual weight loss beyond 510°C is associated with the carbonization of the degraded product or ash. If we compare the overall stability of the ternary hydrogel sample with cellulose microparticles then it is seen that CMP exhibit a total weight loss of 80% while the polymer sample B exhibits a total loss of 60% up to 600°C.

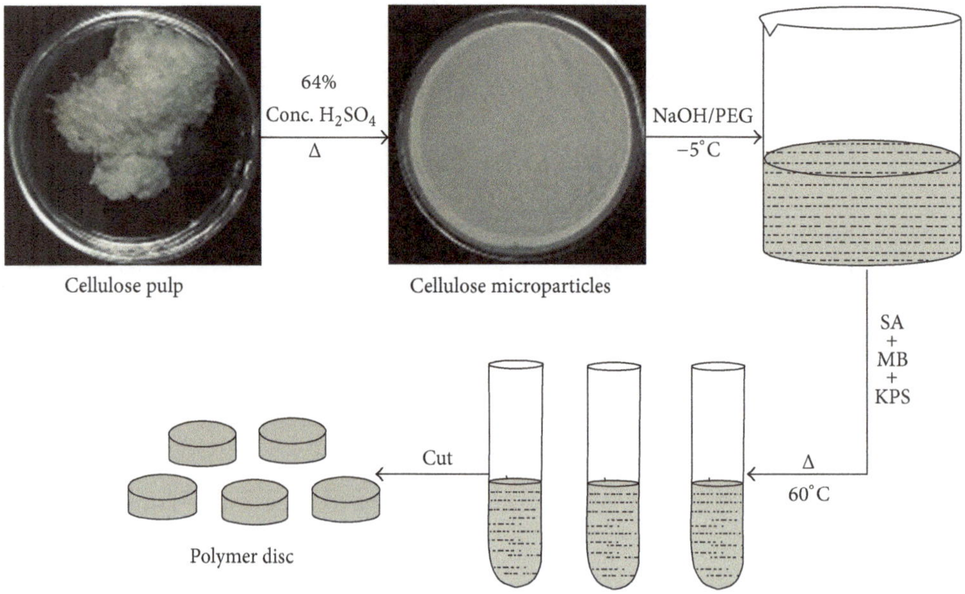

Figure 1: Scheme showing formation of Cell/PEG/poly (SA) semi-IPN hydrogel.

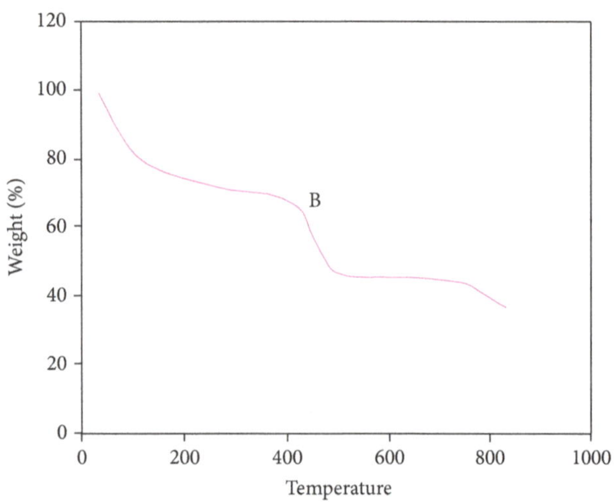

Figure 2: TGA of the semi-IPN sample B.

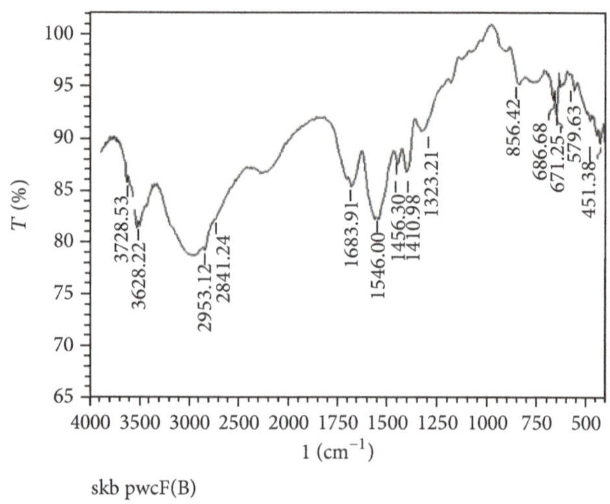

Figure 3: FTIR spectrum of semi-IPN hydrogel sample B.

3.4. FTIR Spectral Analysis. The FTIR spectrum of terpolymer, Cellulose/poly(SA)/PEG, is shown in Figure 3. The presence of poly (sodium acrylate) is indicated by the appearance of sharp peaks at 1411 cm^{-1} and 1546 cm^{-1} due to symmetric and asymmetric stretchings of carboxylate anion. In addition, methylene deformation at 1456 cm^{-1} also appears in the spectrum. Finally –OH stretching vibration in the range of 2400–3400 also appears. There are sufficient spectral evidences about the presence of cellulose in the polymer. The characteristic region of anomeric carbons identifies where a band appears at 856 cm^{-1}, thus confirming the presence of β,1 → 4 linkage. The C–OH out of plane bending mode around 670 cm^{-1} also supports cellulose. Also C–H in-plane bending at 1323 cm^{-1} also confirms cellulose. Finally, a sharp bend at 3628 cm^{-1} corresponds to O–H stretching of alcoholic OH groups from cellulose and poly (ethylene glycol).

3.5. Swelling Behavior of Hydrogels. The polymeric hydrogels, intended to be used for biomedical applications such as drug delivery devices and wound dressings, require a detailed investigation of their swelling capacity, which, in majority of cases, regulate the release of bioactive ingredient from the device. In the present work, the ternary hydrogel systems with varying compositions were investigated for their dynamic water uptake behavior in the swelling media of pH 7.4 at the physiological temperature of 37°C. The results of swelling experiments are discussed below.

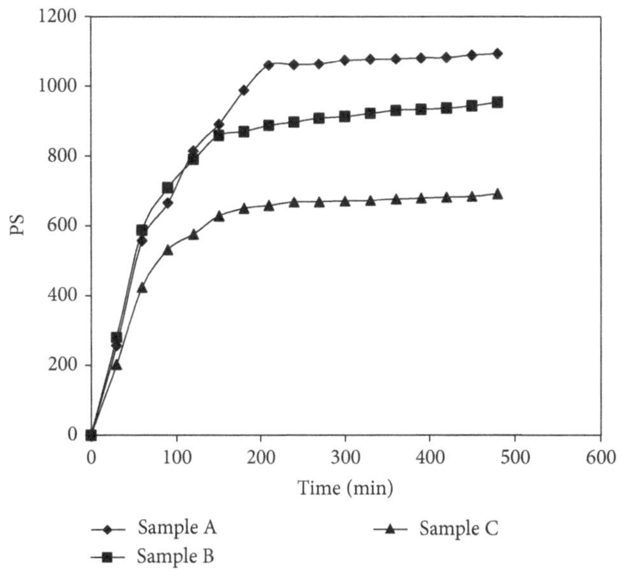

Figure 4: Dynamic water uptake for samples A, B, and C in pH 7.4 at 37°C.

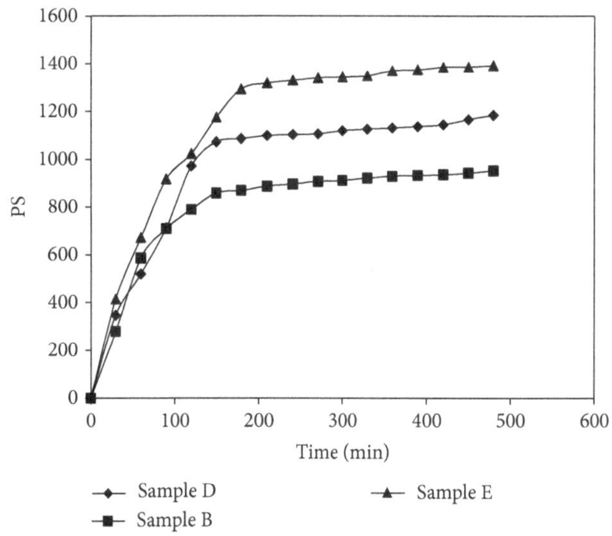

Figure 5: Dynamic water uptake for samples B, D, and E in pH 7.4 at 37°C.

The dynamic water uptake data for hydrogel samples, having varying contents of crosslinker MB (see Table 1), is shown in Figure 4.

It is clear that for all the three samples, namely, A, B, and C, the percent swelling (PS) increases with time and at a given time it decreases with the increase in MB content in the hydrogel. The total PS, exhibited by samples A, B, and C, was 1143, 1003, and 741, respectively. The decrease in PS with MB content is simply because the increased concentration of crosslinker results in an increase in the number of crosslinks junctions between polymeric chains, thus reducing the free space available for accommodation of incoming solvent. In addition, the increased number of crosslinks between polymeric segments renders more rigidity to the chains, thus restricting their relaxation. These two factors ultimately cause a decrease in water uptake [22]. In the present ternary hydrogel system, the three components, namely, Cell, PEG, and poly (SA), are hydrophilic and hence their concentrations in the feed mixture play a crucial role in governing the water absorption behavior of hydrogels. However, the situation is not so simple. Sometimes, the increase in concentration of one or more components results in an uncommon finding. The results of kinetic water uptake for samples D, B, and E that contain increasing amounts of monomer SA in the feed mixture are shown in Figure 5.

The three samples demonstrated total percent swelling of 1235, 1003, and 1443, respectively. It is clear that swelling trend is unusual. As the SA content increases from 6.36 to 10.63 mmol, the percent equilibrium swelling decreases from 1235 to 1003. However, as the SA content is further increased to 14.88 mmol, the equilibrium PS increases to 1443. The observed findings may be explained as follows: initially, when the SA content is increased, the mutual entanglements of poly (SA) chains with those of cellulose and PEG increase, thus producing additional physical crosslinks within the polymer matrix. This lowers the percent swelling of hydrogel. However, as the SA content is increased further to 14.88 mmol, the presence of higher concentration of $-COO^-$ groups along poly (SA) chains not only provides more hydrophilic character to the resulting gel but also promotes chain relaxation process between similar charges. In sample E, the Cell and PEG contents are relatively low as compared to those of SA, and they are not able to produce additional crosslinks with poly (SA) chains via H-bonding interactions (see Table 1). Almost similar types of results have also been reported by Povea et al. [23] who studied swelling behavior of chitosan/poly (acrylic acid-co-acrylamide) hydrogels as a function of initial AAc concentration in the feed mixture. The results were explained based on intersalt complexation between positively charged chitosan molecules and negatively charged $-COO^-$ groups along the copolymeric chains. Moreover, chain relaxation between carboxylate groups was also taken into account. It may be noteworthy to mention here that almost opposite swelling trend is reported by Hosseinzadeh et al. [24] for the carrageenan-g-poly(acrylic acid)/bentonite superabsorbent. In their study percent swelling increased with AAc content, attained a maximum value, and then began to decrease.

The amount of cellulose present in the polymerization mixture also affected the degree of swelling of hydrogels. The results of dynamic uptake of water for samples, B, F, and G containing increasing amounts of cellulose in the feed mixture are shown in Figure 6.

The samples showed equilibrium percent swelling of 1003, 2534, and 3062, respectively. The observed finding may be attributable to the fact that as the cellulose content increases, the number of –OH groups also increases, thus imparting more hydrophilic character to the hydrogels. Thus, cellulose content is a significant parameter to regulate the water absorption behavior of cellulose-based hydrogels. However, in the present case it appears that PEG/NaOH system was

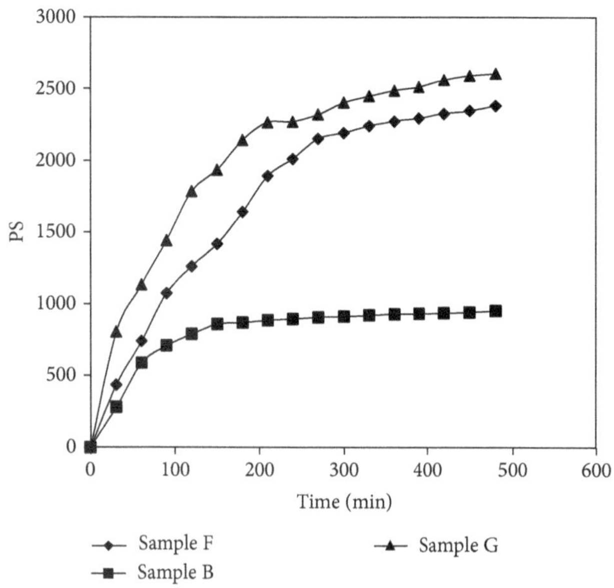

FIGURE 6: Dynamic water uptake for samples B, F, and G in pH 7.4 at 37°C.

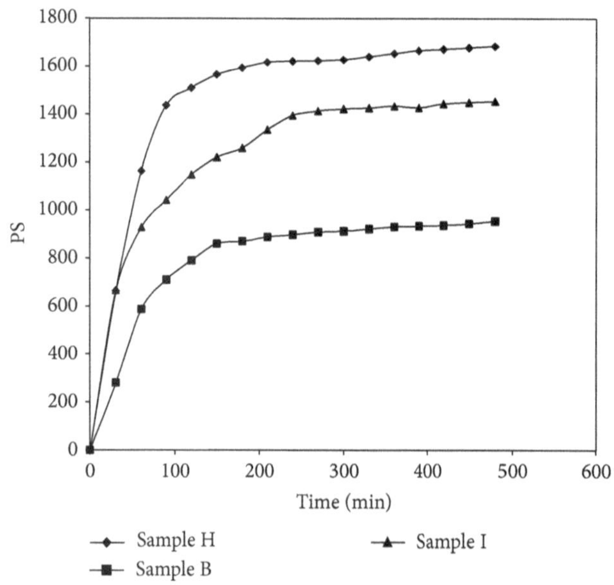

FIGURE 7: Dynamic water uptake for samples B, H, and I in pH 7.4 at 37°C.

quite effective in the breaking of H-bonding interactions of cellulose even at higher cellulose concentrations.

The molar mass of PEG, used as a component of solvent system to dissolve cellulose microparticles (CMP), also affected the equilibrium PS of resulting hydrogels. Samples *B*, *H*, and *I*, prepared by dissolving cellulose in PEG with molar masses of 6000, 4000, and 200, were investigated for their dynamic water uptake behavior at 37°C. The results, as shown in Figure 7, reveal an interesting observation.

The samples exhibited equilibrium percent swelling (EPS) of 1003, 1733, and 1565, respectively. This indicates that as the molar mass decreases from 6000 to 4000, the EPS increases. However, as the molar mass further decreases to 200, the EPS shows a decreasing trend. This may be explained as follows: initially, as the molar mass decreases from 6000 to 4000, the lengths of the PEG chains decrease and this may be associated with decrease in the degree of entanglements with other polymeric chains, namely, those of cellulose and poly (SA). In other words, the extent of mutual entanglements within the polymeric chains decreases. This, in fact, reduces the number of physical crosslinks within the hydrogel and causes an increase in the EPS. However, as the molar mass of PEG is further reduced to 200, there is further decrease in the water absorption of hydrogel. It may be attributed to the fact that with low molar mass of PEG, the average length of polymeric chains is small and therefore chain relaxation process due to invasion of water molecules into the matrix is not so pronounced. In other words, smaller polymeric chains do not relax to a great extent and this results in relatively less equilibrium percent swelling.

3.6. Interpretation of Water Uptake Data. When a glassy polymeric hydrogel is placed in water, the hydrogel undergoes swelling owing to the diffusion of water into hydrogel. This process involves the following steps:

(1) the solvent molecules migrate into preexisting or dynamically formed spaces between polymeric chains within the hydrogel matrix;

(2) the polymeric segments are plasticized if the solvent (usually water) has affinity towards functional groups present along polymeric chains. This results in unfolding of chains. This creates bigger separation between chains, followed by accommodation of water molecules;

(3) if the polymeric chains contain ionizable groups, then the charged groups along the polymeric chains repel each other and enhance the chain relaxation process [25]. This, finally results in extensive swelling of hydrogel.

The fundamental equation used to describe the water penetration mechanism for initial 60% data is

$$F = \frac{M_t}{M_\infty} = kt^n, \quad (2)$$

where F is the fractional water uptake at time t, k is gel characteristic constant, and n is swelling exponent, which indicates Fickian or non-Fickian behavior of the swelling device. Based on the relative rates of diffusion of solvent into hydrogels and relaxation of polymeric chains, the following three types of water sorption mechanisms have been proposed: (1) Fickian diffusion in which the rate of diffusion is much slower than the rate of chain relaxation ($n = 0.5$), (2) Case II diffusion in which rate of relaxation is much slower than the rate of diffusion of solvent ($n > 0.5$), and finally (3) anomalous transport in which the two rates, namely, rate of diffusion and rate of chain relaxation, are comparable ($0.5 < n < 1.0$).

In order to determine the mode of water penetration into hydrogels logarithmic form of (2) was used:

$$\ln F\left(\frac{M_t}{M_\infty}\right) = \ln k + n \ln t. \quad (3)$$

The dynamic water uptake data, obtained for various hydrogel samples and displayed in Figures 4, 5, 6, and 7, were used to draw linear plots between $\ln t$ and $\ln F$ (data not shown). The intercept and slopes of linear plots were used to calculate values of "n" and "k" as given in Table 2. A close look at the various values of "n" given indicates that except for sample A, all the remaining samples exhibit "n" values in the range of 0.45 to 0.65 thus indicating that swelling of all these hydrogel samples follows non-Fickian water transport behavior. This may simply be attributed to the fact that presence of ionized $-COO^-$ groups along the poly (SA) chains causes them to relax owing to the repulsion among similar charges. This is a most commonly observed phenomenon in the case of polymeric hydrogels that contain ionizable groups along their networks, and as far as sample A is concerned, it contains minimum amount of crosslinkers and therefore the polymeric chains are loosely crosslinked. In the swelling medium of pH 7.4, the ionized $-COO^-$ groups repel each other and chains continue to get unfolded or relaxed throughout the swelling process. As a result the mechanism is totally relaxation controlled and this is the case of case II or anomalous transport. Almost similar type of behavior has also been reported by Kim et al. [26] for poly(methacrylic acid-g-polyethylene glycol) hydrogels.

It has been reported that (3) is applicable only for the fractional water uptake value of 0.6. Therefore, some kinetic model is needed to have a complete quantitative interpretation of the swelling data. It has been suggested that for the hydrogels with relatively long duration of swelling process, swelling rate at any time is directly proportional to the quadratic of the swelling capacity before the attainment of equilibrium state. Therefore, the second order equation, proposed by Schott, may be given as [27]:

$$\frac{dM_t}{dt} = k_2(M_\infty - M_t)^2, \quad (4)$$

where M_t is the water uptake at time "t" and "k_2" is the second order rate constant for water absorption process. On integrating the above equation within the limits $t = 0$ $M_t = 0$ and $t = \infty$ $M = M_\infty$, the following equation is obtained:

$$\frac{t}{M_t} = \frac{1}{k_2 M_\infty^2} + \frac{t}{M_\infty} \quad (5)$$

or

$$\frac{t}{M_t} = A + Bt, \quad (6)$$

where A and B are two coefficients whose physical meaning is interpreted as follows: at a long retention time $B_t \gg A$ and therefore $B = 1/M_\infty$; that is, B is reciprocal of the maximum water uptake. On the contrast, at a very short time interval $B_t \ll A$ and so

$$L_t\left(\frac{dM_t}{dt}\right) = \frac{1}{A}$$
$$t \longrightarrow 0. \quad (7)$$

Therefore, the intercept A is reciprocal of initial swelling rate. Finally the Schott kinetic rate constant of swelling, k_2, can be calculated as

$$k_2 = \frac{\text{slope}^2}{\text{Intercept}} = \frac{B^2}{A}. \quad (8)$$

In order to apply second order model on the kinetic data obtained for all the nine samples, in the medium of pH 7.4, plots were obtained between t/M_t and t as shown in Figures 8, 9, 10, and 11, respectively.

TABLE 2: Values of swelling exponent "n" and gel characteristic constant "k" for various samples studied.

Sample code	n	$k \times 10^3$	R^2
A	0.923	10.6	0.975
B	0.609	38.9	0.986
C	0.517	60.1	0.978
D	0.465	33.0	0.987
E	0.573	24.6	0.988
F	0.633	28.0	0.974
G	0.656	87.7	0.988
H	0.455	71.1	0.992
I	0.472	65.6	0.985

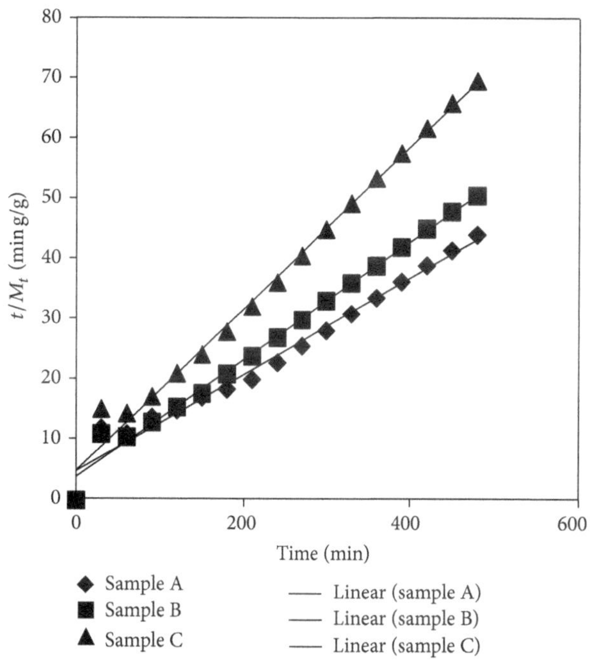

FIGURE 8: Schott kinetic model for samples A, B, and C.

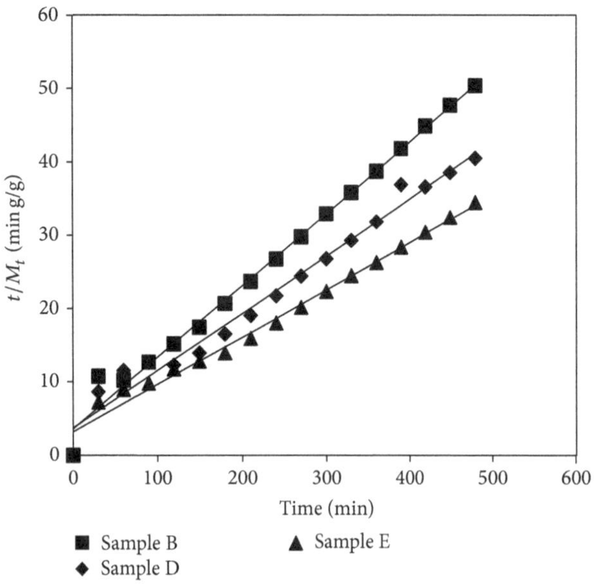

Figure 9: Schott kinetic model for samples B, D, and E.

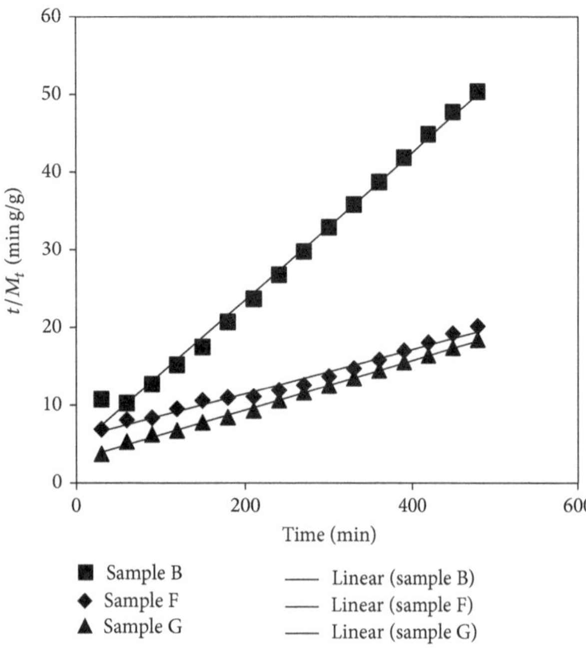

Figure 10: Schott kinetic model for samples B, F, and G.

The slopes and intercepts of linear plots obtained yielded the rate constant k_2, initial swelling rate (i.e., $r_{\text{ini}} = 1/A$), theoretical equilibrium water uptake $M_{\infty(\text{theor})}$, and experimental equilibrium PS $M_{\infty(\text{exp})}$. All the values are displayed in Table 3. It can be seen that for all the linear plots obtained using Schott model, the regression values are large, thus indicating the suitability of this model. In addition, the theoretical and experimental equilibrium water uptakes (expressed as g water/g gel) are quite close to each other. Similar types of results have also been reported elsewhere [28].

3.7. Diffusion Coefficients. Most of the diffusion processes are best interpreted by the Ficks first and second laws of diffusion.

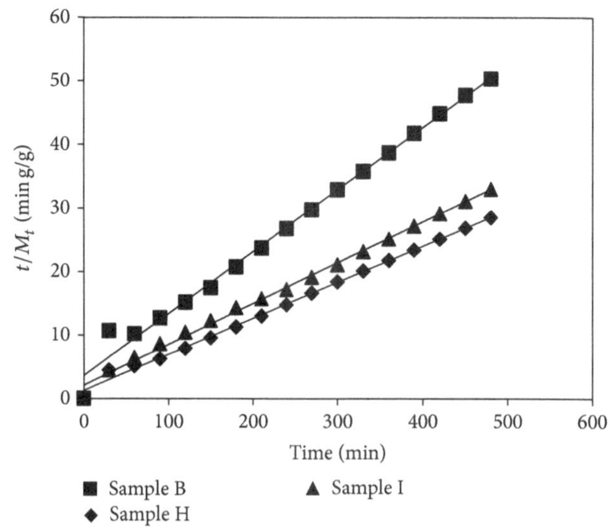

Figure 11: Schott kinetic model for samples B, H, and I.

The initial diffusion coefficient D_I gives an idea about the water penetration into hydrogel matrix in the initial stage of swelling. In order to determine D_I, the following equation was employed [29]:

$$F = \frac{M_t}{M_\infty} = \frac{4\left(D_I t/l^2\right)^{1/2}}{\pi^{1/2}}, \tag{9}$$

where F is fractional uptake and D_I is initial diffusion coefficient. The above equation can be rearranged as

$$F = \frac{4D_I^{1/2}}{l^{1/2} \cdot t^{1/2}}. \tag{10}$$

The slope of linear plot between F and $t^{1/2}$ was used to calculate D_i as

$$D_I = \frac{\text{slope}^2 \cdot l^2 \pi}{16}. \tag{11}$$

The initial 60% of the dynamic swelling data, shown in Figures 4, 5, 6, and 7, were used to draw linear plots between $\ln F$ and $t^{1/2}$. The slopes of the linear plots were used to calculate initial diffusion coefficients D_i.

In order to calculate the average diffusion coefficient D_{ave}, $F = 0.5$ and $t = t_{1/2}$ were substituted in (11), presuming that 50% of the total swelling, attained by hydrogel, can enable us to evaluate D_{ave}. The above substitutions yielded the following expression:

$$D_{\text{ave}} = \frac{0.049 l^2}{t_{1/2}}. \tag{12}$$

To calculate D_{ave}, the dynamic swelling data were used to evaluate the time required for the attainment of 50% of the total swelling of various hydrogel samples.

When a hydrogel disc is placed in swelling medium, the solvent molecules penetrate from all sides into the outer

TABLE 3: Parameters associated with Schott kinetic model for swelling of various hydrogel samples in phosphate buffer of pH 7.4 at 37°C.

Sample code	R^2	$k_2 \times 10^3$	$r_{ini} = (1/A)$ (g water/g gel min)	$M_{\infty(theor)}$ (g water/g gel)	$M_{\infty(exp)}$ (g water/g gel)
A	0.9769	1.384	0.215	12.484	11.435
B	0.9908	2.651	0.278	10.245	10.031
C	0.9908	3.880	0.214	7.434	7.414
D	0.9791	1.641	0.269	12.820	12.352
E	0.9860	1.292	0.311	15.527	14.431
F	0.9263	0.228	0.220	31.055	25.34
G	0.9824	0.494	0.431	29.449	30.62
H	0.9963	2.501	0.778	17.636	17.332
I	0.9954	1.942	0.469	15.552	15.650

TABLE 4: Various diffusion coefficients obtained for swelling of hydrogels with different compositions in buffer media of pH 7.4 at 37°C.

Sample code	$D_I \times 10^6$ (cm^2 min^{-1})	$D_L \times 10^6$ (cm^2 min^{-1})	$D_{ave} \times 10^6$ (cm^2 min^{-1})
A	235.4	9.8	237.2
B	107.5	14.6	69.5
C	1553.2	7.9	64.1
D	1611.4	11.5	57.0
E	3365.8	25.5	274.6
F	8658.7	105.2	306.2
G	8682.8	89.9	253.4
H	7137.2	120.4	554.8
I	7179.8	413.4	745.8

circumference of the disc and cause the outer part to swell. A sharp division between the relaxed, swollen region and interior dry region can be observed. In relation to this phenomenon, an interesting observation was made by Shankar [30], who claimed that a plot between the ln $(1 - M_t/M_\infty)$ and "t" gives a linear plot but with a distinct break in the slope. This is observed experimentally too for both Fickian and non-Fickian diffusions. The break in the slope corresponds to the time when the two advancing boundaries meet each other. Using the slope of the later part, it is possible to calculate D_L. The late time diffusion coefficient D_L was determined using later 60% of the total equilibrium swelling data for various samples using the dynamic swelling data; the ln $(1-M_t/M_\infty)$ values were plotted against t for various samples. The plots obtained were biphasic linear plots as discussed above (for illustration, the ln $(1 - M_t/M_\infty)$ versus t plots for samples A, B, and C are shown in Figure 12).

The slope, obtained for the later part of the swelling process, was used to calculate D_L using the following expression [31]:

$$D_L = -\frac{\left(\text{slope} \times l^2\right)}{\pi^2}. \quad (13)$$

The various types of diffusion coefficients for all the samples, discussed above, are given in Table 4. All the values are expressed in the order of 10^{-6} so that a clear-cut

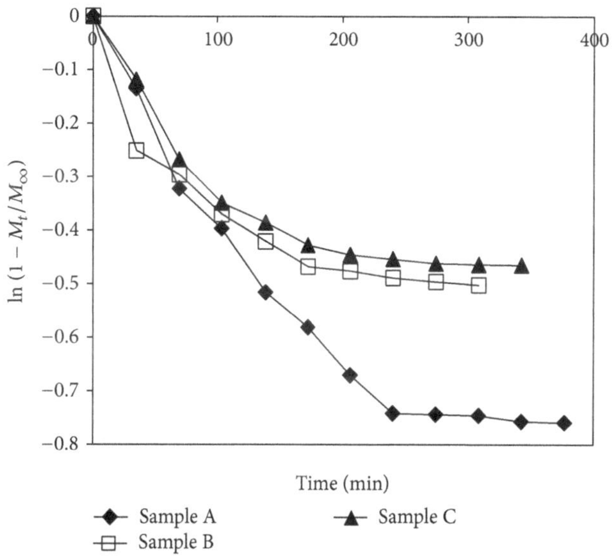

FIGURE 12: ln $(1 - M_t/M_\infty)$ versus t plots for samples A, B, and C.

comparison can be made. A close look at the values of various diffusion coefficients, given, reveal that for all the samples the values of D_I are ten to hundred times greater than those for late time diffusion coefficients D_L. This indicates that solvent penetrates into the hydrogel matrices at a very fast rate in the beginning of the swelling process, and the rate slows down in the later part of the swelling. The various dynamic water uptake data, displayed for different types of hydrogels, also support these findings. In addition, from the ln $(1-M_t/M_\infty)$ versus "t" plots (shown in Figure 12) also indicate the steep slope in the initial stage of swelling.

3.8. pH- and Salt-Dependent Swelling Behavior. The presence of ionizable groups within the hydrogel network renders pH-dependent swelling property to hydrogels. The pH-responsive swelling of such hydrogel systems has been exploited mainly for the targeted delivery of drugs via oral administration [32]. In this work, the representative sample B was equilibrated in buffer solutions of varying pH, in the range of 1 to 11, at 37°C. The results of swelling experiment are displayed in Figure 13.

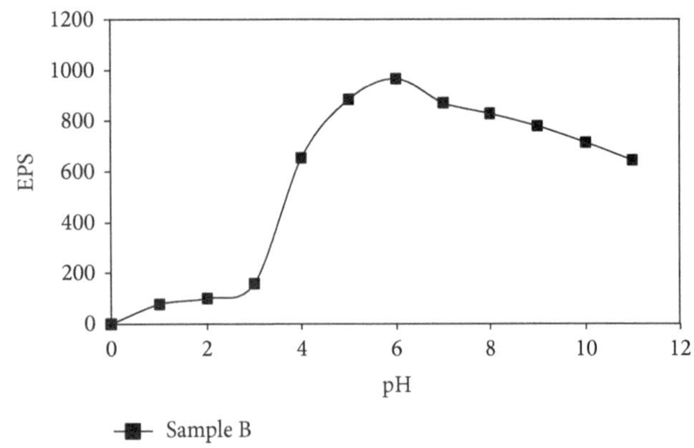

FIGURE 13: Percent equilibrium swelling as a function of pH of the media.

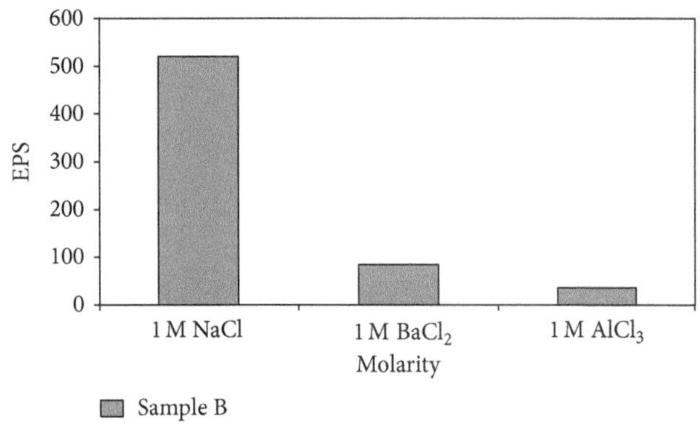

FIGURE 14: Effect of types of salts in swelling media on equilibrium swelling.

It can be seen that hydrogel sample B attains a minimum swelling of 77 percent in the buffer medium of pH 1.0. As the pH of the swelling media increases, the PS also continues to increase and attains a maximum value of 965 at pH 6.0. However, as the pH is further increased the percent swelling shows a slight decreasing trend, attaining 644 percent swelling in the medium of pH 11. The observed findings may be explained as follows. When the pH is quite low, that is, 1.0, the -COOH groups present along the poly (sodium acrylate) chains remain as unionized -COOH groups and produce intermolecular H-bonding interactions, thus imparting compactness to the hydrogel matrix. In addition, the absence of ionization also minimizes the ion osmotic swelling pressure [33] which is also a key factor to enhance the swelling of hydrogel. Thus, hydrogel swells to minimum. When the pH is increased, there starts ionization of -COOH groups to give $-COO^-$ groups along the polymeric chains of poly (SA). These groups begin to repel each other thus causing the polymeric chains to relax or unfold. At the same time, the ionization within the hydrogel matrix also increases ion osmotic swelling pressure. Because of these two factors, the PS continues to increase and attains a maximum value in the range of pH 6-7. In this pH range there is almost complete ionization of -COOH groups, thus resulting in maximum water uptake. However, as the pH is increased further, the PS starts to decrease slightly. This is probably due to the "charge screening effect" of Na^+ ions that are present in excess in the external solution at higher pH. These excess Na^+ ions shield the $-COO^-$ ions and reduce the anion-anion repulsion along the macromolecular chains. Similar types of results are also reported elsewhere [34]. The excellent pH-dependent volume-phase transition opens up the possibilities of using this system for colon targeted drug delivery.

The salt responsiveness of the representative sample B was also investigated by placing preweighed sample B in distilled water and salt solutions of NaCl, $BaCl_2$, and $AlCl_3$, each having a concentration of 1 M. The results, as shown in Figure 14, reveal that there is drastic decrease in the EPS of hydrogels as water is replaced by 1 M salt solutions as swelling media.

The decrease in PES is much more pronounced as the valence of cation increases. The maximum lowering of PES is observed in the case of salt solution of Al^{+++} ions. The observed finding may be attributable to the fact that when salt solutions are taken as swelling media in place of water, there is decrease in the ion osmotic swelling pressure; that is, $\pi = \sum RT(C_{gel} - C_{sol})$, where C_{gel} and C_{sol} are concentrations of mobile ions in the gel and solution phase, respectively [35]. The decrease in π lowers the equilibrium PS. Moreover, the extent of decrease in EPS increases with the valence of counter or mobile ions. This is simply due to facts that as the valence of counter ion increases, the number of ions coming out of gel phase increases. In other words, for entrance of one Al^{+++} ion into the gel phase, expulsion of three monovalent Na^+ ions from the gel phase into swelling medium is required. Therefore, C_{sol} increases in the above expression, thus decreasing the value of π. The present of high valence cations in the gel phase may also produce additional crosslinks between negatively charged $-COO^-$ groups and cations.

4. Conclusion

From the above study, it may be concluded that cellulose pulp can be conveniently dissolved in PEG/NaOH solvent system. When monomer AAc is added in precalculated quantity into cellulose solution and is allowed to undergo free radical polymerization, a semi-IPN hydrogel Cell/PEG/poly (SA) is formed. The swelling property of this hydrogel is governed by parameters like concentration of cellulose, acrylic acid, and crosslinker MB and molecular weight of PEG. The hydrogel exhibits excellent pH-dependent swelling behavior. The presence of salt in the swelling medium also affects the equilibrium percent swelling of hydrogels. The present hydrogel system opens up new possibilities to use it in various biomedical applications including drug delivery and wound healing management.

Conflict of Interests

The authors declare that there is no conflict of interests regarding the publication of this paper.

References

[1] X. Qiu and S. Hu, "'Smart' materials based on cellulose: a review of the preparations, properties, and applications," *Materials*, vol. 6, no. 3, pp. 738–781, 2013.

[2] J. F. Kadla and R. D. Gilbert, "Cellulose structure: a review," *Cellulose Chemistry and Technology*, vol. 34, no. 3-4, pp. 197–216, 2000.

[3] M. Sarvaš, P. Pavlenda, and E. Tákacová, "Effect of hydrogel application on survival and growth of pine seedlings in reclamations," *Journal of Forest Science*, vol. 53, no. 5, pp. 204–209, 2007.

[4] A. Sannino, A. Esposito, L. Nicolais et al., "Cellulose-based hydrogels as body water retainers," *Journal of Materials Science: Materials in Medicine*, vol. 11, no. 4, pp. 247–253, 2000.

[5] M. Patchan, J. L. Graham, Z. Xia et al., "Synthesis and properties of regenerated cellulose-based hydrogels with high strength and transparency for potential use as an ocular bandage," *Materials Science and Engineering C*, vol. 33, no. 5, pp. 3069–3076, 2013.

[6] A. L. Buvanov, I. V. Hofman, A. K. Khripunov, A. A. Tkachenko, and E. E. Ushakova, "High-strength biocompatible hydrogels based on poly(acrylamide) and cellulose: synthesis, mechanical properties and perspectives for use as artificial cartilage," *Polymer Science A*, vol. 55, no. 5, pp. 302–312, 2013.

[7] Z. Sklenar, Z. Vitkova, P. Herdova, K. Horackova, and V. Simunkova, "Formulation and release of alaptide from cellulose-based hydrogels," *Acta Veterinaria*, vol. 81, no. 3, pp. 301–306, 2012.

[8] S. S. Bhattacharya, S. Shukla, Banerjee, S. Choudhary P, and Chakraborty, "Tailored IPN hydrogel bead of sodium carboxymethyl cellulose and sodium carboxymethyl xanthan gum for controlled delivery of diclofenac sodium," *Polymer-Plastics Technology and Engineering*, vol. 52, no. 8, pp. 795–805, 2013.

[9] D. Seliktar, "Extracellular stimulation in tissue engineering," *Annals of the New York Academy of Sciences*, vol. 1047, no. 1, pp. 386–394, 2005.

[10] M. H. Pittler and E. Ernst, "Dietary supplements for body-weight reduction: a systematic review," *The American Journal of Clinical Nutrition*, vol. 79, no. 4, pp. 529–536, 2004.

[11] W. K. Czaja, D. J. Young, M. Kawecki, and R. M. Brown Jr., "The future prospects of microbial cellulose in biomedical applications," *Biomacromolecules*, vol. 8, no. 1, pp. 1–12, 2007.

[12] L. Zhang, D. Ruan, and S. Gao, "Dissolution and regeneration of cellulose in NaOH/Thiourea aqueous solution," *Journal of Polymer Science B: Polymer Physics*, vol. 40, no. 14, pp. 1521–1529, 2002.

[13] R. P. Swatloski, S. K. Spear, J. D. Holbrey, and R. D. Rogers, "Dissolution of cellose with ionic liquids," *Journal of the American Chemical Society*, vol. 124, no. 18, pp. 4974–4975, 2002.

[14] S. Zhang, F.-X. Li, and J.-Y. Yu, "Structure and properties of novel cellulose fibres produced from NaOH/PEG-treated cotton linters," *Iranian Polymer Journal*, vol. 19, no. 12, pp. 949–957, 2010.

[15] F. M. Veronese and G. Pasut, "PEGylation, successful approach to drug delivery," *Drug Discovery Today*, vol. 10, no. 21, pp. 1451–1458, 2005.

[16] X. W. Tan, L. Hartman, K. P. Tan et al., "In vivo biocompatibility of two PEG/PAA interpenetrating polymer networks as corneal inlays following deep stromal pocket implantation," *Journal of Materials Science: Materials in Medicine*, vol. 24, no. 4, pp. 967–977, 2013.

[17] W. Li, R. Wang, and S. Liu, "Nanocrystalline cellulose prepared from softwood kraft pulp via ultrasonic-assisted acid hydrolysis," *BioResources*, vol. 6, no. 4, pp. 4271–4281, 2011.

[18] M. C. I. Mohd Amin, N. Ahmad, N. Halib, and I. Ahmad, "Synthesis and characterization of thermo- and pH-responsive bacterial cellulose/acrylic acid hydrogels for drug delivery," *Carbohydrate Polymers*, vol. 88, no. 2, pp. 465–473, 2012.

[19] A. Zabaniotou, O. Ioannidou, E. Antonakou, and A. Lappas, "Experimental study of pyrolysis for potential energy, hydrogen and carbon material production from lignocellulosic biomass," *International Journal of Hydrogen Energy*, vol. 33, no. 10, pp. 2433–2444, 2008.

[20] A. E. Mucientes, F. Santiago, A. M. Carrero, and B. Talavera, "Superabsorbent hydrogels of poly(sodium acrylate) with crude and exfoliated vermiculites," *Journal of Polymer Engineering*, vol. 33, no. 1, pp. 61–69.

[21] J. Manson, D. Kumar, B. J. Meenan, and D. Dixon, "Polyethylene glycol functionalized gold nanoparticles: the influence of capping density on stability in various media," *Gold Bulletin*, vol. 44, no. 2, pp. 99–105, 2011.

[22] J. N. Hiremath and B. Vishalakshi, "Effect of crosslinking on swelling behaviour of IPN hydrogels of Guar Gum & Polyacrylamide," *Der Pharma Chemica*, vol. 4, no. 3, pp. 946–955, 2012.

[23] M. B. Povea, W. A. Monal, J. V. C. Rodríguez, A. M. Pat, N. B. Rivero, and C. P. Covas, "Interpenetrated chitosan-poly(acrylic acid-co-acrylamide) hydrogels. Synthesis, characterization and sustained protein release studies," *Materials Sciences and Applications*, vol. 2, no. 6, pp. 509–520, 2011.

[24] H. Hosseinzadeh, M. Sadeghzadeh, and M. Badazadeh, "Preparation and properties of carrageenan-g-poly(acrylic acid)/bentonite superabsorbent composite," *Journal of Biomaterials and Nanobiotechnology*, vol. 2, pp. 311–317, 2011.

[25] N. M. Franson and N. A. Peppas, "Influence of copolymer composition on non-fickian water transport through glassy copolymers," *Journal of Applied Polymer Science*, vol. 28, no. 4, pp. 1299–1310, 1983.

[26] B. Kim, K. La Flamme, and N. A. Peppas, "Dynamic swelling behavior of pH-sensitive anionic hydrogels used for protein delivery," *Journal of Applied Polymer Science*, vol. 89, no. 6, pp. 1606–1613, 2003.

[27] H. Schott, "Swelling kinetics of polymers," *Journal of Macromolecular Science: Physics*, vol. 31, no. 1, 1992.

[28] W. Wang and A. Wang, "Synthesis and swelling properties of guar gum-g-poly(sodium acrylate)/Na-montmorillonite superabsorbent nanocomposite," *Journal of Composite Materials*, vol. 43, no. 23, pp. 2805–2819, 2009.

[29] E. Karadag and D. Saraydin, "Influence of some crosslinkers on the swelling of acrylamide-crotonic acid hydrogels," *Turkish Journal of Chemistry*, vol. 21, pp. 151–161, 1997.

[30] V. Shankar, "Calculation of diffusion coefficients of organic vapours from short and longtime sorption data," *Polymer*, vol. 20, no. 2, pp. 254–257, 1979.

[31] P. M. Smith and M. M. Fisher, "Non-Fickian diffusion of water in melamine-formaldehyde resins," *Polymer*, vol. 25, no. 1, pp. 84–90, 1984.

[32] Y. Qiu and K. Park, "Environment-sensitive hydrogels for drug delivery," *Advanced Drug Delivery Reviews*, vol. 53, no. 3, pp. 321–339, 2001.

[33] F. Horkay, I. Tasaki, and P. J. Basser, "Osmotic swelling of polyacrylate hydrogels in physiological salt solutions," *Biomacromolecules*, vol. 1, no. 1, pp. 84–90, 2000.

[34] H. Hosseinzadeh, "Controlled release of diclofenac sodium from pH-responsive carrageenan-g-poly(acrylic acid) superabsorbent hydrogel," *Journal of Chemical Sciences*, vol. 122, no. 4, pp. 651–659, 2010.

[35] O. Okay and S. B. Sariisik, "Swelling behavior of poly(acrylamide-co-sodium acrylate) hydrogels in aqueous salt solutions: theory versus experiments," *European Polymer Journal*, vol. 36, no. 2, pp. 393–399, 2000.

19

Thermal Stability and Kinetic Study of Isotactic Polypropylene/Algerian Bentonite Nanocomposites Prepared via Melt Blending

Fayçal Benhacine, Farida Yahiaoui, and Assia Siham Hadj-Hamou

Laboratoire des Matériaux Polymères, Département de Chimie Macromoléculaire, Faculté de Chimie, Université des Sciences et de la Technologie Houari Boumediene, BP 32, El Alia, 16111 Alger, Algeria

Correspondence should be addressed to Assia Siham Hadj-Hamou; hadjhamouassia@yahoo.fr

Academic Editor: Yves Grohens

Isotactic polypropylene (iPP)/bentonite nanocomposites were prepared via melt blending using bentonite clay originated from Maghnia (Algeria). This clay was, at a first stage, used in its pure form (PBT) and then organically modified by Hexadecyl ammonium chloride (OBT). The effect of Maghnia bentonite dispersion on the iPP matrix was investigated by X-ray diffraction (XRD) and transmission electronic microscopy (TEM). DSC results evidenced that unmodified or organomodified bentonite can act as a nucleating agent increasing the rate of crystallites formation. Moreover, a thermogravimetry analysis confirmed a significant enhanced thermal stability of IPP/clay nanocomposites compared to pure IPP. The Flynn-Wall-Ozawa and Tang methods were applied to determine the activation energy E_a of the degradation process. The apparent activation energy E_α of thermal degradation for IPP/clay nanocomposites was much higher than that of virgin iPP. Comparatively to PBT, results indicate that OBT has an important effect on pure iPP thermal stability. Tensile modulus, tensile strength, and elongation at break were also measured and compared with those of pure iPP.

1. Introduction

Isotactic polypropylene (iPP) is one of the most interesting commodity thermoplastic accounting for about 20% of the total world's polyolefin production, not only for its balance of physical and mechanical properties but also for its environmental friendliness (recyclability) and low cost. It is widely used in many applications, such as fibers, films for food packaging, bottles production, and tubes.

Isotactic polypropylene (iPP)/clay composites are among the pioneer researched nanocomposites because iPP is one of the most widely used polyolefins, and its properties are greatly affected by clay dispersion in the polymer [1–10]. Many researches were reported on preparation and characterization of polypropylene (iPP)/clay nanocomposites, focusing on the dispersed morphology of clay particles [11–14].

Polymer degradation is the purpose of many researches on plastic waste management. The knowledge of degradation kinetics of polymers is important for modeling the degradation process. Several mechanisms were proposed to describe thermal and flame properties of polymer/clay nanocomposites. Generally, the barrier effect of clay layers for heat and volatiles plays a key role in these enhancements. Kinetic data reflect that improvement of thermal stability of polymer/clay nanocomposites is associated with an increase in the activation energy of their degradation; this is explained by the barrier effect developed by clay layers [8, 15–17].

In this survey iPP was mixed in the melt state with pure and organically modified Maghnia bentonite to obtain nanocomposite. Nanoclay dispersion and intercalation mechanisms will be studied on iPP stability and thermal and mechanical properties.

2. Experimental Part

2.1. Materials. Commercial grades isotactic polypropylene (iPP), iPP 05-RF03-0577 C-2807 AQ4, was supplied by Oriental Petrochemicals, Saudi Arabia.

Bentonite (PBT) originated from Maghnia (Algeria) supplied by Bentonite Company of Algeria and analyzed by ENOF Central Laboratory (National Company of Nonferrous Mining Products and Useful Substances). This clay which contains SiO_2 (55–65%), Al_2O_3 (12–18%), Fe_2O_3 (1–3%), Na_2O (1–3%), CaO (1–5%), K_2O (0.76–1.75%), and MgO (2-3%) was initially used in its pure form and then organically modified byHexadecyl ammonium chloride (OBT).

Hexadecyl ammonium chloride was prepared as already stated [18] by mixing adequate amounts of hexadecylamine with concentrated HCl in 300 mL of distilled water added to bentonite clay (OBT) dispersed into hot distilled water at 80°C under vigorous stir for 1 h. The mixture was then filtered and washed with water in ethanol (50/50 vol%) until negative chloride test was obtained using $AgNO_3$.

Hexadecyl ammonium exchanged clay (OBT) was then dried at 80°C for 3 days in a vacuum oven.

2.2. Blending. Melt blending of virgin iPP polymer and 5% in weight of pure bentonite (PBT) or organically modified bentonite (OBT) was conducted using a twin-screw microextruder (5 and 15 microcompounder DSM Xplore) and mixing process parameters (60 rpm screw speed, 10 min mixing time, and temperature profile: 175–180–180°C) were modulated to optimize material final properties.

2.3. Characterizations

2.3.1. X-Ray Diffraction. X-ray diffraction patterns were obtained from a Philips PW3710 diffractometer in 2θ (1.5–50°) range. The monochromatic radiation applied was Cu Kα (1.5406 Å) operating at 35 kV and 25 mA.

2.3.2. Transmission Electronic Microscopy (TEM). As a complementary technique to XRD, TEM was used to investigate the morphology of the elaborated materials using a JEOL 1200EXII microscope operating at 100 kV.

2.3.3. Thermal Characterization

DSC Analysis. Differential scanning calorimetry (DSC) measurements were made on a 204 F1 Phoenix (Netzsch Company). All measurements were carried out in nitrogen atmosphere. The samples were heated from −70 to 200°C, held in the molten state for 2 min to eliminate the influence of thermal history, and then cooled at a rate of 10°C/min to room temperature. Both endothermal flow and exothermal flow were recorded. From these scans the melting temperature (Tmp) and the heat of fusion (ΔH_m) of the composites were measured. The degree of crystallinity χ_c was calculated under the assumption in which the heat of fusion is proportional to the crystalline content, as $\chi = \Delta H_m/(1 - w_{\text{clay}})\Delta H_m^0$, where ΔH_m^0 is the enthalpy of fusion of completely crystalline sample and taken in this work as 146 J/g [19] and w_{clay} is the weight fraction of nanoclay in the nanocomposite.

TGA Analysis. Thermogravimetry analysis (TGA) was conducted on a TA Instruments TGA Q500 using an ultrahigh purity (UHP) nitrogen atmosphere. The samples (films of 10.0 mg) were heated from 25 to 600°C at 5, 10, 20, 30, and 40°C/min heating rates under nitrogen flow.

Mechanical Tests. Tests specimens for tensile measurements were prepared from 1 mm thick plates. Tensile modulus, tensile strength, and elongation at break were measured in a tensile Instron Zwick/Roell Z 100. Tester was set at 200 mm min^{-1} strain rate.

3. Kinetic Models

Many kinetic models were proposed to simplify the general kinetic equation and determine the degradation process parameters.

Reaction rate in TGA studies can be defined as the variation of conversion degree to time or temperature where the degree of conversion (α) is calculated in terms of mass loss according to the following equation:

$$\alpha = \frac{W_0 - W_t}{W_0 - W_\infty} = \frac{\Delta W}{\Delta W_\infty}, \tag{1}$$

where W_0, W_t, and W_∞ are, respectively, weight at the beginning of the degradation step, actual weight at each point of the curve, and the final weight measured after consideration of the specific degradation process. Conversion rate for a kinetic process can be typically indicated as the result of two contributions: the first one is the temperature which affects the kinetic constant and the second one is the concentration of reactive species. Therefore, the kinetic information can be extracted from dynamic experiments by means of various methods. All kinetic studies assume that isothermal conversion rate, $d\alpha/dt$, is a linear function of a temperature-dependent rate constant k and a temperature-independent function of the conversion α; that is,

$$\frac{d\alpha}{dt} = kf(\alpha), \tag{2}$$

where $f(\alpha)$ depends on the particular decomposition mechanism.

According to Arrhenius,

$$k(t) = A\exp\left(-\frac{E_a}{RT}\right), \tag{3}$$

where A is preexponential factor, E_a is the apparent activation energy (kJ/mol), R is the gas constant (8.314 J/mol K), and T is the absolute temperature (K). According to the complexity of the multiple degradation reactions, the function $f(\alpha)$ is generally very complex and it is not possible to solve the differential equation (1) leading to the necessity to introduce some simplifications to determine the kinetic parameters. A simple expression proposed for $f(\alpha)$ is the use of the nth-order model represented in the following equation:

$$f(\alpha) = (1 - \alpha)^n, \tag{4}$$

where n is the reaction order. By inserting (3) and (4) into (2), we can obtain

$$\frac{d\alpha}{dt} = A(1-\alpha)^n \exp\left(-\frac{E_a}{RT}\right). \quad (5)$$

For a dynamic TGA process, introducing the heating rate, $\beta = dT/dt$, into (1), (2) is obtained as follows:

$$\frac{d\alpha}{dT} = \left(\frac{A}{\beta}\right) \exp\left(-\frac{E}{RT}\right)(1-\alpha)^n. \quad (6)$$

Integration of this equation from an initial temperature, T_0, corresponding to a degree of conversion, α_0, to the peak temperature T_p, where $\alpha_0 = \alpha_p$, gives kinetic methods used in evaluating activation energy in this study. Consider the following:

$$\int_{\alpha_0}^{\alpha p} \frac{d\alpha}{(1-\alpha)^n} = \frac{A}{\beta} \int_{T_0}^{Tp} e^{-(Ea/RT)} dT. \quad (7)$$

If T_0 is low, it may be reasonably assumed that $\alpha_0 = 0$ and considering there is no reaction between 0 and T_0, one has

$$\int_{0}^{\alpha p} \frac{d\alpha}{(1-\alpha)^n} = \frac{A}{\beta} \int_{0}^{Tp} e^{-(Ea/RT)} dT. \quad (8)$$

3.1. The Ozawa-Flynn-Wall (OFW) Method (See [20, 21]). This method represents a relatively simple method to determine activation energy directly from weight loss versus temperature data obtained at several heating rates. Equation (7) is integrated using Doyle's approximation, and the result of integration after taking logarithms is

$$\ln \beta = -1.052 \frac{E_a}{RT} + \text{const}, \quad (9)$$

where β, A, E, and T have the known meanings. This is one of the integral methods that can determine activation energy without knowledge of reaction order. It is used to determine activation energy for given values of conversion. Activation energy for different conversion values can be calculated from $\ln \beta$ versus $1/T$ plot.

3.2. The Tang Method (See [22]). New equation for the evaluation of nonisothermal kinetic parameters has been proposed by Tang and it can be written in the following form:

$$\ln \frac{g(\alpha)}{T^{1.89466100}} = -1.00145033 \frac{E_\alpha}{RT_\alpha} + \ln\left[\frac{A_\alpha E_\alpha}{\beta R}\right] + 3.63504095 - 1.89466100 \ln E_\alpha. \quad (10)$$

The corresponding equation has been modified to be applied in an isoconversional way and is given by the following equation:

$$\ln \frac{\beta}{T^{1.89466100}} = 1.00145033 \frac{E_\alpha}{RT_\alpha} + \ln\left[\frac{A_\alpha E_\alpha}{R}\right] - \ln g(\alpha) + 3.63504095 - 1.89466100 \ln E_\alpha, \quad (11)$$

where β is the heating rate, α is the extent of conversion (weight of materials volatilized/initial weight of materials), R is the universal gas constant, and $g(\alpha)$ is a mathematical function whose form is related to the reaction mechanism. E_α, A_α, and T_α are, respectively, the activation energy, the preexponential factor, and the absolute temperature related to a given extent of conversion.

Plotting $\ln[b/T^{1.89466100}]$ versus $1/T$, the activation energy E_α is obtained from the slope.

4. Results and Discussion

4.1. Characterizations of Pure and Organomodified Maghnia Bentonite. Prior to its use as nanofiller in iPP matrix, OBT was first characterized by FTIR spectroscopy and X-ray diffraction to evidence its organic modification.

Figure 1 illustrates FTIR spectra of pure and organically modified bentonite clay.

Pure bentonite clay shows characteristic vibration bands located at 3630 cm^{-1} (OH stretching from lattice hydroxyl), 1032 cm^{-1} (Si–O stretching), and 469 cm^{-1} (Al–O stretching). Those appearing around 920 cm^{-1} and 800 cm^{-1} characterize the tetra- and octahedral layers of bentonite clay [18].

In addition to the characteristic absorption bands of PBT, organomodification of Maghnia bentonite by Hexadecyl ammonium salt is evidenced by the presence of the band at 3625 cm^{-1} and by the appearance of new bands at 2925 cm^{-1}, 2850 cm^{-1}, and 1473 cm^{-1} characteristic of salt [18].

4.2. Morphological Characterization

4.2.1. X-Ray Diffraction. The degree of intercalation of PBT and OBT with iPP matrix was first analyzed by X-ray diffraction. Figure 2 displays XRD patterns of PBT, OBT, iPP/PBT, and iPP/OBT hybrids containing 5% in weight of nanoclay. The XRD pattern of PBT showed a peak at approximately $2\theta = 6^\circ$ corresponding to a basal spacing of 1.47 nm, whereas a strong peak at $2\theta = 3^\circ$ (a basal spacing of 2.94 nm) and a weak peak at $2\theta = 5.40^\circ$ are noticed with OBT. This indicates that clay layers are widely separated by Hexadecyl ammonium ions and that a portion of clay layer was not intercalated.

As shown in Figure 2, the presence of PBT in iPP appeared as a weak peak at a lower angle ($2\theta = 3.18^\circ$) corresponding to a d-spacing of 2.77 nm, which supposes the formation of an intercalated nanocomposite. The weak peak noticed at $2\theta = 6.0^\circ$ ($d = 1.47$ nm) with this nanocomposite also supposes that some clay layers are not totally intercalated by the polymer chains.

Insertion of iPP chains in the interlayer spacing of OBT clay is evidenced from the shift of the characteristic XRD peak of OBT within iPP polymer to a lower angle than the one of OBT clay, resulting in an increase of d-spacing as $2\theta = 2.62^\circ$ corresponding to ($d = 3.37$ nm), due to the large d-spacing of the clay, and enabling intercalation of polymer chains between clay layers.

Furthermore, incorporation of nanoparticles into the iPP matrix can cause changes in polymer crystallization rate

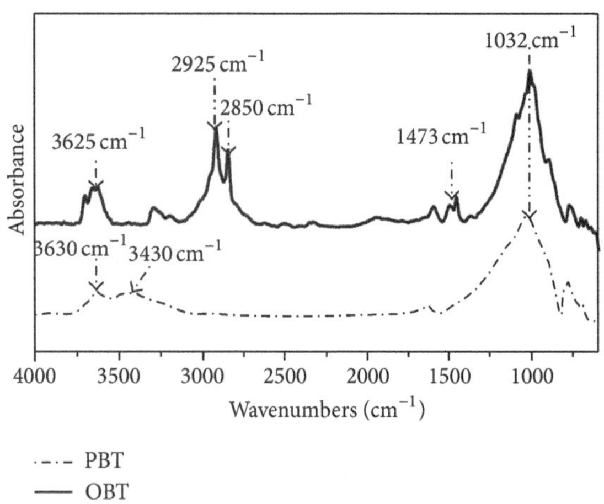

FIGURE 1: FTIR spectra of pure bentonite (PBT) and organomodified bentonite (OBT).

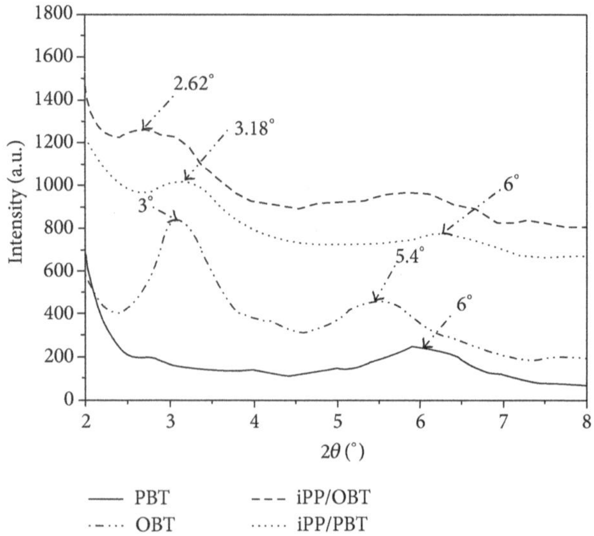

FIGURE 2: Wide angle X-ray diffraction patterns (XRD) of PBT, OBT, iPP/PBT, and iPP/OBT materials.

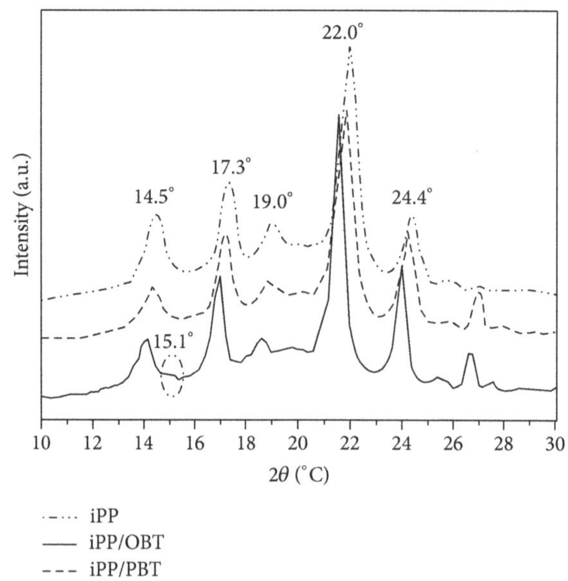

FIGURE 3: XRD curves of pure iPP and its iPP/Maghnia bentonite nanocomposites.

and even induce changes in its crystal structure [23, 24]. To examine this possibility, X-ray diffraction measurements were performed. XRD spectra of pure iPP polymer exhibit distinct α-crystalline diffraction peaks at $2\theta = 14.5°, 17.3°, 19.0°, 22°$, and $24.4°$ (Figure 3). XRD results showed that the α-crystalline form dominates for the two hybrid materials, meaning that the addition of PBT or OBT does not affect the crystal form of the final iPP molecules. Compared with XRD curves of pure iPP, diffraction peaks (040, 130, 111, and 060) for the two iPP/PBT and iPP/OBT systems shifted slightly to lower angles and increased their relative intensity. A possible reason is that the addition of bentonite promotes the nucleation and growth of iPP crystals in an essential way and perfects the α-crystalline form. In addition to the features associated with α -form, only the introduction of 5wt% OBT nanoparticles produces another peak at 2θ value of 15.1° indicating that both monoclinic α and trigonal β-form exist in this composite [25, 26].

4.2.2. Transmission Electronic Microscopy (TEM). A complementary characterization was carried out by TEM. Typical TEM micrograph of iPP/OBT, illustrated in Figure 4(a), confirmed in agreement with XRD the formation of mainly intercalated nanocomposite structures evidenced at high magnification by the presence of stacked layers randomly dispersed in the polymer matrix. However, the micrograph of iPP/PBT (Figure 4(b)) showed the presence of a limited number of intercalated stacks as well as weakly intercalated agglomerates coexisting with a significant fraction of clay agglomerations.

4.3. Thermal Characteristics

4.3.1. DSC Analysis. A DSC analysis was carried out to investigate the effect of PBT or OBT on the melting and crystallization behaviours of iPP polymer. It is known that isotactic polypropylene can crystallize in three forms: the monoclinic form, the trigonal form, and the triclinic form [27]. The α phase is the principal constituent in normal processing conditions.

Figure 5 depicts the DSC melting curves of iPP and its iPP hybrids composites. It is obvious that iPP sample presents a single melting peak at about 160°C attributed to the melting of α-iPP [28].

According to DSC measurements gathered in Table 1, it seems that PBT has no effect on the iPP melting behavior. The melting temperature of the dominant crystal fraction in the iPP/PBT composite remains constant indicating that PBT does not change the iPP crystal organization. However for iPP/OBT nanocomposite the endothermic peak becomes broader and slightly shifts to a high temperature range.

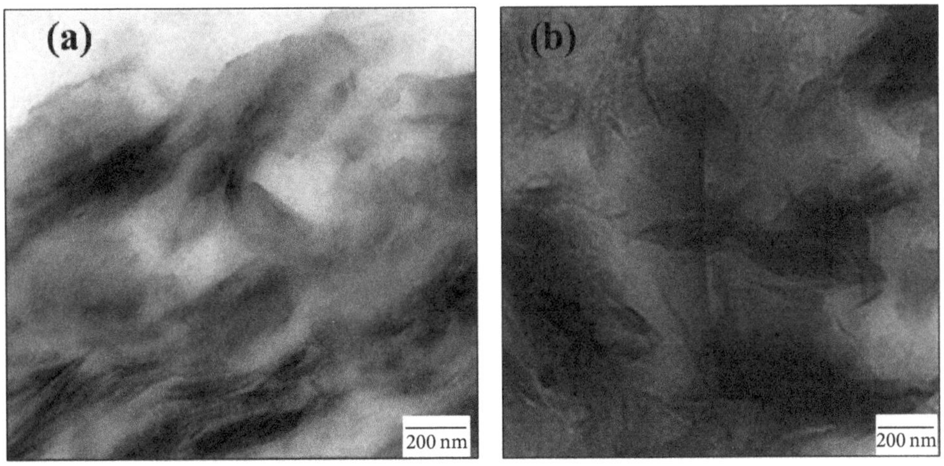

(a) (b)

FIGURE 4: TEM micrographs of (a) iPP/OBT and (b) iPP/PBT.

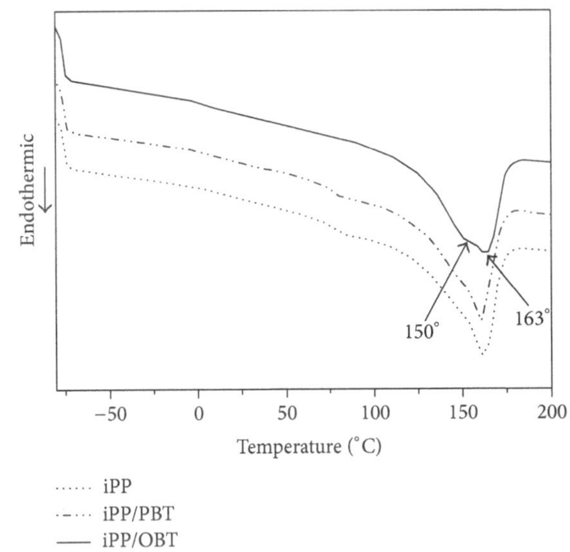

FIGURE 5: DSC melting curves of iPP and its iPP composites.

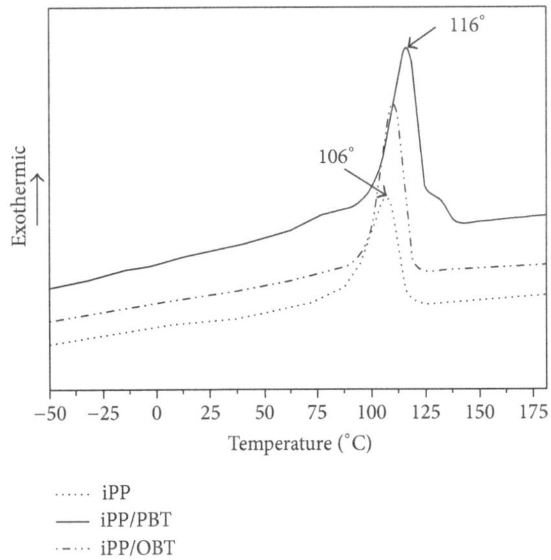

FIGURE 6: DSC crystallization curves of iPP and its iPP composites.

TABLE 1: DSC results of iPP and its iPP/Maghnia bentonite composites.

Samples	T_{mp} (°C)	ΔH_m (J/g)	T_c (°C)	ΔH_m (J/g)	χ_c (%)
iPP	160	76.4	106	75.8	52.3
iPP/PBT	161	77.3	110	81.6	55.7
iPP/OBT	163	86.6	116	85.1	62.4

An endothermic shoulder also appears around 150°C characterizing the β phase of iPP. In this case, the formation of the β phase can be promoted in iPP by the OBT addition. This result is in good agreement with that obtained by XRD analysis.

From Table 1, it can be noticed that the enthalpy of fusion (ΔH_m) for iPP/OBT nanocomposite increases compared to that of pure iPP or iP/PBT composite and therefore the high value of degree of χ obtained for the nanocomposite reflects an increase of iPP crystallinity as a result of the nucleation effect of the OBT [29, 30].

Furthermore, as the aforementioned argument that the organoclay could act as the nucleating agent and accelerate the crystal growth of the iPP polymer, the crystallization temperature of the nanocomposite was clearly enhanced compared with that of the pure iPP or IPP/PBT composite (Figure 6). Indeed, the incorporation of the oganomodified bentonite significantly increases the crystallization temperature of iPP from 106° to 116°C. This suggests that the OBT nanoparticles impart a high efficiency to the heterogeneous nucleation of iPP.

It is known that the organoclay is an effective nucleating agent due to the high surface area of silicate platelets and their chemical affinity for the polymer, which induce a nucleation

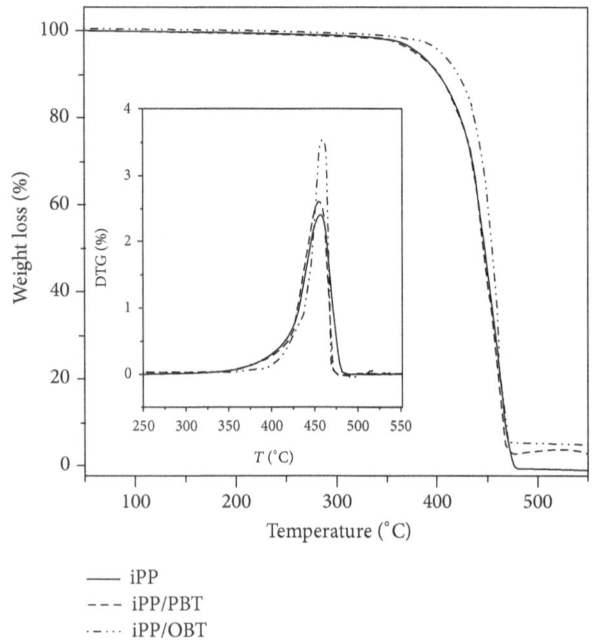

FIGURE 7: TGA and DTG thermograms of iPP, iPP/PBT, and iPP/OBT.

and lamellar ordering effect. For iPP/OBT nanocomposite, 5 wt% of organoclay act as nucleating agent, heighten the crystallization rate, and induce the crystallization at higher temperature compared with that of the pure iPP, whereas, for iPP/PBT composite, according to the XRD and TEM results, PBT layers are not totally intercalated by the polymer chains, prefer to aggregate, which could confine the mobility of the polymer chains to rearrangement and forming of crystal grains, and finally induced the decrease of the crystallization temperature of the iPP/PBT composite but remains higher than that of the pristine iPP.

4.3.2. TGA Analysis. Figure 7 shows the mass loss (TG%) and the derivative mass loss (DTG) curves of iPP and its nanocomposites at 10°C/min heating rate. Typical TGA thermogram of weight loss as a function of temperature for virgin iPP and its nanocomposites can be seen. Degradation of both iPP and iPP/nanoclay materials was performed at one-stage process. It is obvious that iPP and its nanocomposites exhibited a relatively good thermostability, since no remarkable mass loss occurred until 335°C (<0.5%).

Pure iPP begins to degrade at about 425°C. Its thermal degradation initiated primarily by thermal scissions of C–C chain bonds accompanied by transfer of hydrogen at the site of scission [31]. The initial decomposition temperature T_{onset} (calculated by bitangent method from TGA curves) and the peak decomposition temperature at maximum weight loss rate from DTG curves T_{max} and residues (Re) at 580°C are summarized in Table 2.

Temperature T_{onset} of iPP/PBT is about 4°C higher than that of virgin iPP. The amount (5%) of pure clay in iPP does not make a significant enhancement in thermal stability of this hybrid system. This is primarily due to the absence of surfactant on the surface of nanoclay and hence there was a low diffusion of polymer into the nanoclay resulting in agglomeration. However, an increase in the initial decomposition temperature of 13°C was noticed when OBT was added in iPP/OBT nanocomposite. This increase was attributed to the barrier effect of OBT layers dispersed in the isotactic polypropylene matrix.

TABLE 2: Thermogravimetric parameters of iPP/Maghnia bentonite nanocomposites.

Samples	T_{onset} (°C)	T_{max} (°C)	Re (%)
iPP	425	453	0
iPP/PBT	429	455	3
iPP/OBT	438	462	5

As seen from the peak of the first derivative, the temperature, at which iPP decomposition rate is the highest, is T_{max} = 453°C for a 10°C/min heating rate. In the case of iPP/PBT nanocomposites, the main degradation peak remains basically unchanged compared to pure iPP (Table 2). However, an improved thermal stability was obtained with iPP/OBT nanocomposite due to the incorporation of clay particles into iPP matrix, which could enhance thermal stability of iPP [32], as seen from Table 2. A residue (Re) about 5% is obtained with iPP/OBT nanocomposite compared to 0% of its corresponding virgin polymer.

4.3.3. Thermal Degradation of Pure iPP and Its Nanocomposites (iPP/PBT and iPP/OBT). For all samples, degradation was studied through nonisothermal measurements at different heating rates (5, 10, 20, 30, and 40°C). In Figure 8, mass loss at different heating rates for iPP/POBT nanocomposites is shown. As it could be noticed, T_{onset} was gradually shifted to higher values for most heating rates for all samples. This increase in thermal decomposition temperatures of pure iPP is caused by heat hysteresis.

In reference to Yuan et al. works [33], we attempted to eliminate the influence of heating rates on thermal hysteresis, by following the variation of T_{max} as a function of heating rate. As shown in Figure 9, T_{max} of iPP and its nanocomposites increased linearly along with the heating rates (β). The relationship between T_{max} and β can be expressed in the following equation:

$$T_{max} = T_{max}^{0} + A\beta, \quad (12)$$

where T_{max}^{0} is the equilibrium peak thermal decomposition temperature assuming that the heating rate equals 0°C/min and A is the rate constant.

T_{max} and A value of iPP/OBT nanocomposite are higher than those of pure iPP. These results suppose an improvement of OBT barrier effect. The insertion of organomodified OBT in the iPP matrix is likely to enhance its thermal stability.

4.3.4. Determination of Activation Energy. In order to deeply analyze the effect of the incorporation of pure bentonite or organically modified bentonite into iPP matrix on the degradation mechanism of iPP, it is important to evaluate

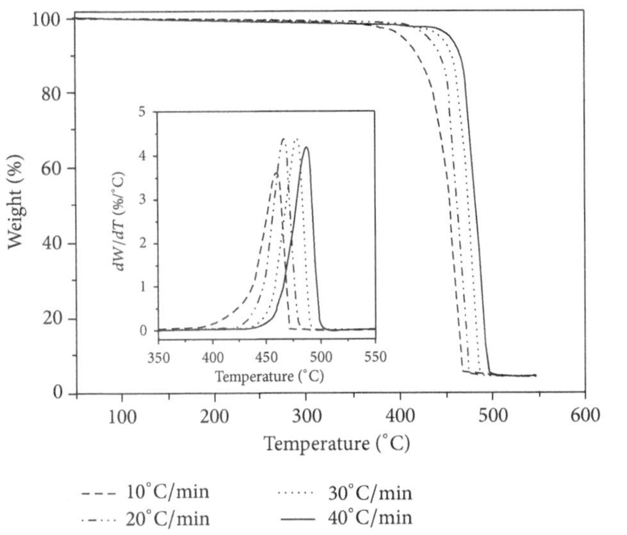

FIGURE 8: Mass loss (%) curves and derivative mass loss curves at different heating rates of iPP/OBT nanocomposite.

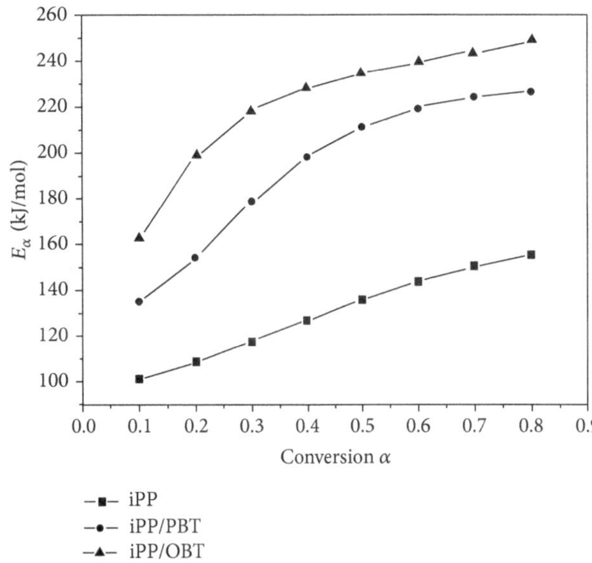

FIGURE 10: Dependence of the activation energy E_a on the degree of the conversion α of the mass loss, as calculated with the OFW method for the different nanocomposites.

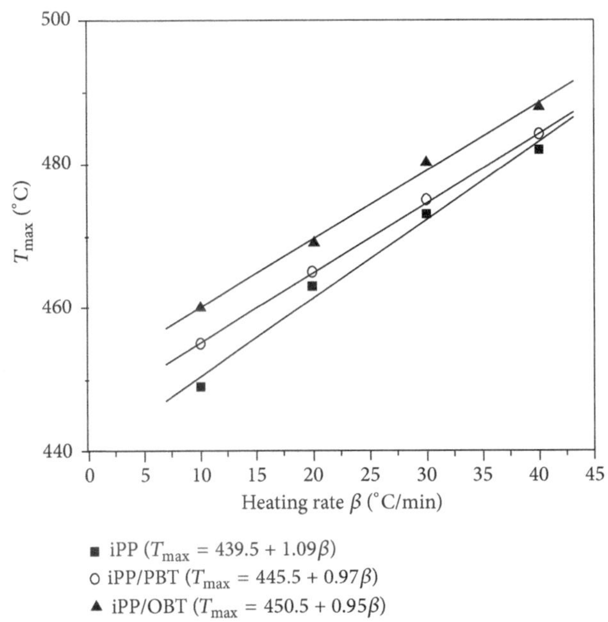

FIGURE 9: Plot of T_{max} versus heating rates of iPP and its iPP/Maghnia bentonite nanocomposites.

the kinetic parameters (activation energy E_a and preexponential factor A).

The activation energy of degradation of pure iPP and iPP/PBT or IPP/OBT nanocomposites was first estimated by the well-known equation of Ozawa, Flynn, and Wall (OFW) and then by the new method developed by Tang.

The activation energy determined from OFW method increased with the degree of conversion for pure iPP and iPP/PBT or iPP/OBT nanocomposites (Figure 10), which should be ascribed to the nature of random scission degradation and the differences in deviation from stationary reaction state at the different heating rates. In addition, activation energies of iPP/PBT or iPP/OBT nanocomposites were much higher than those of pure iPP. This is an indication that the incorporation of nanoclay into the iPP matrix improved iPP thermal stability. These results are in entire agreement determined by Vassiliou et al. [34]. It was reported that the presence of intercalated clay nanoparticles has a stronger stabilizing effect in the early stage. The plate nanoparticles form a more effective diffusion barrier and thus hinder diffusion of formed gasses during iPP thermal degradation [35].

For comparative purpose, activation energy was then calculated using Tang's method. Data of apparent activation energies for pure iPP, iPP/PBT, and iPP/OBT nanocomposites were illustrated in Figure 11.

Activation energies of iPP/Maghnia bentonite nanocomposites are much higher than that of pure iPP and became much more important when organically bentonite was added to iPP.

This is mainly due to the presence of surfactant which widely separated the clay layers and hence allows a high diffusion of polymer into the nanoclay. The plate nanoparticles form a more effective diffusion barrier and thus hinder diffusion of formed gasses during iPP thermal degradation. These results suppose that the shielding effect of the Maghnia bentonite layers could increase the activation energy of the polymer matrix thermal degradation.

4.4. Mechanical Properties. In the last part of this contribution, the effect of the addition of unmodified or organically modified bentonite on the mechanical properties of iPP nanocomposites was also studied and shown in Table 3.

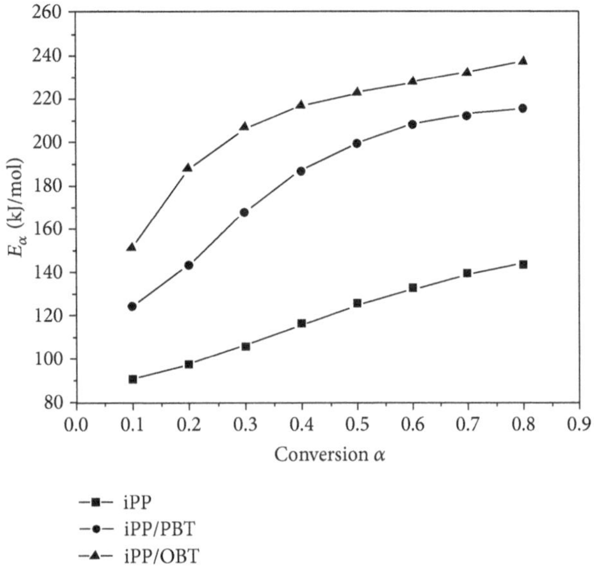

FIGURE 11: Dependence of the activation energy E_α on the degree of the conversion α of the mass loss, as calculated with the Tang method for the different nanocomposites.

TABLE 3: Physical and mechanical properties of iPP and its iPP/Algerian bentonite nanocomposites.

Samples	Young's modulus E (MPa)	Strength σ (MPa)	Elongation at break ε (%)
iPP	1045.0 (±10.4)	25.7 (±0.6)	650.0 (±38.0)
iPP/PBT	1030.0 (±5.5)	26.6 (±0.8)	600.0 (±45.0)
iPP/OBT	1305.5 (±20.2)	30.5 (±1.2)	350.0 (±25.0)

Compared to pure iPP, nanocomposites have higher mechanical properties particularly when OBT was used [36, 37]. This might be due to OBT surface which is much more hydrophobic than PBT surface. Therefore, OBT has good thermodynamical compatibility with the polymer matrix, which leads to improvement in stiffness [38].

5. Conclusion

The purpose of this contribution is the elaboration of nanocomposites by melt intercalation using bentoniteoriginated from Algeria. Higher intercalation level was obtained for the composite prepared using organomodified bentonite compared to the one obtained with pure bentonite. XRD results showed that no significant change induced by the nanofillers incorporation has been noticed on the iPP crystal structure. The α-crystalline form dominated for the two hybrid materials. It should be noted also that in iPP/OBT composite both monoclinic α and hexagonal β forms exist. TEM micrographs of iPP/Maghnia bentonite confirmed in agreement with XRD the formation of mainly intercalated nanocomposite structures.

Both XRD and DSC analyses evidenced an increase in the iPP crystallinity induced by clay incorporation. An improvement of iPP thermal stability was noticed by TGA. Activation energies of both iPP/PBT and iPP/OBT nanocomposites have increased compared to pure iPP, which is due to the incorporation of clay particles into iPP matrix, and consequently improve iPP thermal stability. Tensile tests have shown that nanomaterials have higher mechanical properties particularly when OBT is used for nanocomposites elaboration.

Conflict of Interests

The authors declare that there is no conflict of interests regarding the publication of this paper.

References

[1] A. Vassiliou, D. Bikiaris, K. Chrissafis, K. M. Paraskevopoulos, S. Y. Stavrev, and A. Docoslis, "Nanocomposites of isotactic polypropylene with carbon nanoparticles exhibiting enhanced stiffness, thermal stability and gas barrier properties," *Composites Science and Technology*, vol. 68, no. 3-4, pp. 933–943, 2008.

[2] S. S. Ray and M. Okamoto, "Polymer/layered silicate nanocomposites: a review from preparation to processing," *Progress in Polymer Science*, vol. 28, no. 11, pp. 1539–1641, 2003.

[3] X.-L. Jiang, J.-B. Bao, T. Liu, L. Zhao, Z.-M. Xu, and W.-K. Yuan, "Microcellular foaming of polypropylene/clay nanocomposites with supercritical carbon dioxide," *Journal of Cellular Plastics*, vol. 45, no. 6, pp. 515–538, 2009.

[4] H. Nathani, A. Dasari, and R. D. K. Misra, "On the reduced susceptibility to stress whitening behavior of melt intercalated polybutene-clay nanocomposites during tensile straining," *Acta Materialia*, vol. 52, no. 11, pp. 3217–3227, 2004.

[5] E. Manias, A. Touny, L. Wu, K. Strawhecker, B. Lu, and T. C. Chung, "Polypropylene/montmorillonite nanocomposites. Review of the synthetic routes and materials properties," *Chemistry of Materials*, vol. 13, no. 10, pp. 3516–3523, 2001.

[6] W. Zheng, X. Lu, C. L. Toh, T. H. Zheng, and C. He, "Effects of clay on polymorphism of polypropylene in polypropylene/clay nanocomposites," *Journal of Polymer Science B: Polymer Physics*, vol. 42, no. 10, pp. 1810–1816, 2004.

[7] C. J. Perez and V. A. Alvarez, "Overall crystallization behavior of polypropylene-clay nanocomposites; effect of clay content and polymer/clay compatibility on the bulk crystallization and spherulitic growth," *Journal of Applied Polymer Science*, vol. 114, no. 5, pp. 3248–3260, 2009.

[8] M. A. Pérez, L. B. Rivas, S. M. Rodríguez, Á. Maldonado, and C. Venegas, "Polypropylene/clay nanocomposites. Synthesis and characterization," *Journal of the Chilean Chemical Society*, vol. 55, no. 4, 2004.

[9] M. J. Solomon and A. Somwangthanaroj, "Intercalated polypropylene nanocomposites," Dekker Encyclopedia of Nanoscience and Nanotechnology, 1483.

[10] J.-D. He, M. K. Cheung, M.-S. Yang, and Z. Qi, "Thermal stability and crystallization kinetics of isotactic polypropylene/organomontmorillonite nanocomposites," *Journal of Applied Polymer Science*, vol. 89, no. 12, pp. 3404–3415, 2003.

[11] T. Kashiwagi, E. Grulke, J. Hilding, R. Harris, W. Awad, and J. Douglas, "Thermal degradation and flammability properties of poly(propylene)/carbon nanotube composites," *Macromolecular Rapid Communications*, vol. 23, no. 13, pp. 761–765, 2002.

[12] C. Silvestre, S. Cimmino, M. Raimo, C. Carfagna, V. Capuano, and R. Kotsilkova, "Effect of clay/diamond and clay/carbon

nanosystems on structure-properties relationships of iPP," *Journal of Polymer Science Part B: Polymer Physics*, vol. 42, pp. 3428–3438, 2004.

[13] S. Bouhelal, M. E. Cagiao, S. Khellaf et al., "Nanostructure and microechanical properties of reversibly crosslinked isotactic polypropylene/clay composites," *Journal of Applied Polymer Science*, vol. 115, no. 5, pp. 2654–2662, 2010.

[14] L. Raka, A. Sorrentino, and G. Bogoeva-Gaceva, "Isothermal crystallization kinetics of polypropylene latex-based nanocomposites with organo-modified clay," *Journal of Polymer Science B: Polymer Physics*, vol. 48, no. 17, pp. 1927–1938, 2010.

[15] K. Chrissafis, K. M. Paraskevopoulos, S. Y. Stavrev, A. Docoslis, A. Vassiliou, and D. N. Bikiaris, "Characterization and thermal degradation mechanism of isotactic polypropylene/carbon black nanocomposites," *Thermochimica Acta*, vol. 465, no. 1-2, pp. 6–17, 2007.

[16] L. Wang and J. Sheng, "A kinetic study on the thermal degradation of polypropylene/attapulgite nanocomposites," *Journal of Macromolecular Science B: Physics*, vol. 45, no. 1, pp. 1–11, 2006.

[17] J. Zhu, F. M. Uhl, A. B. Morgan, and C. A. Wilkie, "Studies on the mechanism by which the formation of nanocomposites enhances thermal stability," *Chemistry of Materials*, vol. 13, no. 12, pp. 4649–4654, 2001.

[18] A. Habi, S. Djadoun, and Y. Grohens, "Morphology and thermal behavior of organo-bentonite clay/poly(styrene-co-methacrylic acid)/poly(isobutyl methacrylate-co-4-vinylpyridine) nanocomposites," *Journal of Applied Polymer Science*, vol. 114, no. 1, pp. 322–330, 2009.

[19] B. Monasse and J. M. Haudin, "Growth transition and morphology change in polypropylene," *Colloid & Polymer Science*, vol. 263, no. 10, pp. 822–831, 1985.

[20] T. A. Ozawa, "New method of analyzing thermogravimetric data," *Bulletin of the Chemical Society of Japan*, vol. 38, no. 11, pp. 1881–1886, 1965.

[21] J. H. Flynn and L. A. Wall, "A quick direct method for the determination of activation energy from thermogravimetric data," *Polymer Letters*, vol. 4, pp. 323–328, 1966.

[22] W. Tang, Y. Liu, H. Zhang, and C. Wang, "New approximate formula for Arrhenius temperature integral," *Thermochimica Acta*, vol. 408, no. 1-2, pp. 39–43, 2003.

[23] X. Zhang, M. Yang, Y. Zhao et al., "Polypropylene/montmorillonite composites and their application in hybrid fiber preparation by melt-spinning," *Journal of Applied Polymer Science*, vol. 92, no. 1, pp. 552–558, 2004.

[24] J. Varga, "β-modification of isotactic polypropylene: preparation, structure, processing, properties, and application," *Journal of Macromolecular Science*, vol. 41, no. 4–6, pp. 1121–1171, 2002.

[25] J. Varga and J. Karger-Kocsis, "Rules of supermolecular structure formation in sheared isotactic polypropylene melts," *Journal of Polymer Science B: Polymer Physics*, vol. 34, no. 4, pp. 657–670, 1996.

[26] B. Rozanski, E. Monasse, A. Szkudlarek et al., "Shear-induced crystallization of isotactic polypropylene based nanocomposites with montmorillonite," *European Polymer Journal*, vol. 45, no. 1, pp. 88–101, 2009.

[27] J. Varga, "Supermolecular structure of isotactic polypropylene," *Journal of Materials Science*, vol. 27, no. 10, pp. 2557–2579, 1992.

[28] D. N. Bikiaris, G. Z. Papageorgiou, E. Pavlidou, N. Vouroutzis, P. Palatzoglou, and G. P. Karayannidis, "Preparation by melt mixing and characterization of isotactic polypropylene/SiO_2 nanocomposites containing untreated and surface-treated nanoparticles," *Journal of Applied Polymer Science*, vol. 100, no. 4, pp. 2684–2696, 2006.

[29] C. Hao, Y. Zhao, A. He et al., "Investigation on the melt spinning fibers of PP/organoclay nanocomposites prepared by in-situ polymerization," *Journal of Applied Polymer Science*, vol. 116, no. 3, pp. 1384–1391, 2010.

[30] S. A. Ahmad Ramazani, F. Tavakolzadeh, and H. Baniasadi, "Synthesis of polypropylene/clay nanocomposites using bisupported ziegler-natta catalyst," *Journal of Applied Polymer Science*, vol. 115, no. 1, pp. 308–314, 2010.

[31] S. L. Madorsky, *Thermal Degradation of Organic Polymers*, chapter 4, Wiley-Interscience, New York, NY, USA, 1964.

[32] X. Xu, Y. Ding, F. Wang et al., "Effects of silane grafting on the morphology and thermal stability of poly(ethylene terephthalate)/Clay nanocomposites," *Polymer Composites*, vol. 31, no. 5, pp. 825–834, 2010.

[33] X. Yuan, C. Li, G. Guan, Y. Xiao, and D. Zhang, "Thermal degradation investigation of poly(ethylene terephthalate)/fibrous silicate nanocomposites," *Polymer Degradation and Stability*, vol. 93, no. 2, pp. 466–475, 2008.

[34] A. A. Vassiliou, K. Chrissafis, and D. N. Bikiaris, "In situ prepared PET nanocomposites: effect of organically modified montmorillonite and fumed silica nanoparticles on PET physical properties and thermal degradation kinetics," *Thermochimica Acta*, vol. 500, no. 1-2, pp. 21–29, 2010.

[35] K. Wang, S. Liang, J. Deng et al., "The role of clay network on macromolecular chain mobility and relaxation in isotactic polypropylene/organoclay nanocomposites," *Polymer*, vol. 47, no. 20, pp. 7131–7144, 2006.

[36] J. A. Tarapow, C. R. Bernai, and V. A. Alvarez, "Mechanical properties of polypropylene/clay nanocomposites: effect of clay content, polymer/clay compatibility, and processing conditions," *Journal of Applied Polymer Science*, vol. 111, no. 2, pp. 768–778, 2009.

[37] S. K. Sharma and S. K. Nayak, "Surface modified clay/polypropylene (PP) nanocomposites: effect on physico-mechanical, thermal and morphological properties," *Polymer Degradation and Stability*, vol. 94, no. 1, pp. 132–138, 2009.

[38] M. V. Burmistr, K. M. Sukhyy, V. V. Shilov et al., "Synthesis, structure, thermal and mechanical properties of nanocomposites based on linear polymers and layered silicates modified by polymeric quaternary ammonium salts (ionenes)," *Polymer*, vol. 46, no. 26, pp. 12226–12232, 2005.

20

Edible Polymers: Challenges and Opportunities

Subhas C. Shit and Pathik M. Shah

Central Institute of Plastics Engineering & Technology (CIPET), Plot No. 630, Phase IV, Vatva, G.I.D.C., Ahmedabad 382 445, India

Correspondence should be addressed to Subhas C. Shit; cipetahmd@gmail.com

Academic Editor: Jean-Marc Saiter

Edible polymers have established substantial deliberation in modern eons because of their benefits comprising use as edible materials over synthetic polymers. This could contribute to the reduction of environmental contamination. Edible polymers can practically diminish the complexity and thus improve the recyclability of materials, compared to the more traditional non-environmentally friendly materials and may be able to substitute such synthetic polymers. A synthetic hydrogel polymer unlocked a new possibility for development of films, coatings, extrudable pellets, and synthetic nanopolymers, particularly designed for medical, agricultural, and industrial fields. Edible polymers offer many advantages for delivering drugs and tissue engineering. Edible polymer technology helps food industries to make their products more attractive and safe to use. Novel edible materials have been derived from many natural sources that have conventionally been regarded as discarded materials. The objective of this review is to provide a comprehensive introduction to edible polymers by providing descriptions in terms of their origin, properties, and potential uses.

1. Introduction

The increasing mandate for high quality, ready-to-eat food products with a long shelf life contributes to the expansion of new processing technologies which ensure that the product's natural properties and appearance were not dramatically transformed. Food wrapping contributes to a lot of wastes in our homes and it is something that most of us just accept. Edible polymers were the polymeric material which can be easily consumed by human beings or lower animals in whole or part via the oral cavity and given harmless effect to the health. Edible polymers may be applied directly on the surface as additional protection to preserve product quality and stability. The requirements imposed on edible polymers were determined by the specific properties of the product and changes in these properties during production and storage [1]. There were several reasons for investigating edible polymers. One of them was the introduction of new food product categories, such as safe, convenient, and high quality products. They protect food against the loss of nutrients. In practice, edible coatings which control the rate of transport of the product's molecular components from the inside to the outside of the packaging may slow down adverse reactions which were responsible for undesirable changes in food products [1]. The efficiency of edible polymer was determined by their mass transport properties [2]. Technologist develops and formulates implantable drug delivery systems made of bioresorbable polymers that release precise quantities of therapeutic agents gradually over time [3]. Polymer drug conjugates were products of nanomedicine enrolled in cancer diagnosis and treatment [4].

Natural polymers can be an alternative source for packaging development due to their precise taste and biodegradability. Edible polymers have appeared as a substitute for synthetic plastic for food applications and have received significant attention in recent years because of their advantages over synthetic polymer. The main advantage of edible polymer over traditional synthetics was that they can be consumed with the products. There is no package to dispose and even if the films are not consumed, they could still contribute to the reduction of environmental effluence. The edible polymers were produced exclusively from renewable, edible ingredients and therefore were anticipated to degrade more readily than other polymeric materials [5]. Edible polymer can expand the organoleptic properties of packaged foods, provided that they contain various components like

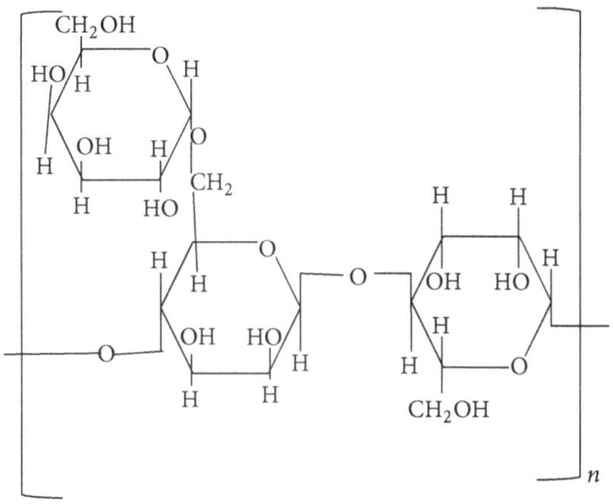

Scheme 1: Common structure of hydrocolloids.

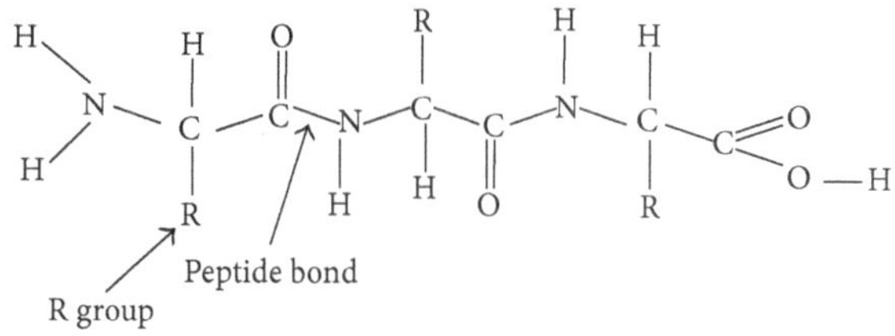

Scheme 2: Common structure of polypeptide.

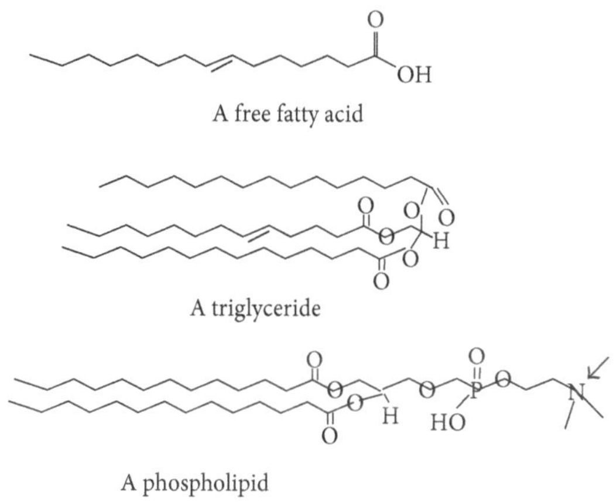

Scheme 3: Common structure of lipids.

flavorings, colorings, and sweeteners. Their use based on natural polymers and food grade additives has been continually increasing in the food industry and medical industries [6]. The edible polymer can be produced with a variety of natural products such as polysaccharides, proteins, and lipids, with the addition of plasticizers and surfactants. The functionality and performance of edible polymer mainly depend on their barrier, mechanical, and color properties, which in turn depend on film composition and its formation process. Food products were usually coated by dipping or spraying, forming a thin film on the food surface that acts as a semipermeable membrane, which in turn control the moisture loss or/and suppress the gas transfer [5]. The edible polymer also functions as carriers for antimicrobial and antioxidant agents. Production of edible polymer causes less waste and pollution; however, their permeability and mechanical properties were generally poorer than synthetic polymer.

At present, around 150 million tons of plastic were produced yearly all over the world, and production and consumption continue to upturn [7]. Most of these plastics were crude oil based, and any increase in their production results in an increase of oil use and this causes serious environmental contamination, due to wasted and undegraded polymer. Edible and biodegradable polymers must be considered as a substitute to more traditional recycling procedures and this has stimulated researchers to synthesize new polymers that can be returned to the biological cycle after practice. Therefore, the use of agricultural polymers that were easily biodegradable would not only solve these problems, but also provide a potential new use for excess farm products. In the food packaging sector, starch-based material has received great attention due to its biodegradability, edible, commonly available, abundant, low cost, nonallergic, easy to use and thermoprocessable [8]. Components used for the preparation of edible polymer can be classified into four categories: hydrocolloids, polypeptides, lipids, and composites. Hydrocolloid films have good barrier properties of oxygen, carbon dioxide, and lipids (Scheme 3) but not to water vapor (Scheme 1). Most hydrocolloid polymers also possess superb mechanical properties (Tables 1 and 2), which were quite useful for fragile food products. Among them, protein-based edible polymers were the most attractive ones. These polymers have impressive gas barrier properties compared with those prepared from lipids and polysaccharides. When being not moist, the O_2 permeability of soy protein-based film was 500, 260, 540, and 670 times lower than that of low density polyethylene, methylcellulose, starch, and pectin, respectively [9].

The mechanical properties of some protein-based edible polymers like rapeseed protein blended with gelatin (tensile strength—53.45 MPa) were better than polysaccharide and fat-based films [13]. Protein-based edible polymer can form bonds at different positions and offer high potential for forming numerous linkages (Scheme 2). Low value fish, such as lizard fish, were usually rejected from surimi manufacturing because of poor surimi gel strength. Hence, they have hitherto has mostly been used as animal feed and/or sold for low prices because of the lack of techniques for using them as foodstuffs. In order to obtain effective utilization of low value fish, including lizard fish, there was a need for more information on polymer produced from low value fish meat, their mechanical properties, and their applications. Nowadays there was a proecological tendency of dropping the use of synthetic materials. One of the problem-solving strategies was to replace the plastic packaging with new, biodegradable materials. Edible polymers seem to be a good alternative to the plastic foil. They were defined as material which can be consumed and provides a good barrier to moisture, oxygen, and solute movements for the food [14]. Because edible polymers were considered a packaging as well as a food component, they should fulfill a number of necessities, such as high barrier and mechanical efficiencies

Table 1: Tensile strength and elongation at break properties of hydrocolloid polymers [10, 11].

Polymer	Tensile strength (TS) MPa	Elongation at break (*E* %)
Cellulose derivatives	44–65	10–50
Collagen	1–70	10–70
Chitosan	10–100	20–80
Gelatin	25–140	7–22
Starch	35–46	1.7–3.4
Soy protein	3.7–4.5	152–160
Lentil protein	4-5	58–70
Whey protein	2.5–3.0	15–18
Peanut protein	3-4	147–150
Mung Bean	5.70–6.51	32–40
Low density polyethylene	16–18	>1000
Oriented polypropylene	50–60	73–100
Polyethylene terephthalate	81–85	19–25
Polyvinylidene chloride	65–75	18–23

and biochemical, physicochemical, and microbial stabilities; they should be nontoxic, nonpolluting, and of low cost [15].

2. Categories of Edible Polymer

The edible polymers can be comprised in any of the four categories:

(1) hydrocolloids, (2) polypeptides, (3) lipids, (4) synthetic and composite edible polymers.

2.1. Hydrocolloids. Hydrocolloids are often called hydrophilic polymers, of vegetable, animal, microbial, or synthetic origin, that generally contain many hydroxyl groups (polysaccharides) and may be polyelectrolytes. Polysaccharides were commercially available for use in food and nonfood industries as stabilizers, thickening and gelling agents, crystallization inhibitors, and encapsulating agents [16]. Many hydrocolloids were polyelectrolytes, for example, starch alginate, carrageenan, carboxymethylcellulose, gum arabic, chitosan, pectin, and xanthan gum.

Starch consists of two kinds of molecules, amylose (normally 20–30%) and amylopectin (normally 70–80%), which was primarily derived from cereal grains and tubers like corn (maize), wheat, potato, tapioca, and rice. Amylose was accountable for the film forming capacity of starch [17]. High amylose starch films have been made that were flexible, oxygen impermeable, oil resistant, heat sealable, and water soluble. Polymer of high-amylose corn starch or potato starch was more stable during aging [18]. Starch-based edible polymer exhibits physical characteristics similar to plastic polymer in which they were odorless, tasteless, colorless, nontoxic, biologically absorbable, semipermeable to carbon dioxide, and resistant to passage of oxygen. Since the water activity was critical for microbial, chemical, and enzymatic activities, edible starch-based films can retard microbial growth by lowering the water activity within the package. The ability of polysaccharide-based (starch, carrageenan, and chitosan) coatings to extend the shelf life of strawberry fruit (*Fragaria ananassa*) was studied by Ribeiro et al., mainly for industrial applications [19]. Films, made with oxidized starch, showed improved tensile properties and fracture elongation percentage, with respect to those of the native starch [20]. *Kaempferia rotunda* and *Curcuma xanthorrhiza* are essential oils enriched in cassava starch-based edible coating which could extend patin fillet's shelf life and could be used as alternative fish preservation [21]. Alginates were derived from seaweeds and possess good film-forming properties that make them particularly useful in food applications. Alginate has a potential to form biopolymer film or coating component because of its unique colloidal properties, which include thickening, stabilizing, suspending, film forming, gel production, and emulsion stabilizing [22]. Alginate was used in various pharmaceutical preparations such as gaviscon, bisodol, and asilone. Gaviscon is a nonprescription medicine, which is taken by mouth to treat heartburn and gastroesophageal reflux disease. It was also observed that raft-forming chewable tablets prepared using an optimum amount of sodium alginate, calcium carbonate, and sodium bicarbonate could be an efficient dosage form in the treatment of gastroesophageal reflux disease [23]. 5-fluorouracil encapsulated with alginate beads was for the treatment of breast cancer [24]. Alginate based phenobarbitone tablets can be easily prepared for the treatment of chronic seizures in terms of controlled release of drug to obtain the effect over long time period [25]. Alginate was used extensively as an impression-making material in dentistry, prosthetics, life casting, and occasionally for creating positives for small-scale casting. It was also used in the food industry, for thickening soups and jellies. Edible polymer prepared from Alginates forms strong films and exhibits poor water resistance because of their hydrophilic nature [26].

Carrageenans were water-soluble polymers with a linear chain of partially sulfated galactans. These sulfated polysaccharides were extracted from the cell walls of various red seaweeds. Carrageenan film formation includes a gelation mechanism during moderate drying, leading to a three-dimensional network formed by polysaccharide-double helices and to a solid film after solvent evaporation [27]. It is used in desserts, ice cream, cream, milkshakes, salad dressings, sweetened condensed milks, and sauces, which is gel to increase viscosity. It is also used in beer, toothpaste, fruit gushers, soy milk, and so forth [28]. Recently, carrageenan films were also found to be less opaque than those made of starch [29].

Only four cellulose derivative forms are used for edible coatings or films: hydroxypropyl cellulose (HPC), hydroxypropyl methylcellulose (HPMC), carboxymethylcellulose (CMC), or methylcellulose (MC). Cellulose derivatives exhibit thermogelation; therefore when suspensions are heated they form a gel, whereas they return to original consistency when cooled [30]. The films casting from aqueous solutions of MC, HPMC, HPC, and CMC tend to have moderate strength, are resistant to oils and fats, and are flexible, transparent, flavorless, colorless, tasteless, water-soluble, and moderate barriers to oxygen. MC is more

TABLE 2: Water vapor permeability (WVP) properties of different polymers [12].

Film formulation	WVP ($g\,m^{-1}\,s^{-1}\,Pa^{-1}$)
Corn zein	5.35×10^{-10}
Corn zein plasticized with glycerol	8.90×10^{-10}
Fish skin gelatin	2.59×10^{-10}
Whey protein plasticized with sorbitol	7.17×10^{-10}
Wheat gluten plasticized with glycerol	7.00×10^{-10}
Gelatin (obtained from pigskin) plasticized with sorbitol	1.6×10^{-10}
Amylose	3.8×10^{-10}
Corn starch plasticized with glycerol	2.57×10^{-10}
Corn starch plasticized with sorbitol	1.75×10^{-10}
Amylomaize starch plasticized with sorbitol	1.21×10^{-10}
Hydroxypropyl methylcellulose withplasticizer and oil	1.90×10^{-10}
Amylomaize starch with sorbitol andsunflower oil	9.7×10^{-11}
Methylcellulose	8.70×10^{-11}
Methylcellulose 3%	$8.4–12.1 \times 10^{-11}$
Chitosan 2% (unknown source)	$3.66–4.80 \times 10^{-11}$
Chitosan 3%	$6.19–15.27 \times 10^{-11}$
Cellophane	8.4×10^{-11}
PVDC	2.22×10^{-13}
LPDE (low density polyethylene)	9.14×10^{-13}
HDPE (high density polyethylene)	2.31×10^{-13}

resistant to water and it is the lowest hydrophilic cellulose derivatives. However, cellulose derivative films are poor water vapor barriers because of the inherent hydrophilic nature of polysaccharides and they possess poor mechanical properties. A way to improve the moisture barrier would be the incorporation of hydrophobic compounds, such as fatty acids, into the cellulose ether matrix to develop a composite film [31]. As another example, Galus et al. [32] developed bilayers by adding a mixture of lipids (paraffin oil, paraffin wax, or a mixture of hydrogenated palm oil and triolein) onto a methylcellulose layer. Edible coatings made of CMC, MC, HPC, and HPMC have been applied to some fruits and vegetables for providing barriers to oxygen, oil, or moisture transfer.

Microcrystalline cellulose and powdered cellulose were used as inactive fillers in drug tablets and as thickeners and stabilizers in processed foods [33]. Cellulose powder was, for example, used in Kraft's Parmesan Cheese to prevent caking inside the tube. Methylcellulose (MC) and hydroxypropyl methylcellulose (HPMC) were used in coating formulations to reduce oil uptake in deep-fat frying potato strips and dough discs. MC coatings were more effective in reducing oil uptake than HPMC ones [34].

One method of enhancing the moisture barrier would be by incorporation of hydrophobic compounds such as fatty acids into the cellulose ether matrix to develop a composite film [5], for example, hydroxypropyl methylcellulose (HPMC). It was found that the moisture barrier was improved by stearic acid incorporation into the film forming solution; however, there were difficulties in preparing a homogenous composite film with both hydrophobic and hydrophilic compounds. For these reasons, homogeneous packaging film was prepared from HMPC by chemical modification allowing an increase in hydrophobicity. Cross-linking of HPMC can be considered a useful approach to prepare nonwater soluble cellulose derivatives with interesting moisture barrier properties, which could be used as biodegradable packaging materials [35].

Pectin was a group of plant-derived polysaccharides that appear to work well with low moisture foods but was poor moisture barriers. Pectin was a heterogeneous grouping of acidic structural polysaccharides, found in fruit and vegetables, and mainly prepared from citrus peel and apple pomace. The conventional application was given the jellylike consistency to jams or marmalades and jelly beans, which would otherwise be sweet juices. In medicine, pectin increases the viscosity and volume of stool so that it is used against constipation and diarrhea [36].

The agar was hydrophilic colloids that have the ability to form reversible gels simply by cooling a hot aqueous solution. Agar exhibits features that make it convenient for cutting meats. It forms strong gels characterized by melting points far above the initial gelation temperature [37]. Agar has been used comprehensively as a gelling agent in the food industry. However, despite its biodegradability and its enormous gelling power, agar has not been used widely due to poor aging. Both photodegradation and fluctuations in ambient temperature and humidity alter agar crystalline, leading to the formation of microfractures and polymer embrittlement [38]. The effect of agar on the structure and the functional properties of emulsified edible films have been recently studied by Phan et al. [39].

Chitosan was one of the auspicious natural polymers with characteristics such as biodegradability, chemical inertness, biocompatibility, high mechanical strength, good film-forming properties, and low cost [40, 41]. Additionally, chitosan was a nontoxic cellulose-like polyelectrolyte polymer hydrogel that was suitable for the production of artificial muscles, as this material undergoes a large volume change in response to changes in pH, temperature, or solvent composition [42]. Modern studies have suggested that chitin may play a role in a probable pathway in human allergic disease [43]. Chitosan was the second most abundant natural and nontoxic polymer in nature after cellulose. Some desirable properties of chitosan were that it forms films without the addition of additives and exhibits good oxygen and carbon dioxide permeability, as well as excellent mechanical properties and antimicrobial activity against bacteria, yeasts, and molds [44, 45]. Chitosan products were highly viscous, resembling natural gums. Chitosan can form translucent films to enhance the quality and extend the storage life of food products [46]. Pure chitosan films were generally cohesive and compact and the film surface has a smooth contour without pores or cracks. Chitosan was used in a wide range of applications such as wastewater treatment [47], separation membranes [48], food packaging drug delivery systems [49], and biosensors [50]. However, in sensor applications, the poor electrical conductivity of hydrogen results in a poor response time and a high operating voltage limits its applicability in devices. Hence, composites have been attempted by incorporating a rigid conducting polymer (such as PANI) into a flexible matrix (such as chitosan) to combine the good possibility of the matrix and the electrical conductivity of the conductive polymer [51]. The addition of tannic acid as a cross-linking agent of chitosan matrix has improved the mechanical properties and those related to the water affinity such as solubility, permeability, and contact angle [52].

Gums in edible-polymer preparation were used for their texturizing capabilities. In edible-forming preparations, guar gum was used as a water binder, stabilizer, and viscosity builder. Gum arabic, owing to its solubility in hot or cold water, was the least viscous one of the hydrocolloid gums. Xanthan gum was readily dispersed in water; hence high consistency was obtained rapidly in both hot and cold systems. A blend of guar gum, gum Arabic, and xanthan gum provided uniform coatings with good cling and improved adhesion in wet batters. The mesquite gum forms films with excellent water vapor (Tables 1 and 2) barrier properties when small amounts of lipids were added in their formulation [52].

2.2. Polypeptides. Polypeptide- (protein-) based edible polymer can be used for the individual packaging of small portions of food, particularly products that were not currently individually packaged for practical reasons, such as beans, nuts, and cashew nuts. Additionally, polypeptide-based edible polymer can function as carriers for antimicrobial and antioxidant agents. Additional possible application of polypeptide-based edible polymer could be their use of multilayer food packaging materials together with nonedible polymer. In this case, the protein-based edible polymer would be the internal layers in direct contact with food materials. Through functions relating to mechanical and barrier properties, protein-based edible polymer may be able to substitute synthetic polymer films.

Polypeptides were anticipated to be good oxygen barriers at low relative humidity. However, the fact that proteins were not totally hydrophobic and contain predominantly hydrophilic amino acid residues limits their moisture-barrier properties. Collagens were the major structural proteins of connective tissue, such as bone, hide, tendons cartilage, and ligaments. The main application of collagen films was as barrier membrane [53]. Films with a thickness of 0.01–0.5 mm are formed by air-drying a casted collagen preparation similar to ophthalmological shields and made of biodegradable materials. The drugs can be loaded into collagen membranes by hydrogen bonding, covalent bonding, or simple entrapment. They can be sterilized and become made flexible. This was used in ophthalmology in delivering drugs to yes and also used over wounds to make healing rapid and uniform [54]. Collagen polymer offers several advantages; it was biocompatible and nontoxic to most tissues; it has well-documented structural, physical, chemical, and immunological properties; it can be processed into a variety of forms, and it is readily isolated and purified in large quantities [55, 56]. Gelatin was unique in forming a thermoreversible substance with a melting point close to body temperature, which was mainly significant in edible and pharmaceutical applications. Fundamentally, gelatin was obtained by controlling hydrolysis from the fibrous insoluble protein, collagen, which was widely found in nature as the major constituent of skin, bones, and connective tissue. The characteristic features of gelatin were the high content of the amino acids glycine, proline, and hydroxyproline. Gelatin also has a mixture of single and double unfolded chains of a hydrophilic character. At approximately 40°C, gelatin aqueous solutions were in the sol state and form physical, thermoreversible gels on cooling. Gelatin was used to encapsulate low moisture or oil phase food constituents and pharmaceuticals [6]. Such encapsulation provides fortification against oxygen and light, as well as defining the amount of ingredient or drug dosage. In addition, gelatin films have been formed as coatings on meats to reduce oxygen, moisture, and transport of oil [57]. Addition, gelatin was competent to form clear and strong films and was used for microencapsulation and capsule coatings in food and pharmaceutical manufacturing. Microencapsulation is the process of surrounding or enveloping one substance within another substance on a very small scale, yielding capsules ranging from less than one micron to several hundred microns in size. Substances may be microencapsulated with the intention that the core material is confined within capsule walls for a specific period of time. Alternatively, core materials may be encapsulated so that the core material will be released either gradually through the capsule walls, known as controlled release or diffusion, or when external conditions trigger the capsule walls to rupture, melt, or dissolve [58]. Short chain gelatin peptides are effective in the treatment of stomach ulcers.

Zein was a prolamin protein and therefore dissolves in 70–80% ethanol [59]. Zein was a relatively hydrophobic and thermoplastic material [60]. Zein coating has also shown an ability to reduce moisture and loss of firmness and delay color alteration in fresh fruit like apples [61]. Furthermore, zein may similarly take part in the coating of conventional packaging plastics. Zein's properties make it valuable in processed foods and pharmaceuticals, in competition with insect shellac. It was now used as a crust for candy, nuts, fruit, pills, and other encapsulated foods and drugs [62]. Wheat gluten was a water insoluble protein of wheat flour which was comprised of a mixture of polypeptide molecules and considered to be globular proteins. The cohesiveness and elasticity of the gluten give reliability to wheat dough and expedites film formation. The poor resistance of wheat gluten films to water vapor (Tables 1 and 2) was due to the hydrophilic nature of the protein and the substantial amount of hydrophilic plasticizer added to impart acceptable film flexibility. The properties of wheat gluten films can be upgraded by using a cross-linking agent such as glutaraldehyde or heat curing. Wheat gluten films can be abundantly biodegraded after 36 days in aerobic fermentation and within 50 days in farmland soil without liberating toxic products. It can be used in a wide selection of applications including milling, bakery products, meats, pasta, and breads [63]. Soy protein from soybeans has been extensively used as a food ingredient in nearly every food product available to the consumer, since it contains high nutrition and excellent functional properties. Since soy proteins were abundant, inexpensive, biodegradable, and nutritional, they show the potential to be developed as edible and biodegradable films [64, 65]. Milk proteins can be classified into two types: casein and whey proteins. The acid casein can be converted to functional soluble caseinates by neutralization through the addition of alkali. Edible protein films based on various caseinates can be obtained by solubilization in water followed by casting and drying. Laminated films that comprised casein did protect dried fruit and vegetables from moisture absorption and oxidation [12]. Caseinate-lipid emulsion coatings were fruitful in plummeting moisture loss from skinned carrots and zucchini [66].

Mung beans were of interest as a potential component of biopolymeric films because of their high protein content [67]. Bourtoom [68] synthesized and investigated the films from mung bean protein. It was found that the mechanical properties (tensile strength and elongation at break) of mung bean protein films had superior mechanical properties and water vapor barrier properties (Tables 1 and 2). These were better than other protein sources such as casein, soy protein isolate, wheat gluten, peanut proteins, and water-soluble fish protein films. However, the mung bean protein films still showed substantially lower mechanical and water vapor barrier properties (Tables 1 and 2), compared with some synthetic polymers (high density polyethylene, polyvinyl chloride, cellulose acetate, and polyester).

2.3. Lipids. Lipid compounds exploited as edible polymer consist of acetylated monoglycerides, natural wax, and surfactants. The most operative lipid substances were paraffin wax and beeswax. The primary utility of a lipid was to block passage of moisture due to their relative low polarity. In contrast, the hydrophobic characteristic of lipid forms thicker and more brittle films. Accordingly, they must be associated with film-forming agents such as proteins or cellulose derivatives. Generally, water vapor permeability decreases when the concentration of hydrophobicity phase rises. Lipid-based films were often maintained on a polymer structure matrix, usually a polysaccharide, to provide mechanical strength.

Paraffin wax was a plagiarist from the distillate portion of crude petroleum and consists of a mixture of solid hydrocarbon resulting from ethylene catalytic polymerization. Paraffin wax was permitted for use on raw fruit and vegetable and cheese. Carnauba wax was an exudate from palm tree leaves. Beeswax was produced from honey bees. Candelilla was obtained from candelilla plant. Mineral oil consists of a mixture of liquid paraffin and naphthenic hydrocarbon. Various types of wax were used as barrier films to gas and moisture (skin on fresh fruits) and to improve the surface appearance of various foods (e.g., the sheen on sweet). If applied as a thick layer, they must be removed before consumption (certain cheese); when used in thin layers, they were considered edible. Various types of wax (notably paraffin, carnauba, candellila, and beeswax) were the most efficient edible compounds providing a humidity barrier [69]. Acetylation of glycerol monosterate by its reaction with acetic anhydride yields 1-stearodiacetin. This acetylated monoglyceride displays the exclusive characteristic of solidifying from the molten state into a flexible, wax-like solid. Most lipids in the solid state can be stretched to only about 102% of their original length before rupturing. Acetylated glycerol monostearate, however, can be stretched up to 800% of its original length; water vapor permeability of this film was much less than that of polysaccharide film with the exception of methylcellulose or ethylcellulose. Acetylated monoglyceride coatings have been used on poultry and meat cuts to retard moisture loss during storage [6].

Shellac resins were a secretion by the insect *Laccifer lacca* and were composed of a complex mixture of aliphatic alicyclic hydroxyl acid polymers. Shellac was not generally recognized as safe substance; it was only permitted as an indirect food additive in food coatings and adhesives. It was mostly used in coatings for the pharmaceutical industry and only a few studies have been reported on foods [70]. Rosin and its derivatives were widely used in coating for citrus and other fruits. These coatings were designed primarily to impart high gloss at the time of inspection by the buyer, usually after application of the coating. When coatings were applied to fruit, they form an additional barrier through which gases must pass. Because coatings differ in gas permeance and ability to block openings in the skin, they have different effects on gas exchange. Citrus fruits with shellac- and wood resin-based coatings generally have lower internal O_2, higher internal CO_2, and higher ethanol content than fruits with wax coatings. High ethanol content, in turn, is an indication of off flavor. Shellac- and wood resin-based coatings also tend to increase the prevalence of postharvest pitting [6].

2.4. Synthetic and Composite Edible Polymer. Edible polymers may be diverse in nature, consisting of a blend of polysaccharides, protein, and/or lipids. This approach empowers one to utilize the distinct functional characteristics of each class of film former. The combination between polymers to form films could be from proteins and carbohydrates, proteins and lipids, carbohydrates and lipids, or synthetic polymers and natural polymers. The main objective of producing composite polymer was to improve the permeability or mechanical properties as dictated by the need of a specific application. These heterogeneous films were applied either in the form of an emulsion, suspension, or dispersion of the nonmiscible constituents, or in successive layers, or in the form of a solution in a common solvent. The method of application affects the barrier properties of the films obtained. Kamper and Fennema [71] introduced the emulsion films from methylcellulose and fatty acids to improve water vapor barrier of cellulose films. Recently, many scientists have extensively discovered the development of composite polymer based on the work of Kamper and Fennema [71]. Examples of these studies were using lipid and hydroxypropyl methylcellulose, methylcellulose and lipid, methylcellulose and fatty acid, corn zein, methylcellulose and fatty acid, whey isolate and lipids, casein and lipids, gelatin and soluble starch, hydroxypropyl starch and gelatin, corn zein and corn starch [72], gelatin and fatty acid, soy protein isolates and gelatin [73], and soy protein isolate and polylactic acid [74].

Polyvinyl acetate was a nontoxic commercially important polymer prepared through emulsion polymerization which has been investigated as a coating film containing fungicides for protection of diverse foods and as a coating for pharmaceutical products. Biopolymers such as polysaccharides and proteins or even biocompatible synthetic polymers such as polymethyl methacrylate (PMMA) could be better alternatives for combinations with antimicrobial polymers aiming at antimicrobial chemotherapy in vivo. The use of antimicrobial macromolecular agents, however, was not restricted to in vivo applications; they may also be used ex vivo for water disinfection, food packing and preservation, and many antifouling applications [75]. For more than 50 years, techniques such as compression spray and dip coating and encapsulation have been used in the pharmaceutical industry to incorporate bioactive agents with polymers. Such polymers have largely included cellulose derivatives, polyethylene glycol (PEG), and poly (N-vinylpyrrolidone). Peppas and coworkers have initiated the use of pH-responsive complexation hydrogels of poly (methacrylic acid) grafted with PEG, referred to as P (MAA-g-EG), for oral protein delivery. Micelle-forming polymers, such as block copolymers of poly (ethylene oxide) and poly (propylene oxide), or Pluronics, have been thoroughly studied in drug delivery. Polyplexes formed by cooperative electrostatic interactions between polyethyleneimine (PEI) and DNA were widely studied for gene delivery [76].

An anionic copolymer based on methacrylic acid and methyl methacrylates was used for the coating of tablets and pills which resistance to gastric juice and improve the keeping property protection against moisture, light, and air coatings stable under tropical conditions [77]. Polymer spheres with a sugar coating on the outside and plastic coating on the inside have been made by European scientists. This gives them dual functionality to target and deliver drugs. Helmut Schlaad, from the Max Planck Institute of Colloids and Interfaces, Potsdam, and colleagues from Germany and Switzerland made the spheres by dissolving glycosylated polybutadiene-poly (ethylene oxide) block copolymers in water. When dissolved, the copolymers spontaneously formed hollow colloids called vesicles with a glucose coating on the outside and a poly (ethylene oxide) coating on the inside. The polymer vesicles could be used as living cell mimics or drug delivery vessels. They could be used to target drugs and biomolecules to injured or cancerous tissues [78].

3. Special Applications of Edible Polymers

Heightened consumer demand for enhanced keeping quality and freshness of foods has given rise to the concept of active packaging—a type of packaging that alters conditions surrounding the food to maintain product quality and freshness, improve sensory properties, or enhance product safety and shelf life. To meet the growing demand of recyclable or natural packaging materials and consumer demands for safer and better quality foods, new and novel food-grade packaging materials or technologies have been and continue to be developed. Examples of these packaging materials include biobased polymers, bioplastic or edible polymer packaging products made from raw materials originating from agricultural or marine sources [79]. The application of nanocomposites promises to expand the use of edible and biodegradable polymers [80]. It will help to reduce the packaging waste associated with processed foods and will support the preservation of fresh foods, extending their shelf life. Polymer composites were mixtures of polymers with inorganic or organic additives having certain geometries (fibers, flakes, spheres, and particulates). The use of nanoscale fillers was leading to the development of polymer nanocomposites and represents a radical alternative to these conventional polymer composites [81]. Pharmaceutical and food products were often coated with edible shellac, which must be of a sufficiently reduced viscosity. The application of polyvinyl acetate (PVA) was assessed as a potential protective edible coating in round-type tomato fruit on the green stage of maturity. The edible polymeric coating did not significantly affect brightness, weight loss, and RR; however, environmental conditions at which the fruits were maintained had significant effects. Despite the statistically nonsignificant effects, results showed that the PVA coating was associated with slight protection of the fruit compared to uncoated fruits [82]. Gelatin capsules [83] were small shells made from gelatin that were used to enclose various medications and supplements. Capsules can make it easier to take certain drugs and supplements, especially those in a powder or liquid form. The gelatin was almost odorless and tasteless, and it typically does not cause digestive problems for most people. Gelatin also usually was not interacting with other drugs. Gelatins from alternatives to mammalian species were gaining prominence, especially gelatins from marine fish species. Because of their good film-forming abilities, fish gelatins may be a good alternative to synthetic plastics for making films preserve foodstuffs [84].

Edible shellac coatings preserve fruits and vegetables during their long trek from the farm to your local produce stand.

To improve these coatings, Agricultural Research Service (ARS) scientists developed a new coating made of food-grade polyvinyl acetate that is cheaper to use and more effective than shellac at preventing postharvest fruit decay without discoloring the fruit. ARS has applied for a patent for this breakthrough [85]. It was cheaper to use and more effective than shellac at preventing postharvest fruit decay without discoloring the fruit. The coating can be applied to fruits and vegetables by dipping, spraying, or brushing on. Polyvinyl acetate (PVA) was a synthetic polymer that was used as an ingredient in chewing gum [86]. In addition to the polyvinyl acetate, the coating for fresh produce can include plasticizers, surfactants that aid coverage, gloss-enhancing additives, and other ingredients for specific uses, such as coating candy and baked goods. The new coating has several advantages over shellac, which slows fruit respiration and keeps the fruit firm [85]. Shellac tends to whiten or "blush" when it was wide open to moisture. This often occurs when apples were moved from cold storage to a humid environment. Another problem was that citrus fruits and some apple varieties develop "off" flavors when coated with shellac. And the fact that shellac coatings are primarily composed of insect exudates has made them objectionable to some consumers.

Several examples of snacks of orally soluble edible polymer, including their compositions and manufacturing techniques, were disclosed [87]. The snacks may include one or more layers of film that was orally soluble and fragments quickly upon placement in a human mouth without leaving substantial residue that can be felt by the human tongue or which needs to be swallowed or ejected from the mouth. An edible snack unit of tasty chocolate-flavored orally soluble polymer is comprised of a base medium formed in the shape of an ingestible snack, said base medium being formed at least in part from an orally soluble polymer, and said snack being orally soluble such that it disintegrates relatively slowly when placed on the roof of a human mouth and licked [88]. These film-forming edible polymers can include pullulan, hydroxypropyl methylcellulose, hydroxyethyl cellulose, hydroxypropyl cellulose, polyvinyl pyrrolidone, carboxymethyl cellulose, polyvinyl alcohol, sodium alginate, polyethylene glycol, xanthan gum, tragacanth gum, guar gum, acacia gum, gum arabic, polyacrylic acid, methyl methacrylate copolymer, carboxyvinyl polymer, amylose, high amylose starch, hydroxypropylated high amylose starch, dextrin, pectin, chitin, chitosan, levan, elsinan, collagen, gelatin, zein, gluten, soy protein isolate, whey protein isolate, casein, polysaccharides, natural gums, polypeptides, polyacrylates, starch, gum karaya, gelatin, mixtures thereof, and others. Chewing gum consists of a gum base, sugar, corn syrup, softeners, and flavorings. The gum base was the insoluble part left in the mouth while chewing. The gum base was made of resins from trees, latexes, or the milky juices from plants, and man-made polymers. Thomas Adams desired to use the chicle as a rubber substitute, but instead, rubber was used to make chewing gum—unvulcanized styrene-butadiene synthetic rubber (SBR), others utilize polyvinyl acetate polymer (PVA) [89]. Modern chewing gum was made with a synthetic rubber, polyisobutylene, which not only makes it elastic, but also contributes to the obstinate, sticky quality. US 3761286 [90] reveals a chewing gum composition comprising chewing gum and a water insoluble hydrophilic polymer of a hydroxyalkyl acrylate or methacrylate having controllably, releasable adsorbed there in a flavor, wherein the hydrophilic polymer can be poly-(2-hydroxy ethyl methacrylate). Gebreselassie et al. [91] invented swellable gum compositions which increase in volume upon chewing. The gum compositions include a super absorbent hydrophilic polymer in combination with a gum base or as a component of a gum base.

Researchers around the globe were trying to find ways of improving therapeutic efficacy of drugs by modifying the formulation technique, polymeric systems, and so forth. The use of novel edible polymers offers benefits. Care should be taken to properly select polymers while designing a delivery system. The eventual goal was to introduce cost effective, biocompatible, multifunctional, and less toxic polymers so that the delivery systems pass through the various phases of clinical trials and benefit the society. It was believed that the advances in polymer sciences will transfigure the design, development, and performance of polymer based drug delivery systems. PCL and PLA were chemically bonded onto starch and can be used directly as thermoplastics or compatibilizer [92]. Starch-g-PVA behaves good properties of both components such as processability, hydrophilicity, biodegradability, and gelation ability. Starch was a natural polymer which has widespread application ranging from a simple filler or binder to a more functional ingredient in the formulation of capsules, coatings, subcutaneous implants, and tablets [93]. It was easy to purify and soluble in water, which makes it convenient for delivery by injection and thus was considered as an ideal candidate for nanoparticle preparation. Protein based nanoparticles have the advantage of greater stability during storage and were easy to scale up as compared to other delivery systems [94]. It was a versatile protein carrier which was used in drug targeting for achieving better pharmacokinetic profile of peptide or protein based medicines.

Hyaluronic acid (HA), a natural polyanionic polysaccharide, was distributed widely in the extracellular matrix and the joint liquid of mammalians and approved for injections by the Food and Drug Administration (FDA) [95]. It was nontoxic, biocompatible mucoadhesive polysaccharide having negative charge and was biodegradable. It was mainly distributed in the connective tissue, eyes, intestine, and lungs. Guar gum was used as a binder, disintegrated in tablet formulations. It also acts as a stabilizer, emulsifier, thickening, and suspending agent in liquid formulations [96]. It has been widely used for colonic drug delivery applications. The swelling ability of guar gum was used in the delay of drug release from the dosage forms. Its utility as a carrier for colon specific drug delivery was based on its degradation by colonic bacteria. Chitin and chitosan derivatives were also used in the treatment of industrial effluents because of their affinity to metal ions. N-carboxymethyl chitosan has been used widely in pharmaceutical areas for achieving controlled release of drugs, orthopedic devices, and connective tissue. Pectin hydrogels can be used as a binder in tablet formulations and have been used in controlled-release matrix tablet formulations [97]. Using an extruder/spheronizer, spherical pellets containing calcium pectate were prepared. These were then coated with pectin solution resulting in the formation of insoluble calcium pectinate gel around the pellets. The use of pectin to develop other oral controlled release drug delivery systems has been reported by many authors. Polysaccharides, for instance, contain hydroxyl groups that allow for direct reaction to drugs with carboxylic acid functions, thereby producing ester linkages that are biodegradable and thus facilitate the release of the drug in the body.

Finally, another relevant application of drug delivery systems was vaccines. Vaccines were classified into two groups, namely, proteins and nucleic acids. In both cases, they require polymeric carriers because they are susceptible to degradation by peptidases or nucleases. The DNA vaccine was a newly developed system. The DNA in this vaccine system encodes a protein antigen of interest that induces activation of the immune system. This DNA can be encapsulated into polymeric carriers, thereby protecting it from degradation [98]. It is then released into the phagosome, thus allowing it to reach the cell nucleus and express the foreign protein. Polymeric carriers of antineoplastic drugs can passively accumulate in cancerous tissues because of differences in the biochemical and physiological features of healthy and malignant tissues, while they can actively accumulate in the same tissues because they have been conjugated to targeting moieties. Scientists in the US have designed an ingestible electronic device that was composed entirely of edible materials and produces its own electric current [99]. Here the team designed and fabricated an ingestible current source consisting of flexible polymer electrodes and a sodium ion electrochemical cell. Having flexible polymer electrodes means they can be folded into an edible capsule, while the sodium ion cell serves as an on-board energy supply. The idea was for the patient to consume a pill that encapsulates the device. Polyesters based on PLA, PGA, and their copolymers PLGA and poly (ε-caprolactone) (PCL) have been extensively employed because of their biocompatibility and biodegradability. Polyesters have been used for the encapsulation of many types of therapeutic agents like cancer, bacterial, and parasitic infections [100].

4. Future Trends

A noble group of edible polymers was under enlargement, with the goal of allowing for the incorporation and/or controlled release of active compounds using nanotechnological solutions such as nanoencapsulation and multifaceted systems. Nowadays, nanotechnologies were being used to enhance the nutritional features of food by means of nanoscale additives and nutrients and nanosized delivery systems for bioactive polymeric compounds. Nanocomposites perception represents a motivating route for creating new and innovative materials, also in area of edible polymers. Materials with a large variety of properties have been realized, and even more due to be realized. Micro- and nanoencapsulation of active compounds with edible polymer coatings may help to control their release under specific conditions, thus protecting them from moisture, heat, or other extreme conditions and enhancing their stability and viability. Coating foods with nanolaminates involves either dipping them into a series of solutions containing substances that would be adsorbed to a food's surface or spraying substances onto the food surface. These nanolaminate coatings could be elaborated entirely from food-grade ingredients and could include various functional agents such as antimicrobials, antibrowning agents, antioxidants, enzymes, flavorings, and colorant. In fact, the layer-by- layer electrodeposition technique [101] could be used to coat highly hydrophilic food systems such as fresh-cut fruits and vegetables including further vitamins and antimicrobial agents. The nanocomposite materials obtained by mixing natural, edible polymers, and sheets of crystalline solid layered offer a great variety of property profile. They were even able to compete, both in price and in performance, with synthetic polymeric materials. Consumer demands were driving research and development for alternatives to petroleum-based packaging materials including those with recyclable or edible properties, as well as those materials made from renewable/sustainable agricultural products. Edible films, gels, or coatings were considered biopolymers with numerous desirable properties and may be made from a variety of materials, including polysaccharides, lipids, and proteins, alone or in combination with other components. Edible biopolymers also have been developed from other sources and applied to foods, including fungal exopolysaccharides (pullan) or fermentation byproducts (polylactic acid).

5. Conclusions

The practice of edible polymers on many food products continues to develop. The potential paybacks of edible polymers as carriers of antimicrobial agents, flavors, antioxidants, coloring agents, vitamins, probiotics, and nutraceutical excuse continued research in this field of active packaging. A new trend in the storage of foods and in the food industry, nowadays, was the use of recyclable, renewable agricultural products for the production of packages, edible films, and coatings. Edible polymer technology plays an energetic role in drug discovery and was improving outcomes for patients today and addressing unmet therapeutic needs in the future. Vaccine drug delivery systems were now being proven to be patient friendly as they avoid the need to administer booster doses and provide a long term therapy in small doses. Edible polymer vaccines on the other hand open an attractive avenue for the oral delivery of vaccines. The development of new technologies to improve the delivery properties of edible polymer was a major issue for future research. At that moment, most studies on food applications have been conducted at a laboratory scale. However, further research should be focused on a commercial scale with the purpose of providing more accurate information that can be used to commercialize fresh cut products coated with edible polymers. Food industries were looking for edible polymers that could be used on a broad spectrum of foods and add value to their products while increasing their shelf life.

When the active ingredients (antimicrobials, antioxidants, and nutrients) were added to edible polymer, mechanical, sensory, and even functional properties can be dramatically affected. Scientists have researched new production methods to make coatings edible, safe, and fully functional. According to a broad range of research findings, edible polymer may be applied to prolong the shelf life of food products, control material exchange, and improve the products sensory properties, nutritive value, and attractiveness. Edible biopolymers were competitive as both commodity polymers and specialty applications where a specific functionality was valued. In a scenario where petroleum-based polymers were replaced by biopolymers, edible biopolymers from food crops would primarily be used in food applications, whereas other biopolymers could well cover the demand for commodity polymers. A continued development of functionality and processing was needed for a commercial breakthrough.

Conflict of Interests

The authors declare that there is no conflict of interests regarding the publication of this paper.

References

[1] S. Kokoszka and A. Lenart, "Edible coatings—formation, characteristics and use—a review," *Polish Journal of Food and Nutrition Sciences*, vol. 57, no. 4, pp. 399–404, 2007.

[2] G. G. Buonocore, M. A. del Nobile, A. Panizza, S. Bove, G. Battaglia, and L. Nicolais, "Modeling the lysozyme release kinetics from antimicrobial films intended for food packaging applications," *Journal of Food Science*, vol. 68, no. 4, pp. 1365–1370, 2003.

[3] R. C. Eberhart, S. Su, K. T. Nguyen et al., "Bioresorbable polymeric stents: current status and future promise," *Journal of Biomaterials Science, Polymer Edition*, vol. 14, no. 4, pp. 299–312, 2003.

[4] D. Bikiaris, "Nanomedicine in cancer treatment: drug targeting and the safety of the used materials for drug nanoencapsulation," *Biochemical Pharmacology*, vol. 1, article E122, 2012.

[5] A. Dhanapal, P. Sasikala, L. Rajamani, V. Kavitha, G. Yazhini, and M. S. Banu, "Edible films from polysaccharides," *Food Science and Quality Management*, vol. 3, pp. 9–18, 2012.

[6] T. Bourtoom, "Edible films and coatings: characteristics and properties," *International Food Research Journal*, vol. 15, no. 3, pp. 237–248, 2008.

[7] *World Economic Situation and Prospects 2012*, Sales no. E.12.Ii.C.2, United Nations Publication, 2012.

[8] E. Salleh, I. I. Muhamad, and N. Khairuddin, "Structural characterization and physical properties of antimicrobial (AM) starch-based films," *World Academy of Science, Engineering and Technology*, vol. 3, pp. 7–25, 2009.

[9] Z. Akbari, T. Ghomashchi, and S. Moghadam, "Improvement in food packaging industry with biobased nanocomposites," *International Journal of Food Engineering*, vol. 3, no. 4, article 3, 2007.

[10] S. Saremnezhad, M. H. Azizi, M. Barzegar, S. Abbasi, and E. Ahmadi, "Properties of a new edible film made of faba bean protein isolate," *Journal of Agricultural Science and Technology*, vol. 13, no. 2, pp. 181–192, 2011.

[11] O. Skurtys, C. Acevedo, F. Pedreschi, J. Enronoe, F. Osorio, and J. M. Aguilera, *Food Hydrocolloid Edible Films and Coatings*, Food Science and Technology, Nova Publisher, 2010.

[12] M. E. Embuscado and K. C. Huber, *Edible Films and Coatings for Food Applications*, Springer, London, UK, 2009.

[13] S. A. Jang, G. O. Lim, and K. B. Song, "Preparation and mechanical properties of edible rapeseed protein films," *Journal of Food Science*, vol. 76, no. 2, pp. C218–C223, 2011.

[14] P. Mokrejs, F. Langmaier, D. Janacova, M. Mladek, K. Kolomaznik, and V. Vasek, "Thermal study and solubility tests of films based on amaranth flour starch-protein hydrolysate," *Journal of Thermal Analysis and Calorimetry*, vol. 98, no. 1, pp. 299–307, 2009.

[15] P. Bergo, P. J. A. Sobral, and J. M. Prison, "Effect of glycerol on physical properties of cassava starch films," *Journal of Food Processing and Preservation*, vol. 34, no. 2, pp. 401–410, 2010.

[16] A. M. Stephen, G. O. Phillips, and P. A. Williams, *Food Polysaccharides and Their Applications*, CRC Press, Taylor & Francis Group, 2nd edition, 2006.

[17] C. A. Romero-Bastida, L. A. Bello-Pérez, M. A. García, M. N. Martino, J. Solorza-Feria, and N. E. Zaritzky, "Physicochemical and microstructural characterization of films prepared by thermal and cold gelatinization from non-conventional sources of starches," *Carbohydrate Polymers*, vol. 60, no. 2, pp. 235–244, 2005.

[18] K. Krogars, J. Heinämäki, M. Karjalainen, J. Rantanen, P. Luukkonen, and J. Yliruusi, "Development and characterization of aqueous amylose-rich maize starch dispersion for film formation," *European Journal of Pharmaceutics and Biopharmaceutics*, vol. 56, no. 2, pp. 215–221, 2003.

[19] C. Ribeiro, A. A. Vicente, J. A. Teixeira, and C. Miranda, "Optimization of edible coating composition to retard strawberry fruit senescence," *Postharvest Biology and Technology*, vol. 44, no. 1, pp. 63–70, 2007.

[20] E. Argüello-García, J. Solorza-Feria, J. R. Rendón-Villalobos, F. Rodríguez-González, A. Jiménez-Pérez, and E. Flores-Huicochea, "Properties of edible films based on oxidized starch and zein," *International Journal of Polymer Science*, vol. 2014, Article ID 292404, 9 pages, 2014.

[21] R. Utami, Kawiji, E. Nurhartadi, A. Y. T. Putra, and I. Setiawan, "The effect of cassava starch-based edible coating enriched with Kaempferia rotunda and Curcuma xanthorrhiza essential oil on refrigerated patin fillets quality," *International Food Research Journal*, vol. 21, no. 1, pp. 413–419, 2014.

[22] J. Rhim, "Physical and mechanical properties of water resistant sodium alginate films," *Lebensmittel-Wissenschaft und Technologie*, vol. 37, no. 3, pp. 323–330, 2004.

[23] S. T. Prajapati, A. P. Mehta, I. P. Modhia, and C. N. Patel, "Formulation and optimisation of raft-forming chewable tablets containing H_2 antagonist," *International Journal of Pharmaceutical Investigation*, vol. 2, pp. 176–182, 2012.

[24] B. Arica, S. Çaliş, H. S. Kaş, M. F. Sargon, and A. A. Hincal, "5-fluorouracil encapsulated alginate beads for the treatment of breast cancer," *International Journal of Pharmaceutics*, vol. 242, no. 1-2, pp. 267–269, 2002.

[25] R. Malviya, "Swelling and erosion based formulations for the treatment of chronic seizures using (3)2 factorial design," *Middle-East Journal of Scientific Research*, vol. 11, no. 1, pp. 77–84, 2012.

[26] W. Borchard, A. Kenning, A. Kapp, and C. Mayer, "Phase diagram of the system sodium alginate/water: a model for

biofilms," *International Journal of Biological Macromolecules*, vol. 35, no. 5, pp. 247–256, 2005.

[27] T. Karbowiak, F. Debeaufort, D. Champion, and A. Voilley, "Wetting properties at the surface of iota-carrageenan-based edible films," *Journal of Colloid and Interface Science*, vol. 294, no. 2, pp. 400–410, 2006.

[28] V. D. Prajapati, P. M. Maheriya, G. K. Jani, and H. K. Solanki, "Carrageenan: a natural seaweed polysaccharide and its applications," *Carbohydrate Polymers*, vol. 105, pp. 97–112, 2014.

[29] T. Karbowiak, H. Hervet, L. Léger, D. Champion, and F. Debeaufort, "Effect of plasticizers (water and glycerol) on the diffusion of a small molecule in iota-carrageenan biopolymer films for edible coating application," *Biomacromolecules*, vol. 7, no. 6, pp. 2011–2019, 2006.

[30] G. O. Phillips and P. A. Williams, *Handbook of Hydrocolloids*, CRC Press, Cambridge, UK, 2009.

[31] Y. S. Pathare, V. S. Hastak, and A. N. Bajaj, "Polymers used for fast disintegrating oral films: a review," *International Journal of Pharmaceutical Sciences Review*, vol. 21, no. 1, pp. 169–178, 2013.

[32] S. Galus, H. Mathieu, A. Lenart, and F. Debeaufort, "Effect of modified starch or maltodextrin incorporation on the barrier and mechanical properties, moisture sensitivity and appearance of soy protein isolate-based edible films," *Innovative Food Science and Emerging Technologies*, vol. 16, pp. 148–154, 2012.

[33] H. Patel, V. Shah, and U. Upadhyay, "New pharmaceutical excipients in solid dosage forms—a review," *International Journal of Pharmacy and Life Sciences*, vol. 2, no. 8, pp. 1006–1019, 2011.

[34] M. A. García, C. Ferrero, N. Bértola, M. Martino, and N. Zaritzky, "Edible coatings from cellulose derivatives to reduce oil uptake in fried products," *Innovative Food Science and Emerging Technologies*, vol. 3, no. 4, pp. 391–397, 2002.

[35] V. Comaa, I. Sebtia, P. Pardonb, F. H. Pichavantb, and A. Deschampsa, "Film properties from crosslinking of cellulosic derivatives with a polyfunctional carboxylic acid," *Carbohydrate Polymers*, vol. 51, pp. 265–271, 2003.

[36] P. Srivastava, "Sources of pectin, extraction and its applications in pharmaceutical industry—an overview," *Indian Journal of Natural Products and Resources*, vol. 2, no. 1, pp. 10–18, 2011.

[37] N. Natrajan and B. W. Sheldon, "Efficacy of nisin-coated polymer films to inactivate *Salmonella typhimurium* on fresh broiler skin," *Journal of Food Protection*, vol. 63, no. 9, pp. 1189–1196, 2000.

[38] Y. Freile-Pelegrin, D. Robledo, M. J. Chan-Bacab, and B. O. Ortega-Morales, "Antileishmanial properties of tropical marine algae extracts," *Fitoterapia*, vol. 79, no. 5, pp. 374–377, 2008.

[39] T. D. Phan, F. Debeaufort, D. Luu, and A. Voilley, "Functional properties of edible agar-based and starch-based films for food quality preservation," *Journal of Agricultural and Food Chemistry*, vol. 53, no. 4, pp. 973–981, 2005.

[40] M. Mucha, K. Wańkowicz, and J. Balcerzak, "Analysis of water adsorption on Chitosan and its blends with hydroxypropylcellulose," *E-Polymers*, vol. 16, pp. 1–10, 2007.

[41] B. Wang, J. Zhang, G. Cheng, and S. Dong, "Amperometric enzyme electrode for the determination of hydrogen peroxide based on sol-gel/hydrogel composite film," *Analytica Chimica Acta*, vol. 407, no. 1-2, pp. 111–118, 2000.

[42] S. J. Kim, S. R. Shin, S. M. Lee, I. Y. Kim, and S. I. Kim, "Electromechanical properties of hydrogels based on chitosan and poly(hydroxyethyl methacrylate) in NaCl solution," *Smart Materials and Structures*, vol. 13, no. 5, pp. 1036–1039, 2004.

[43] C. G. Lee, "Chitin, chitinases and chitinase-like proteins in allergic inflammation and tissue remodeling," *Yonsei Medical Journal*, vol. 50, no. 1, pp. 22–30, 2009.

[44] J. Vartiainen, R. Motion, H. Kulonen, M. Rättö, E. Skyttä, and R. Ahvenainen, "Chitosan-coated paper: effects of nisin and different acids on the antimicrobial activity," *Journal of Applied Polymer Science*, vol. 94, no. 3, pp. 986–993, 2004.

[45] S. Y. Park, K. S. Marsh, and J. W. Rhim, "Characteristics of different molecular weight chitosan films affected by the type of organic solvents," *Journal of Food Science*, vol. 67, no. 1, pp. 194–197, 2002.

[46] C. Ribeiro, A. A. Vicente, J. A. Teixeira, and C. Miranda, "Optimization of edible coating composition to retard strawberry fruit senescence," *Postharvest Biology and Technology*, vol. 44, no. 1, pp. 63–70, 2007.

[47] G. Crini, "Non-conventional low-cost adsorbents for dye removal: a review," *Bioresource Technology*, vol. 97, no. 9, pp. 1061–1085, 2006.

[48] W. Won, X. Feng, and D. Lawless, "Pervaporation with chitosan membranes: separation of dimethyl carbonate/methanol/water mixtures," *Journal of Membrane Science*, vol. 209, no. 2, pp. 493–508, 2002.

[49] J. Nunthanid, M. Laungtana-Anan, P. Sriamornsak et al., "Characterization of chitosan acetate as a binder for sustained release tablets," *Journal of Controlled Release*, vol. 99, no. 1, pp. 15–26, 2004.

[50] Y. Tsai, S. Chen, and H. Liaw, "Immobilization of lactate dehydrogenase within multiwalled carbon nanotube-chitosan nanocomposite for application to lactate biosensors," *Sensors and Actuators B: Chemical*, vol. 125, no. 2, pp. 474–481, 2007.

[51] X. H. Xu, G. L. Ren, J. Cheng, Q. Liu, D. G. Li, and Q. Chen, "Self-assembly of polyaniline-grafted chitosan/glucose oxidase nanolayered films for electrochemical biosensor applications," *Journal of Materials Science*, vol. 41, no. 15, pp. 4974–4977, 2006.

[52] R. Díaz-Sobac, H. García, C. I. Beristain, and E. J. Vernon-Carter, "Morphology and water vapor permeability of emulsion films based on mesquite gum," *Journal of Food Processing and Preservation*, vol. 26, no. 2, pp. 129–141, 2002.

[53] R. Dimitriou, G. I. Mataliotakis, G. M. Calori, and P. V. Giannoudis, "The role of barrier membranes for guided bone regeneration and restoration of large bone defects: current experimental and clinical evidence," *BMC Medicine*, vol. 10, article 81, 2012.

[54] R. Khan and M. H. Khan, "Use of collagen as a biomaterial: an update," *Journal of Indian Society of Periodontology*, vol. 17, no. 4, pp. 539–542, 2013.

[55] K. Dybka and P. Walczak, "Collagen hydrolysates as a new diet supplement," *Food Chemistry and Biotechnology*, vol. 73, pp. 83–92, 2009.

[56] T. Wittaya, "Protein-based edible films: characteristics and improvement of properties," in *Structure and Function of Food Engineering*, chapter 3, InTech, 2012.

[57] M. I. Khan, M. N. Adrees, M. R. Tariq, and M. Sohaib, "Application of edible coating for improving meat quality: a review," *Pakistan Journal of Food Sciences*, vol. 23, no. 2, pp. 71–79, 2013.

[58] J. Sris, A. Seethadevi, K. Suria Prabha, P. Muthuprasanna, and P. Pavitra, "Microencapsulation: a review," *International Journal of Pharma and Bio Sciences*, vol. 3, no. 1, pp. 509–521, 2012.

[59] T. Wittaya, "Protein-based edible films: characteristics and improvement of properties," in *Structure and Function of Food Engineering*, chapter 3, InTech.

[60] R. Shukla and M. Cheryan, "Zein: the industrial protein from corn," *Industrial Crops and Products*, vol. 13, no. 3, pp. 171–192, 2001.

[61] J. Bai, V. Alleyne, R. D. Hagenmaier, J. P. Mattheis, and E. A. Baldwin, "Formulation of zein coatings for apples (Malus domestica Borkh)," *Postharvest Biology and Technology*, vol. 28, no. 2, pp. 259–268, 2003.

[62] H. Dhanya and S. Divia, "Development of Zein-Pectin nanoparticle as drug carrier," *International Journal of Drug Delivery*, vol. 4, pp. 147–152, 2012.

[63] M. M. Moore, T. J. Schober, P. Dockery, and E. K. Arendt, "Textural comparisons of gluten-free and wheat-based doughs, batters, and breads," *Cereal Chemistry*, vol. 81, no. 5, pp. 567–575, 2004.

[64] S. Y. Cho and C. Rhee, "Mechanical properties and water vapor permeability of edible films made from fractionated soy proteins with ultrafiltration," *Lebensmittel-Wissenschaft und Technologie*, vol. 37, no. 8, pp. 833–839, 2004.

[65] J. Zhang, P. Mungara, and J. Jane, "Mechanical and thermal properties of extruded soy protein sheets," *Polymer*, vol. 42, no. 6, pp. 2569–2578, 2001.

[66] Bisson, "Preparation of a Casein-Based Puffed Product," United States Patent. Patent Number: 4,744,99, May 1988.

[67] S. Keereekasetsuk and T. Bourtoom, "Influence of plasticizers on the properties of edible film from Mung bean protein," in *Proceedings of the 14th World Congress of Food Science and Technology*, Shanghai, China, October 2008.

[68] T. Bourtoom, "Factors affecting the properties of edible film prepared from mung bean proteins," *International Food Research Journal*, vol. 15, no. 2, pp. 167–180, 2008.

[69] F. F. Shih, K. W. Daigle, and E. T. Champagne, "Effect of rice wax on water vapour permeability and sorption properties of edible pullulan films," *Food Chemistry*, vol. 127, no. 1, pp. 118–121, 2011.

[70] S. Berg, M. Bretz, E. M. Hubbermann, and K. Schwarz, "Influence of different pectins on powder characteristics of microencapsulated anthocyanins and their impact on drug retention of shellac coated granulate," *Journal of Food Engineering*, vol. 108, no. 1, pp. 158–165, 2012.

[71] S. L. Kamper and O. N. Fennema, "Water vapor permeability of an edible, fatty acid, bilayer film," *Journal of Food Science*, vol. 49, pp. 1482–1485, 1984.

[72] S. Y. Ryu, J. W. Rhim, H. J. Roh, and S. S. Kim, "Preparation and physical properties of zein-coated high-amylose corn starch film," *LWT—Food Science and Technology*, vol. 35, no. 8, pp. 680–686, 2002.

[73] N. Cao, Y. Fua, and Y. He, "Preparation and physical properties of soy protein isolate and gelatin composite films," *Lwt—Food Science and Technology*, vol. 35, pp. 680–686, 2002.

[74] J. W. Rhim, A. Gennadios, A. Handa, C. L. Weller, and M. A. Hanna, "Solubility, tensile, and color properties of modified soy protein isolate films," *Journal of Agricultural and Food Chemistry*, vol. 48, no. 10, pp. 4937–4941, 2000.

[75] A. M. Carmona-Ribeiro and L. D. D. M. Carrasco, "Cationic antimicrobial polymers and their assemblies," *International Journal of Molecular Sciences*, vol. 14, no. 5, pp. 9906–9946, 2013.

[76] S. Yamano, J. Dai, S. Hanatani et al., "Long-term efficient gene delivery using polyethylenimine with modified Tat peptide," *Biomaterials*, vol. 35, no. 5, pp. 1705–1715, 2014.

[77] H. U. Petereit, C. Meier, and E. Roth, United States Patent. Patent no. : Us 7,160,558 B2, January 2007.

[78] H. Schlaad, L. You, R. Sigel et al., "Glycopolymer vesicles with an asymmetric membrane," *Chemical Communications*, no. 12, pp. 1478–1480, 2009.

[79] D. S. Cha and M. S. Chinnan, "Biopolymer-based antimicrobial packaging: a review," *Critical Reviews in Food Science and Nutrition*, vol. 44, no. 4, pp. 223–237, 2004.

[80] L. Lagarón Cabedo, D. Cava, J. Feijoo, R. Gavara, and E. Gimenez, "Improving packaged food quality and safety, part 2: nanocomposites," *Food Additives and Contaminants*, vol. 22, no. 10, pp. 994–998, 2005.

[81] E. W. Gacitua, A. A. Ballerini, and J. Zhang, "Polymer nanocomposites: synthetic and natural fillers a review," *Maderas: Ciencia y Tecnología*, vol. 7, no. 3, pp. 159–178, 2005.

[82] G. Y. Cortez-Mazatán, L. A. Valdez-Aguilar, R. H. Lira-Saldivar, and R. D. Peralta-Rodríguez, "Polyvinyl acetate as an edible coating for fruits. Effect on selected physiological and quality characteristics of tomato," *Revista Chapingo Serie Horticultura*, vol. 17, no. 1, pp. 15–22, 2011.

[83] A. Lodha, A. Patel, J. Chaudhuri, P. Jadia, T. Joshi, and J. Dalal, "Formulation and evaluation of transparent ibuprofen soft gelatin capsule," *Journal of Pharmacy and Bioallied Sciences*, vol. 4, pp. 95–97, 2012.

[84] M. C. Gómez-Guillén, M. Pérez-Mateos, J. Gómez-Estaca, E. López-Caballero, B. Giménez, and P. Montero, "Fish gelatin: a renewable material for developing active biodegradable films," *Trends in Food Science and Technology*, vol. 20, no. 1, pp. 3–16, 2009.

[85] K. Grohmann and R. D. Hagenmaier, "Edible Food Coatings Containing Polyvinyl Acetate," United States Patent. Patent Number: 6,162,475, December 2000.

[86] P. R. Gunjal, T. S. Kalmegh, M. Gadhave, and S. A. Jadhav, "Review on medicated chewing gum as a novel drug delivery system," *International Journal of Universal Pharmacy and Life Sciences*, vol. 2, no. 4, pp. 19–36, 2012.

[87] T. M. Pearce and T. V. Pearce, "Snacks of Orally Soluble Edible Films," United States Patent. Pub. no. : Us 2003/0224090 A1, December 2003.

[88] T. Pearce, "Delivery Units of Thick Orally Soluble Polymer," United States Patent Application Publication. Pub. no. : Us 2004/0247746 A1, December 2004.

[89] E. John, *Vanity, Vitality, and Virility*, pp. 189–197, Oxford University Press, New York, NY, USA, 2004.

[90] T. H. Shepherd and F. E. Gould, "Chewing Gum Composition Containing Flavored Homopolymers," United States Patent Office. Patent no. 3,761,286, September 1973.

[91] P. Gebreselassie, N. Boghani, and G. Visscher, "Edible Compositions Containing Swellable Polymers," United States Patent Application Publication. Pub. no. : Us 2006/0153949 A1, July 2006.

[92] D. R. Lu, C. M. Xiao, and S. J. Xu, "Starch-based completely biodegradable polymer materials," *Express Polymer Letters*, vol. 3, no. 6, pp. 366–375, 2009.

[93] V. G. Kadajji and G. V. Betageri, "Water soluble polymers for pharmaceutical applications," *Polymers*, vol. 3, pp. 1972–2009, 2011.

[94] Y. Luo and Q. Wang, "Zein-based micro- and nano-particles for drug and nutrient delivery: a review," *Journal of Applied Polymer Science*, 2014.

[95] J. Necas, L. Bartosikova, P. Brauner, and J. Kolar, "Hyaluronic acid (hyaluronan): a review," *Veterinarni Medicina*, vol. 53, no. 8, pp. 397–411, 2008.

[96] V. T. Chamle, P. S. Misal, S. G. Shep, and R. A. Hajare, "A potential natural gum as a polymer used in NDDS: recent investigations," *International Journal of Current Trends in Pharmaceutical Research*, vol. 1, no. 1, pp. 67–73, 2013.

[97] I. J. Ogaji, E. I. Nep, and J. D. Audu-Peter, "Advances in natural polymers as pharmaceutical excipients," *Pharmaceutica Analytica Acta*, vol. 3, no. 1, article 146, 2012.

[98] G. Vilar, J. Tulla-Puche, and F. Albericio, "Polymers and drug delivery systems," *Current Drug Delivery*, vol. 4, pp. 367–394, 2012.

[99] Y. J. Kim, S. Chun, J. Whitacre, and C. J. Bettinger, "Self-deployable current sources fabricated from edible materials," *Journal of Materials Chemistry B*, vol. 1, no. 31, pp. 3781–3788, 2013.

[100] B. Mattix, T. R. Olsen, T. Moore et al., "Accelerated iron oxide nanoparticle degradation mediated by polyester encapsulation within cellular spheroids," *Advanced Functional Materials*, vol. 24, no. 6, pp. 800–807, 2014.

[101] P. Jochenweiss and D. Julianmcclements, "Functional materials in food nanotechnology," *Journal of Food Science*, vol. 71, no. 9, pp. 10–116, 2006.

21

Main Chain Noncentrosymmetric Hydrogen Bonded Macromolecules Incorporating Aniline, Alkanol, and Alkanoic Acid Hydrogen Bond Donors

Jeremy R. Wolf[1,2]

[1] *Southern Illinois University Carbondale, Carbondale, IL 62901, USA*
[2] *ABC Laboratories, Columbia, MO 65202, USA*

Correspondence should be addressed to Jeremy R. Wolf; wolfchem@hotmail.com

Academic Editor: Jean-Marc Saiter

The syntheses and characterization of three noncentrosymmetric main chain hydrogen bonded macromolecules which incorporate aniline, alkanoic acid, and alkanol hydrogen bond donor units are reported. These macromolecules participate in weak intermolecular hydrogen bonding as demonstrated using attenuated total reflectance (ATR) FTIR. The phase transitions of these macromolecules depend on the identity of the hydrogen bond donor.

1. Introduction

Poled organic thin films are technologically useful materials [1–9] which have several advantages over traditional inorganic crystals including ease of processing [10], larger second order susceptibilities [11], an intrinsically low dielectric constant [10], and significantly faster responses [12–15]. Our approach to poled organic films involves surface-induced polar alignment of main chain noncentrosymmetric hydrogen bonded liquid crystalline macromolecules [16]. We have previously synthesized several main chain noncentrosymmetric hydrogen bonded macromolecules (Table 1) [16], and while several of these macromolecules possess liquid crystalline phases, the melting points of these macromolecules are very high and they exhibit poor solubilities in organic solvents [16–20].

In several of our studies, we found that replacing a benzoic acid hydrogen bond donor group with a weaker hydrogen bond donor (such as a OH group) in our macromolecules resulted in a large melting point depressions and increased solubility in organic solvents [19, 20] presumably due to the weaker donor-acceptor hydrogen bonding. Alkanoic acids, alkanols, and aniline are all weaker hydrogen bond donors than their benzoic acid or phenol analogs [21], therefore incorporating these hydrogen bond donors into our noncentrosymmetric hydrogen bonded macromolecules should result in lower melting points and increased solubility. In this paper, we report the synthesis and characterization of main chain noncentrosymmetric hydrogen bonded macromolecules incorporating alkanoic acid, alkanol, and aniline hydrogen bond donors and stilbazole acceptors.

2. Experimental Section

2.1. Materials. All chemicals were purchased from Fisher or Aldrich chemical company and used as received. Tetrahydrofuran (THF) was distilled over sodium-benzophenone under an argon atmosphere. Chromatography was performed using Sorbent Technology 60 angstrom, 63–200 μm mesh silica (10940-25). Thin layer chromatography was performed using Whatman flexible plates with 250 μm layer of fluorescent silica gel (UV_{254}) or EM Science glass TLC plates (60 F_{265}). All final products were dried at appropriate temperatures (below their melting and/or decomposition temperatures) in a Napco E-series 5831 vacuum oven prior to analysis.

2.2. Model Mixtures. A standard procedure for making model mixtures was as follows. One component was weighted

Table 1: First generation noncentrosymmetric main chain hydrogen bonded macromolecules.

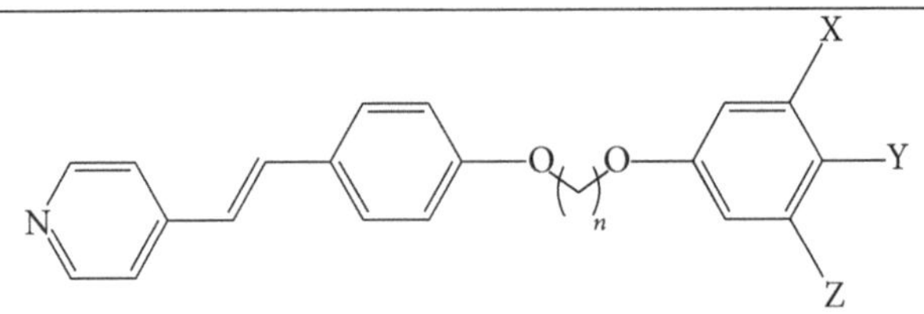

Polymer	n	X	Y	Z	m.p. (°C)
1	6	H	CO_2H	H	279*
2	10	H	CO_2H	H	244*
3	10	H	CO_2H	CH_3	198*
4	10	CH_3	CO_2H	CH_3	183
5	10	H	OH	H	188

*Polymer possesses a liquid crystalline phase.

into a glass vial. The required amount of the other compound (to obtain the desired mole ratio) was weighted into the same vial. The content of the vial was heated in a silicon oil bath until both components were visibly melted. The vial was removed from the oil bath and the mixture was allowed to crystallize (at room temperature). The melting/crystallization sequence was repeated two additional times and the resulting mixture was analyzed.

2.3. Measurements. Proton (300 MHz) and carbon 13 (75 MHz) nuclear magnetic resonance spectra were acquired using a Varian VXR 300 wide bore instrument and processed using either VNMR or Mestrec software. Infrared spectroscopy was performed on Thermo-Nicolet Nexus 670-FTIR using an Avatar multibounce HATR accessory; spectra were processed using OMNIC software. Mass spectra were acquired by the staff at the Mass Spectroscopy Lab at University of Illinois, Urbana, IL. Differential scanning calorimetry was performed on a Mettler Toledo DSC 821e equipped with a Julabo FT 900 cooling unit using heating and cooling rates of 10°C/min unless otherwise stated; all reported transition temperatures are from the second cycle of a DSC scan unless otherwise stated. All DSC transition temperatures reported are the midpoints of the transition; the enthalpies of the transitions are reported in parentheses following the transition temperature. Polarized optical microscopy was performed on an Olympus BXP polarizing microscope equipped with and Instec HCS400 heating stage; heating and cooling rates used varied between 1 and 5°C/min.

2.4. Experimental Procedures for Synthesizing All Macromolecules and Intermediates. The syntheses of macromolecules **1–5** were described by our group previously [17–20]. Compounds **6**, **8**, **12**, and **15** were synthesized according to the literature procedures [21–24].

*tert-Butyl [4-(10-Chlorodecyloxy)-phenyl]-carbamate (**7**).* Sodium hydride (1.085 g, 47.2 mmol) was placed in a round-bottomed flask equipped with a condenser and argon purge. Anhydrous DMF (20 mL) was added via syringe. Solid compound **6** (3.002 g, 14.3 mmol) was slowly added to the flask; this mixture was stirred for 30 minutes. 1,10-Dichlorodecane (9.0 mL, 42.6 mmol) was added to the flask via syringe and the mixture was stirred at room temperature for 30 minutes. Water (50 mL) was added to the reaction mixture and the solution was extracted with ethyl acetate (5 × 50 mL). The extract was washed with brine and dried over $MgSO_4$, and the crude product was adsorbed onto silica. The product was purified via column chromatography (85/15 Hex/EtOAc); evaporation of the solvent yielded 0.751 g (14%) of **7** as a white solid: R_f = 0.39 (85/15 Hex/EtOAc); ^{1}H NMR (300 MHz, $CDCl_3$) δ 1.44 (m, 22H), 1.74 (m, 2H), 3.53 (t, J = 6.83, 2H), 3.91 (t, J = 6.27, 2H), 6.34 (bs, 1H), 6.83 (d, J = 9.41, 2H), and 7.26 (d, J = 10.25, 2H); ^{13}C NMR (75 MHz, $CDCl_3$) δ 25.98, 26.84, 28.34, 28.84, 29.25, 29.31, 29.35, 29.41, 29.69, 32.61, 45.16, 68.26, 114.83, 120.53, 131.21, and 155.20; Mass Spec M^+ = 383.3 Calc.: 383.2227.

*tert-Butyl (4-10-[4-(2-Pyridin-4-yl-vinyl)-phenoxy]-decyloxy-phenyl)-carbamate (**9**).* Compound **7** (0.674 g, 1.8 mmol), compound **8** (0.311 g, 1.6 mmol), and K_2CO_3 (0.764 g, 5.5 mmol) were placed in a round-bottomed flask equipped with a condenser and argon purge. Anhydrous DMF (20 mL) was added via syringe and the reaction was heated to 120°C for 7 hours; the mixture was cooled; and water (100 mL) was added. The solution was extracted with EtOAc (7 × 50 mL); the extract was washed with brine and dried over $MgSO_4$; and the crude product was adsorbed onto silica. The product was purified by column chromatography (60/40 EtOAc/Hex); evaporation of the solvent yielded 0.412 g (48%) of **9** as a yellow solid: R_f = 0.49 (60/40 EtOAc/Hex); ^{1}H NMR (300 MHz, DMSO-d_6) δ 1.38 (m, 21H), 1.66 (m, 4H), 3.86 (t, J = 6.80, 2H), 3.98 (t, J = 6.45, 2H), 6.79 (d, J = 8.63, 2H), 6.95 (d, J = 8.78, 2H), 7.06 (d, J = 16.74, 2H), 7.50 (m, 6H), 8.49 (d, J = 6.02, 2H), and 9.11 (bs, 1H); ^{13}C NMR (75 MHz, $CDCl_3$) δ 25.47, 28.16, 28.62, 28.74, 28.92, 67.50, 78.61, 114.38, 114.74, 119.62, 120.57, 123.40, 128.53, 128.60, 132.46, 132.68, 144.64, 152.91, 153.86, and 159.19.

*4-{10-[4-(2-Pyridin-4-yl-vinyl)-phenoxy]-decyloxy}-phenylamine (**10**).* Compound **9** (0.175 g, 0.3 mmol), THF (18 mL), and MeOH (10 mL) were combined in a round-bottomed flask. Concentrated HCl (10 mL) was slowly added to the flask and the mixture was allowed to stir for 16 hr at room temperature. The solution was neutralized (pH = 7) with solid $NaHCO_3$ and the precipitate was collected by vacuum filtration; the solid so obtained was dried in the oven affording 0.098 g (66%) of **10** as a tan solid: ^{1}H NMR (300 MHz, DMSO-d_6) δ 1.32 (m, 12H), 1.71 (m, 4H), 3.82 (d, J = 6.74, 2H), 4.02 (d, J = 6.30, 2H), 4.41 (bs, 2H), 6.59 (m, 4H), 6.96 (m, 3H), 7.43 (m, 5H), and 8.50 (d, J = 5.31, 2H); ^{13}C NMR (75 MHz, DMSO-d_6) δ 26.11, 26.21, 29.31, 29.37, 29.48, 29.66, 68.64, 69.27, 71.18, 115.77, 116.70, 116.53, 121.27, 124.43, 129.15, 129.61, 133.49, 142.62, 145.48, 150.17, 150.53, and 160.11; Mass Spec M^+ = 445.2 Calc.: 444.2777.

*10-[4-(2-Pyridin-4-yl-vinyl)-phenoxy]-decan-1-ol (**11**).* Compound **8** (0.303 g, 1.5 mmol), K_2CO_3 (0.644 g, 4.7 mmol),

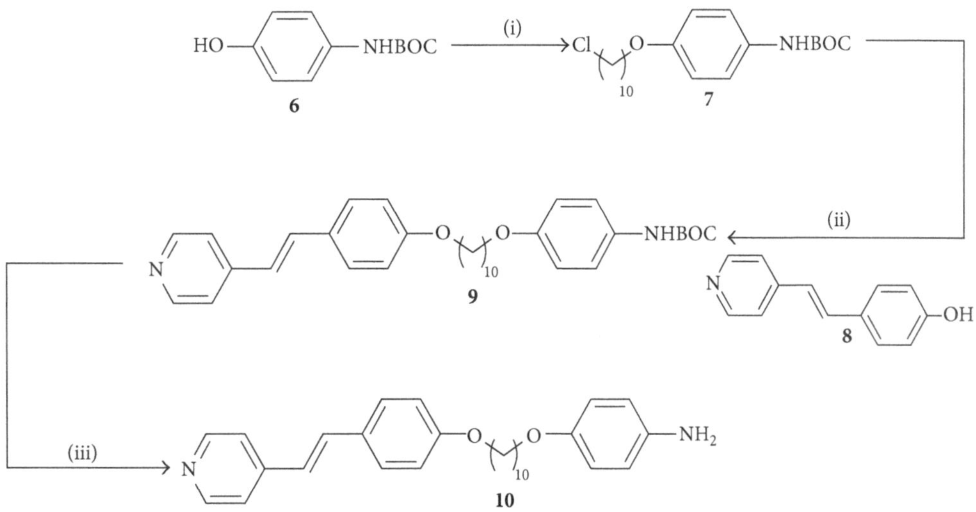

SCHEME 1: Synthetic scheme for obtaining macromolecule **10**: (i) NaH, 1,10-dichlorodecane; (ii) K_2CO_3, **3**; (iii) HCl, THF.

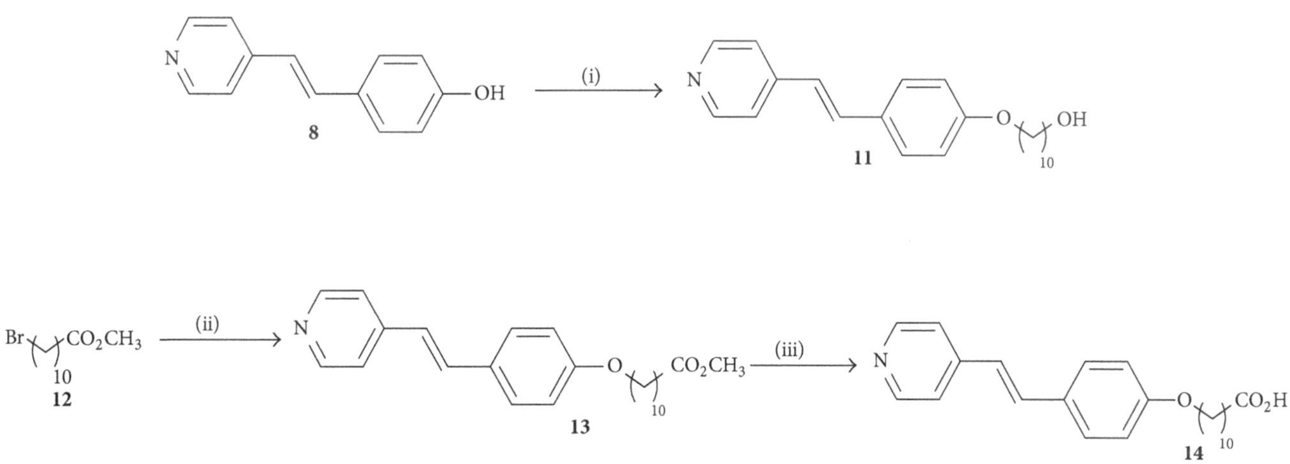

SCHEME 2: Synthetic schemes for obtaining macromolecules **6** and **9**: (i) 10-chlorodecanol, K_2CO_3; (ii) **8**, K_2CO_3; (iii) NaOH, THF, MeOH.

and 1-chloro-10-decanol (0.28 mL, 1.4 mmol) were combined in a round-bottom flask equipped with a condenser and argon purge. Dry DMF (10 mL) was added via syringe and the solution was heated to 120°C for 24 hours and then cooled to room temperature. Water (100 mL) was added; the solution was extracted with ethyl acetate (7 × 50 mL); the acetate layer was dried over $MgSO_4$; and the crude product was evaporated onto silica gel. The product was purified via column chromatography using graduated elutions from 60/40 to 40/60 (Hex/EtOAc); evaporation of solvent afforded pale yellow solid. The solid so obtained was recrystallized in methanol to afford 80 mg (15%) of **11** as a white powder: R_f = 0.43 (60/40 Hex/EtOAc); ^{1}H NMR (300 MHz, DMSO-d_6) δ 1.36 (m, 16H), 1.71 (t, J = 6.77, 2H), 3.99 (t, J = 11.07, 2H), 4.33 (t, J = 4.80, 1H), 6.96 (d, J = 8.60, 2H), 7.08 (d, J = 16.61, 1H), 7.50 (m, 3H), 7.58 (d, J = 8.86, 2H), and 8.51 (d, J = 5.86, 2H); ^{13}C NMR (75 MHz, DMSO-d_6) δ 25.45, 28.54, 28.59, 28.72, 28.92, 28.99, 32.49, 114.66, 120.49, 123.30, 128.45, 128.52, 132.60, 144.56, 149.39, and 149.85; Mass Spec M^+ = 353.3 Calc.: 353.2355.

*Methyl-11-[4-(2-pyridin-4-yl-vinyl)-phenoxy]-undecanoate (**13**).* Compound **12** (1.915 g, 6.9 mmol), K_2CO_3 (2.578 g, 18.7 mmol), and compound **8** (1.200 g, 6.1 mmol) were placed in a dried round-bottomed flask equipped with a condenser and argon purge; this mixture was heated to 70°C for 24 hours and 110°C for an additional 24 hours. The mixture was cooled to room temperature, water (100 mL) was added, and the mixture was extracted with EtOAc (7 × 50 mL); the acetate extract was washed with brined and dried over $MgSO_4$, and the crude product was adsorbed onto silica. The product was purified via column chromatography using elutions from 50/50 to 70/30 (EtOAc/Hex); evaporation of the solvent afforded 0.579 g (24%) of compound **13** as an impure bright-yellow solid: R_f = 0.55 (70/30 EtOAc/Hex); ^{1}H NMR (300 MHz, DMSO-d_6) δ 1.38 (m, 14H), 1.69 (t, J = 6.04, 2 h), 2.26 (t, J = 7.42, 2H), 3.56 (s, 3H), 3.98 (t, J = 7.15, 2H), 6.94 (d, J = 8.26, 2H), 7.05 (d, J = 16.76, 1H), 7.42 (s, 1H), 7.49 (d, J = 4.94, 2H), 7.56 (d, J = 8.40, 2H), and 8.48 (d, J = 4.36, 2H).

*11-[4-(2-Pyridin-4-yl-vinyl)-phenoxy]-undecanoic Acid (**14**).* Compound **13** (0.579 g, 1.5 mmol), KOH (1.636 g, 32.7 mmol),

TABLE 2: FTIR stretches of model compounds, model mixtures, and macromolecules.

Sample	υ Acid C=O (cm^{-1})	Sample	υ Amine N–H (cm^{-1})	υ O–H (cm^{-1})
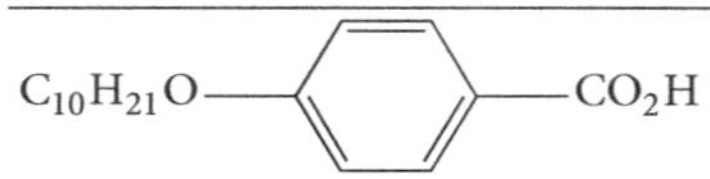 **(18)**	1666	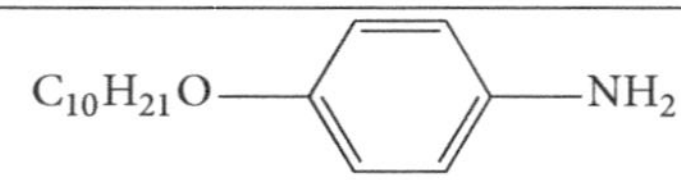 **(15)**	3436^{+} 3355^{++}	N/A
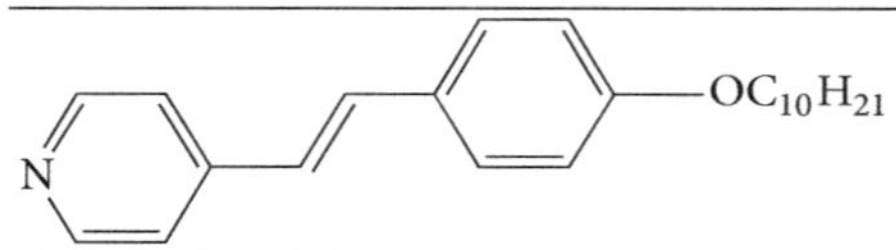 **(16)**	N/A	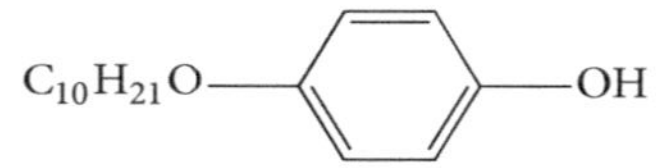 **(17)**	N/A	3358
Lauric acid	1694	Decanol	N/A	3330
1 : 1 (mol : mol) Lauric acid : **16 (mixture C)**	1708	1 : 1 (mol : mol) **15 : 16 (Mixture A)**	3431^{+} 3353^{++}	N/A
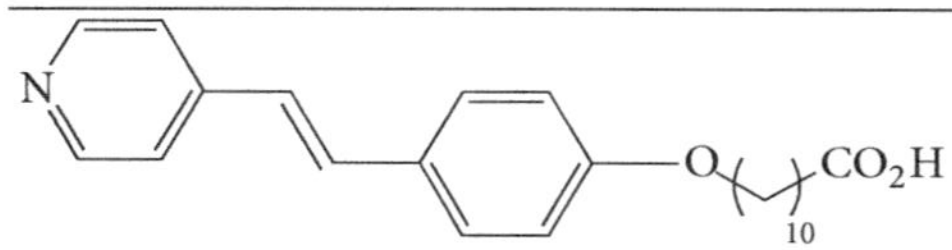 **(14)**	1708	1 : 1 (mol : mol) Decanol : **16 (Mixture B)**	N/A	3320
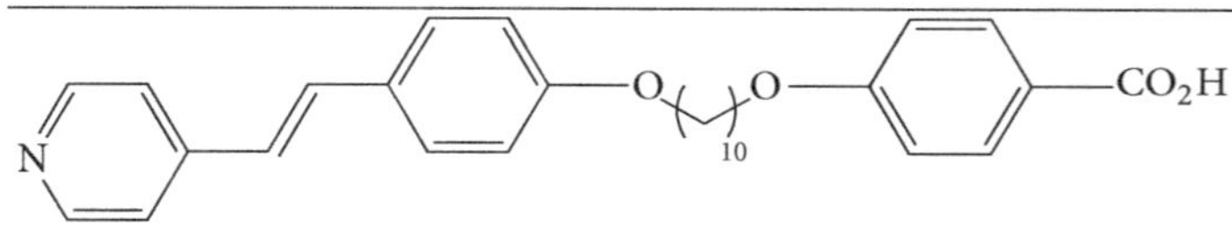 **(2)**	1709	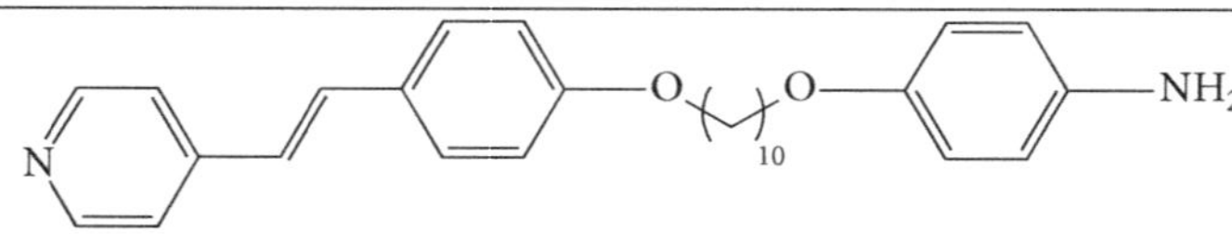 **(10)**	3302^{*}	NA
		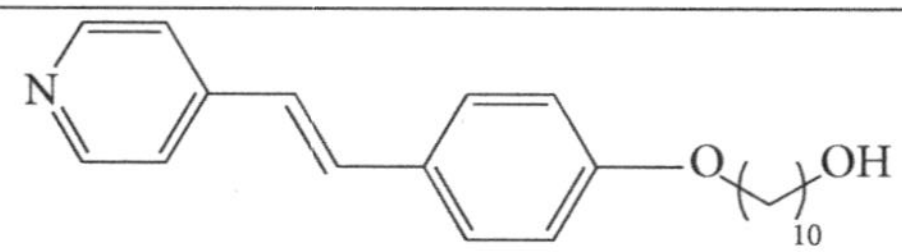 **(11)**	N/A	3301
		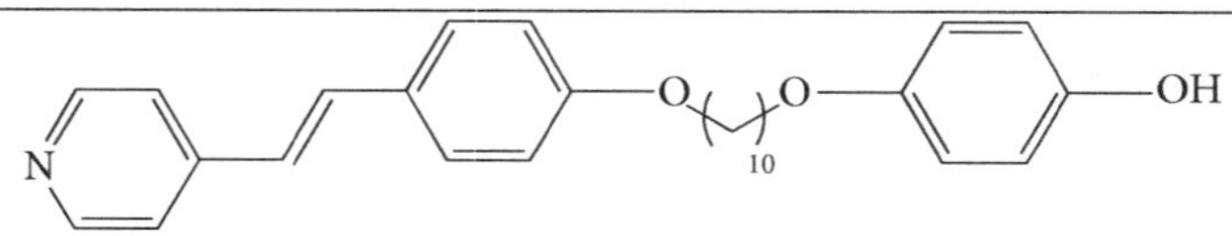 **(5)**	3078^{*}	N/A

*Center of a broad peak spanning 30–50 cm^{-1}; $^{+}$asymmetric N–H stretch; $^{++}$symmetric N–H stretch.

5N NaOH (22 mL), THF (30 mL), and MeOH (20 mL) were placed in a round-bottomed flask equipped with a condenser; this mixture was refluxed for 2 days. The mixture was cooled to room temperature and neutralized (pH = 7) with 5 N HCl; the mixture was cooled in the refrigerator. The solution was vacuum filtered and the solid so obtained was washed with water and acetone; this solid was dried in vacuum to afford 0.495 g (89%) of **14** as an off-white solid: R_f = 0.33 (EtOAc); ^{1}H NMR (300 MHz, DMSO-d_6) δ 1.36 (m, 14H), 1.69 (t, J = 6.38, 2H), 2.16 (t, J = 7.34, 2H), 3.96 (t, J = 6.36, 2H), 6.94 (d, J = 8.96, 2H), 7.07 (d, J = 16.44, 1H), 7.43 (s, 1H), 7.49 (d, J = 5.98, 2H), 7.57 (d, J = 8.48, 2H), and 8.48 (d, J = 5.47, 2H); ^{13}C NMR (75 MHz, DMSO-d_6) δ; Mass Spec M^+ = 381.3 Calc.: 381.2304.

*4-[2-(4-Decyloxy-phenyl)-vinyl]-pyridine (**16**).* Compound **8** (1.504 g, 7.6 mmol), K_2CO_3 (2.618 g, 18.9 mmol), decylbromide (1.60 mL, 7.8 mmol), and dry DMF (20 mL) were combined in an oven dried round bottomed flask equipped with a condenser and an argon purge; this mixture was heated to 120°C for 18 hours and then cooled to room temperature. Water (100 mL) was added and the solution was extracted with EtOAc (5 × 100 mL). The combined organic layers were washed with brine and dried over $MgSO_4$. The crude product was purified via column chromatography using elutions from 60/40 to 50/50 (Hex/EtOAc). Evaporation of the solvent afforded an yellow solid, which was recrystallized twice from hexanes to afford 0.902 g (35%) of **16** as a pale yellow solid: R_f = 0.37 (50/50 Hex/EtOAc); ^{1}H NMR (300 MHz, $CDCl_3$) δ 0.88 (t, J = 6.36, 3H), 1.27 (m, 14H), 1.79 (t, J = 6.83, 2H), 3.98 (t, J = 6.92, 2H), 6.90 (m, 3H), 7.26 (d, J = 16.49, 1H), 7.34 (d, J = 6.34, 2H), 7.46 (d, J = 8.86, 2H), and 8.54 (d, J = 6.32, 2H); ^{13}C NMR (75 MHz, $CDCl_3$) δ 14.09, 22.66, 26.00, 29.20, 29.23, 29.37, 29.55, 31.87, 67.16, 68.12, 114.82, 120.60, 123.56, 128.35, 132.78, 145.62, 150.11, 154.68, and 159.82; IR (ATR) 2918, 2851, 1590, 1514, 1256, and 1168 cm^{-1}; MS (EI+) *m/e* 337 (197, 180, 168, 115, 91, and 62).

3. Results and Discussion

3.1. Synthesis of Macromolecules. The synthesis of macromolecule 10 is shown in Scheme 1. Compound 6 was reacted with 1,10-dichlorodecane to afford O-alkylated compound 7. The identity of compound 7 (as the O-alkylated product) was confirmed using a nuclear overhauser effect (NOE) NMR experiment, while the yield of compound 7 was low, extending the reaction time or increasing the temperature during the reaction resulted in an inseparable mixture of N-alkylated and O-alkylated products which contained little to no compound 7. Alkylation of 7 with stilbazole 8 afforded compound 9 which was deprotected using HCl to afford macromolecule 10 which precipitated from solution upon neutralization of the cleavage medium. Macromolecule 10 is soluble in hot DMSO (120°C) and hot (120°C) DMF but insoluble in less polar organic solvents.

The syntheses of macromolecules 11 and 14 are shown in Scheme 2. Macromolecule 11 was synthesized in one step via basic coupling of compound 8 and 10-chlorodecanol. Compound 12 was alkylated with stilbazole 8 to afford compound 13. Basic hydrolysis of 13 afforded macromolecule 14, which precipitated upon neutralization of the basic cleavage medium. Macromolecule 11 is soluble in DMSO, DMF, and alcohols at room temperature, whereas the solubility of 14 is similar to that of macromolecule 10.

3.2. Hydrogen Bonding in Our Macromolecules. Solid state FTIR was used to analyze the hydrogen bonding in our noncentrosymmetric macromolecules. Select FTIR stretches for model compounds, model mixtures (which mimic the hydrogen bonding in our macromolecules), and our macromolecules are shown in Table 2. The amine N–H peaks in model mixture A and macromolecule 10 are red shifted with respect to the N–H shifts in 4-decyloxyaniline which suggest strong intermolecular hydrogen bonding in 10. In macromolecule 10, the asymmetric and symmetric N–H stretches merge together into one broad peak which can be attributed to the numerous possible orientations of the N–H (i.e., donor-donor or donor-acceptor) hydrogen bonds in the macromolecule. The O–H stretches in macromolecule 11 and mixture B are red shifted by 10–20 cm^{-1} with respect to those in 1-decanol. The magnitude of this red shift is small compared to the approximately 250 cm^{-1} red shift in the O–H stretch observed when phenols form hydrogen bonds with stilbazoles [19, 20], which suggests relatively weak intermolecular hydrogen bonding in macromolecule 11 compared to its phenol analogue (compound 5). The C=O stretch in macromolecule 14 is blue shifted 14 cm^{-1} relative to that in lauric acid. This shift is smaller than the 43 cm^{-1} shift between the C=O between 13 and macromolecule 2 which suggests that the intermolecular hydrogen bonds in macromolecule 14 are weaker than those of inmacromolecule 2.

3.3. Phase Transitions of Noncentrosymmetric Hydrogen Bonded Macromolecules. The phase transitions of the macromolecules synthesized in this paper were characterized using Polarized Optical Microscopy (POM) and Differential Scanning Calorimetry (DSC) and are summarized in Figure 1. The melting point of macromolecule 10 is 31°C lower than that of the phenol analog (macromolecule 5) and 87°C lower than that of its benzoic acid analogue (macromolecule 2). This melting point depression can be attributed to the relatively weaker amine-stilbazole hydrogen bonding. Macromolecules 11 and 14 have crystal-crystal transitions on the heating and cooling cycles but no liquid crystalline phases. The melting points of 11 and 14 are comparable to that of 10 suggesting that (like macromolecule 10) macromolecules 11 and 14 have relatively weak intermolecular hydrogen bonding (relative to those in macromolecules 2 and 5).

4. Conclusions

The synthesis of three new main chain noncentrosymmetric hydrogen bonded macromolecules incorporating alkanol, alkanoic acid, and aniline donors is described. Relatively weak intermolecular hydrogen bonding is present in macromolecules **10**, **11**, and **14**. Macromolecules **11** and **14**

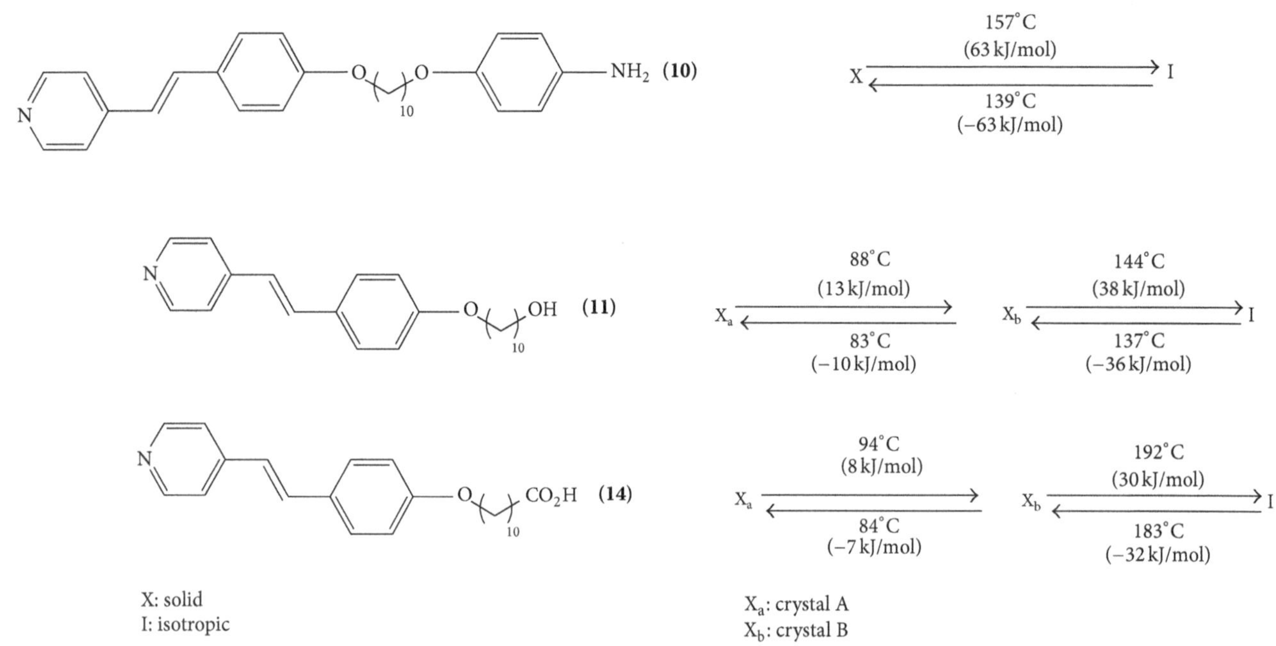

FIGURE 1: Phase transitions of macromolecules **10**, **11**, and **14**.

both possess enantiotropic crystal → crystal transitions in addition to melting and crystallization temperatures, while macromolecule **10** only has melting and crystallization phase transitions. The phase transition temperatures of **10**, **11**, and **14** are lower than that of the phenol and benzoic acid analogues due to weaker hydrogen bonding.

Conflict of Interests

The author declares that there is no conflict of interests regarding the publication of this paper.

Acknowledgments

This paper is dedicated to the memory of Dr. Daniel J. Dyer, Professor of chemistry at Southern Illinois University, Carbondale. The author would like to thank the National Science Foundation under Grant CHE_0094195, Oak Ridge Associated Universities, the Materials Technology Center at Southern Illinois University Carbondale, and the Office of Research Development and Administration at Southern Illinois University for support of this research.

References

[1] R. A. Hann and D. Bloor, "Organic materials for non-linear optics II," in *Proceedings of the 2nd International Symposium on Organic Materials for Non-Linear Optics Organised by the Applied Solid State Chemistry Group of the Dalton Division of the Royal Society of Chemistry*, D. Bloor and R. A. Hann, Eds., Oxford, UK, September 1990.

[2] P. N. Butcher and D. Cotter, *The Elements of Nonlinear Optics*, Cambridge University, New York, NY, USA, 1990.

[3] D. V. Skobeltsyn, *Nonlinear Optics*, Consultants Bureau, New York, NY, USA, 1970.

[4] G. C. Baldwin, *An Introduction to Nonlinear Optics*, Plenum, New York, NY, USA, 1969.

[5] R. Schwödiauer, G. S. Neugschwandtner, K. Schrattbauer et al., "Preparation and characterization of novel piezoelectric and pyroelectric polymer electrets," *IEEE Transactions on Dielectrics and Electrical Insulation*, vol. 7, no. 4, pp. 578–586, 2000.

[6] M. C. Petty, "Molecular electronics: prospects for instrumentation and measurement science," *Measurement Science and Technology*, vol. 7, no. 5, pp. 725–735, 1996.

[7] J. Tsibouklis, M. Petty, M. C. Petty, and W. J. Feast, "Pyroelectric polymer structures," *International Journal of Electronics*, vol. 78, no. 6, pp. 1101–1105, 1995.

[8] G. Eberle, H. Schmidt, and W. Eisenmenger, "Piezoelectric polymer electrets," *IEEE Transactions on Dielectrics and Electrical Insulation*, vol. 3, no. 5, pp. 624–646, 1996.

[9] Q. X. Chen and P. A. Payne, "Industrial applications of piezoelectric polymer transducers," *Measurement Science and Technology*, vol. 6, no. 3, pp. 249–267, 1995.

[10] C. H. Bosshard, M. Bösch, I. Liakatas, M. Jäger, and P. Günter, "Second-order nonlinear optical organic materials: recent developments," in *Nonllnear Optical Effects and Materials*, P. Günter, Ed., vol. 72 of *Springer Series in Optical Sciences*, chapter 3, pp. 163–299, Springer, 2000.

[11] A. Carella, A. Castaldo, R. Centore, A. Fort, A. Sirigu, and A. Tuzi, "Synthesis and second order nonlinear optical properties of new chromophores containing 1,3,4-oxadiazole and thiophene rings," *Journal of the Chemical Society, Perkin Transactions 2*, no. 11, pp. 1791–1795, 2002.

[12] Y. Cui, G. Qian, L. Chen, Z. Wang, J. Gao, and M. Wang, "Enhanced thermal stability of dipole alignment in inorganic-organic hybrid films containing benzothiazole chromophore," *Journal of Physical Chemistry B*, vol. 110, no. 9, pp. 4105–4110, 2006.

[13] M. Jazbinsek, L. Mutter, and P. Günter, "Photonic applications with the organic nonlinear optical crystal DAST," *IEEE Journal on Selected Topics in Quantum Electronics*, vol. 14, no. 5, pp. 1298–1311, 2008.

[14] J. Chen, S. Marturunkalkul, L. Li, J. Kumar, and S. K. Tripathy, "Second order nonlinear optical materials," in *Handbook of Conducting Polymers*, pp. 727–743, CRC Press, New York, NY, USA, 1998.

[15] E. Chen and A. Murphy, *Broadband Optical Modulators*, CRC Press, New York, NY, USA, 2011.

[16] C. Landorf, J. Wolf, C. Li et al., "Design of polar organic liquid crystalline thin films," in *Organic Thin Films for Photonic Applications*, vol. 64 of *Trends in Optics and Photonics Series*, pp. 145–152, Optical Society of America, Washington, DC, USA, 2002.

[17] C. Landorf, J. Simpson, J. Jacobsen, and D. J. Dyer, "Design and synthesis of noncentrosymmetric organic thin films with Hydrogen bonded main chain polymers," *Polymeric Materials: Science and Engineering*, vol. 83, pp. 172–173, 2000.

[18] D. J. Dyer, J. Wolf, C. Li et al., "Design and synthesis of noncentrosymmetric hydrogen bonded main chain liquid crystalline polymers: towards polar order in organic thin films," *Polymer Preprints*, vol. 44, no. 1, p. 578, 2003.

[19] J. R. Wolf, *Synthesis and characterization of noncentrosymmetric Hydrogen bonded liquid crystalline polymers [Ph.D. thesis]*, Southern Illinois University, 2005.

[20] J. R. Wolf, T. Zhao, C. Landorf, and D. J. Dyer, "Synthesis and characterisation of laterally substituted noncentrosymmetric main chain hydrogen-bonded polymers," *Liquid Crystals*, vol. 41, no. 5, pp. 721–730, 2014.

[21] L. Jongen, B. Goderis, I. Dolbnya, and K. Binnemans, "Influence of the chain length on the thermal behavior of lanthanide(III) 4-alkoxybenzoates," *Chemistry of Materials*, vol. 15, no. 1, pp. 212–217, 2003.

[22] A. R. Lapucha, "Heterocyclic analogs of stilbenes: reactions of (E)-4-azastilbenes with polymethylene dibromides," *Polish Journal of Chemistry*, vol. 61, no. 4–6, pp. 563–567, 1987.

[23] D. S. Clyne and L. Weiler, "The synthesis of 14-membered macrocyclic ethers," *Tetrahedron*, vol. 55, no. 48, pp. 13659–13682, 1999.

[24] R. M. Meudtner, M. Ostermeier, R. Goddard, C. Limberg, and S. Hecht, "Multifunctional "clickate" as versatile extended heteroaromatic building blocks: efficient synthesis via click chemistry, conformational preferences, and metal coordination," *Chemistry—A European Journal*, vol. 13, no. 35, pp. 9834–9840, 2007.

Permissions

The contributors of this book come from diverse backgrounds, making this book a truly international effort. This book will bring forth new frontiers with its revolutionizing research information and detailed analysis of the nascent developments around the world.

We would like to thank all the contributing authors for lending their expertise to make the book truly unique. They have played a crucial role in the development of this book. Without their invaluable contributions this book wouldn't have been possible. They have made vital efforts to compile up to date information on the varied aspects of this subject to make this book a valuable addition to the collection of many professionals and students.

This book was conceptualized with the vision of imparting up-to-date information and advanced data in this field. To ensure the same, a matchless editorial board was set up. Every individual on the board went through rigorous rounds of assessment to prove their worth. After which they invested a large part of their time researching and compiling the most relevant data for our readers. Conferences and sessions were held from time to time between the editorial board and the contributing authors to present the data in the most comprehensible form. The editorial team has worked tirelessly to provide valuable and valid information to help people across the globe.

Every chapter published in this book has been scrutinized by our experts. Their significance has been extensively debated. The topics covered herein carry significant findings which will fuel the growth of the discipline. They may even be implemented as practical applications or may be referred to as a beginning point for another development. Chapters in this book were first published by Hindawi Publishing Corporation; hereby published with permission under the Creative Commons Attribution License or equivalent.

The editorial board has been involved in producing this book since its inception. They have spent rigorous hours researching and exploring the diverse topics which have resulted in the successful publishing of this book. They have passed on their knowledge of decades through this book. To expedite this challenging task, the publisher supported the team at every step. A small team of assistant editors was also appointed to further simplify the editing procedure and attain best results for the readers.

Our editorial team has been hand-picked from every corner of the world. Their multi-ethnicity adds dynamic inputs to the discussions which result in innovative outcomes. These outcomes are then further discussed with the researchers and contributors who give their valuable feedback and opinion regarding the same. The feedback is then collaborated with the researches and they are edited in a comprehensive manner to aid the understanding of the subject.

Apart from the editorial board, the designing team has also invested a significant amount of their time in understanding the subject and creating the most relevant covers. They scrutinized every image to scout for the most suitable representation of the subject and create an appropriate cover for the book.

The publishing team has been involved in this book since its early stages. They were actively engaged in every process, be it collecting the data, connecting with the contributors or procuring relevant information. The team has been an ardent support to the editorial, designing and production team. Their endless efforts to recruit the best for this project, has resulted in the accomplishment of this book. They are a veteran in the field of academics and their pool of knowledge is as vast as their experience in printing. Their expertise and guidance has proved useful at every step. Their uncompromising quality standards have made this book an exceptional effort. Their encouragement from time to time has been an inspiration for everyone.

The publisher and the editorial board hope that this book will prove to be a valuable piece of knowledge for researchers, students, practitioners and scholars across the globe.

List of Contributors

Jyotishkumar Parameswaranpillai and Abhilash George
Department of Polymer Science and Rubber Technology, Cochin University of Science and Technology, Cochin, Kerala 682022, India

Jürgen Pionteck
Leibniz-Institute for Polymer Research Dresden, Hohe Strare 6, 01069 Dresden, Germany

Sabu Thomas
School of Chemical Sciences, Mahatma Gandhi University, Priyadarshini Hills, Kottayam, Kerala 686560, India
Centre for Nanoscience & Nanotechnology, Mahatma Gandhi University, Priyadarshini Hills, Kottayam, Kerala 686560, India

Sandeep Rai and Snehal Lokhandwala
Shroff S. R. Rotary Institute of Chemical Technology, BlockNo. 402, At & Post Vataria, District Bharuch, Gujarat 393002, India

Anjali Bajpai
Department of Chemistry, Government Autonomous Model Science College, Jabalpur 482001, India

P. J. Jandas, S. Mohanty and S. K. Nayak
Laboratory forAdvancedResearch inPolymericMaterials (LARPM), Central Institute of Plastics Engineering andTechnology (CIPET), Bhubaneswar, Orissa 751024, India

Bhaskar Patham
General Motors Technical Centre India Pvt. Ltd., Vehicle Performance Centers, Vehicle CAE Methods Group, International Tech Park Ltd. Whitefield Road, Bangalore 560 066, India

M. P. Poornendu Thejaswini
General Motors Technical Centre India Pvt. Ltd., Vehicle Performance Centers, Materials and IMDS Group, International Tech Park Ltd. Whitefield Road, Bangalore 560 066, India

Eun-Soo Park
Youngchang Silicone Co., Ltd., 481-7 Gasan-Dong, Kumchun-Gu, Seoul 153-803, Republic of Korea

M. A. El-Sheikh, S. M. El-Rafie, E. S. Abdel-Halim and M. H. El-Rafie
National Research Centre, Textile Research Division, El-Behooth Street, Dokki, P.O. 12311, Cairo, Egypt

Renuka R. Gonte and P. C. Deb
Department of Applied Chemistry, Defence Institute of Advanced Technology, Girinagar, Pune 411 025, India

K. Balasubramanian
Department of Materials Engineering, Defence Institute of Advanced Technology, Girinagar, Pune 411 025, India

Hyeyoung Shin
Kumho Petrochemical R&D Center, Hwaam-Dong 57-1, Yuseong, Daejeon 305-348, Republic of Korea

Chong Geng and Qingfeng Yan
Department of Chemistry, State Key Laboratory of New Ceramics and Fine Processing, Tsinghua University, Beijing 100084, China

Fan Li
Department of Chemistry, State Key Laboratory of New Ceramics and Fine Processing, Tsinghua University, Beijing 100084, China
Department of Material Science and Engineering, Tsinghua University, Beijing 100084, China

Elijah I. Nep, Patricia O. Odumosu, Ndidi C. Ngwuluka, Patrick O. Olorunfemi, and Nelson A. Ochekpe
Biomaterials and Drug Delivery Research Group, Faculty of Pharmaceutical Sciences, University of Jos, PMB 2084, Jos 930001, Nigeria

Kamlesh Pandey, Nidhi Asthana and Mrigank Mauli Dwivedi
National Center of Experimental Mineralogy and Petrology, University of Allahabad, Allahabad, UP 211002, India

S. K. Chaturvedi
Department of Env. Science, Chitrakoot Gramodaya Vishwavidyalay, Chitrakoot, MP 485334, India

C. P. Rejisha, S. Soundararajan, N. Sivapatham and K. Palanivelu
Department of Plastics Technology, Central Institute of Plastics Engineering and Technology (CIPET), Chennai, Tamil Nadu 600 032, India

Sharief ud Din Khan and M. A. Wahab
Department of Physics, Jamia Millia Islamia, New Delhi 110025, India

Manju Arora and Parveen Saini
CSIR-National Physical Laboratory, Dr. K.S. Krishnan Marg, New Delhi 110012, India

Thejas Gopal Krishne Urs, Mahadevaiah Dasaiah and Rudrappa Somashekar
Department of Physics, University of Mysore, Manasagangotri, Mysore 570006, India

Mohd Hafizuddin Ab Ghani, Mohd Nazry Salleh, Ruey Shan Chen, Sahrim Ahmad, Mohd Rashid Yusof Hamid, Ismail Hanafi, and Nishata Royan Rajendran Royan
Material Science Programme, School of Applied Physics, Faculty of Science and Technology, The National University of Malaysia, 43600 Bangi, Malaysia

Emi Govorlin BajsiT and Vesna Rek
Department of Polymer Engineering and Organic Chemical Technology, Faculty of Chemical Engineering and Technology, University of Zagreb, Maruli´cev trg 19, 10 000 Zagreb, Croatia

Ivana SosiT
Department of Industrial Ecology, Faculty of Chemical Engineering and Technology, University of Zagreb, Maruli´cev trg 19, 10 000 Zagreb, Croatia

P. U. Belsare, A. B. Zade and M. S. Dhore
Department of Chemistry, Laxminarayan Institute of Technology, Rashtrasant Tukadoji Maharaj Nagpur University, Nagpur, Maharashtra 440010, India

P. P. Kalbende
Department of Applied Chemistry, Datta Meghe Institute of Engineering, Technology and Research, Sawangi (Meghe), Wardha 442001, India

S. K. Bajpai
Polymer Research Laboratory, Department of Chemistry, Government Model Science College, Jabalpur, Madhya Pradesh 482001, India

M. P. Swarnkar
Department of Chemistry, Kymore Science College, Kymore, Katni, Madhya Pradesh 483880, India

Fayçal Benhacine, Farida Yahiaoui and Assia Siham Hadj-Hamou
Laboratoire des Mat´eriaux Polym`eres, D´epartement de Chimie Macromol´eculaire, Facult´e de Chimie, Universite des Sciences et de la Technologie Houari Boumediene, BP 32, El Alia, 16111 Alger, Algeria

Subhas C. Shit and Pathik M. Shah
Central Institute of Plastics Engineering & Technology (CIPET), Plot No. 630, Phase IV, Vatva, G.I.D.C., Ahmedabad 382 445, India

Jeremy R. Wolf
Southern Illinois University Carbondale, Carbondale, IL 62901, USA
ABC Laboratories, Columbia, MO 65202, USA